Nonlinear Systems and Complexity

Volume 34

Series Editor

Albert C. J. Luo, Southern Illinois University, Edwardsville, IL, USA

Nonlinear Systems and Complexity provides a place to systematically summarize recent developments, applications, and overall advance in all aspects of nonlinearity, chaos, and complexity as part of the established research literature, beyond the novel and recent findings published in primary journals. The aims of the book series are to publish theories and techniques in nonlinear systems and complexity; stimulate more research interest on nonlinearity, synchronization, and complexity in nonlinear science; and fast-scatter the new knowledge to scientists, engineers, and students in the corresponding fields. Books in this series will focus on the recent developments, findings and progress on theories, principles, methodology, computational techniques in nonlinear systems and mathematics with engineering applications. The Series establishes highly relevant monographs on wide ranging topics covering fundamental advances and new applications in the field. Topical areas include, but are not limited to: Nonlinear dynamics Complexity, nonlinearity, and chaos Computational methods for nonlinear systems Stability, bifurcation, chaos and fractals in engineering Nonlinear chemical and biological phenomena Fractional dynamics and applications Discontinuity, synchronization and control.

More information about this series at https://link.springer.com/bookseries/11433

Jiazhong Zhang
Editor

Dynamics and Fault Diagnosis of Nonlinear Rotors and Impellers

Springer

Editor
Jiazhong Zhang
School of Energy and Power Engineering
Xi'an Jiaotong University
Shaanxi Province, China

ISSN 2195-9994 ISSN 2196-0003 (electronic)
Nonlinear Systems and Complexity
ISBN 978-3-030-94300-4 ISBN 978-3-030-94301-1 (eBook)
https://doi.org/10.1007/978-3-030-94301-1

This Springer imprint is published by the registered company Springer Nature Switzerland AG
The registered company address is: Gewerbestrasse 11, 6330 Cham, Switzerland

Preface

Over the decades after 1980s, some of the abstract terminologies in nonlinear dynamics, such as fractal, attractor, bifurcation, and chaos, have evolved to common ones, and nonlinear approach has become a main and fascinating stream recently to study complex phenomena in large-scale engineering systems. Indeed, there exists a rich variety of nonlinear phenomena in engineering, and nonlinearity in engineering is a direct application of nonlinear dynamics. On the other hand, a key understanding of the nature and nonlinear phenomena of practical systems can develop the corresponding mathematical theory, enabling further application of nonlinear design to engineering.

This contributed volume, entitled Dynamics and Fault Diagnosis of Nonlinear Rotors and Impellers, presents recent developments in nonlinear dynamics applied to engineering, including 11 chapters. Specifically, the authors stress stability and bifurcation in large-scale, complex rotor dynamic systems; periodic motions and their bifurcations in nonlinear circuit systems; fault diagnosis of complex engineering systems with nonlinear approaches; singularities in fluid machinery and bifurcation analysis; nonlinear behaviors in rotor dynamic systems with multi-mistuned blades; mode localization induced by mistuning in impellers with periodical and cyclic symmetry; and nonlinear behaviors in fluid–structure interaction and their control. These new results will maximize readers' understanding on the recent progress in nonlinear dynamics applied to large-scale engineering systems in general and nonlinear rotors and impellers. That is, the readers could have, in a single volume, the opportunity to understand the state of the art or progress in the areas of nonlinear dynamics applied to engineering system with large-scale or infinite dimension.

The editor has handled some edited books and monograph in recent years with a selection of relevant studies in the areas of complexity and chaos in nonlinear systems. The materials presented in this edited book will enrich the new developments in nonlinear dynamics applied to engineering systems, which are mostly governed by nonlinear evolutionary partial differential equations, rather than ordinary differential equations. Additionally, the numerical and analytical methods for such systems are presented, such as model reduction method on AIMs for N-

S equations, nonlinear dynamic system with infinite dimension and composed of continuum mechanics, multilevel finite element method, and hierarchical basis, so that the nonlinear dynamics could be introduced to such reduced systems with high computing efficiency. In particular, the studies on mode or vibration localization induced by material and aerodynamic mistuning in centrifugal impellers with cyclic symmetry are presented, and further a theoretical explanation for such kind of fatigue and the controlling strategy are given. Consequently, this book might well be used as a source for supplementary material for traditional courses in nonlinear dynamics, dynamic system, etc.

I would also like to take this opportunity to express my gratitude to the authors for their support and contributions to chapters, and particular thanks go to Professor Albert C. J. Luo for constant support and inspiration. I hope that the results presented will be useful for other specialists in science and engineering.

Xi'an, Shaanxi Province, China Jiazhong Zhang

Contents

Chapter 1
Periodic Motions to Chaos in a Nonlinear Rotor System

Yeyin Xu and Albert C. J. Luo

1.1 Introduction

Rotor dynamics has been of great interest for over a century. In 1919, Jeffcott [1] studied the unbalance effects of the eccentric mass on the whirling motions in a single disc rotor with rigid support. In 1933, Smith [2] investigated the transverse vibration of a rotor system with flexible shaft and flexibly-supported bearings. In 1955, Yamamoto [3] observed a subharmonic oscillation of order-1/2 in an experiment of a rotor system with angular clearances of supported bearings. The theoretical studies of a rotor system with the bearing radial clearance was provided by Yamamoto in [4]. The radial clearance was modeled as a nonlinear term. The stability criterion was presented, and jumping phenomenon of periodic motions was obtained. In 1964, Ehrich [5] analyzed the instability of a rotor system concerning the internal damping of the shaft. The unstable speed range was found to be governed by the ratio of the external to internal damping. The whirling mode in the unstable region had speeds close to half of the rotating speed. In 1969, Gunter [6] found a lower support stiffness with high damping could sufficiently reduce bearing forces. Barrent et al. [7] developed an optimal method for computing the optimum bearing damping. In 1982, Childs [8] studied the asymmetric bearing-shaft clearance effect on a Jeffcott rotor. The asymmetric bearing clearance was modeled as a discontinuous stiffness model. Periodic motions with the 1/2 and 1/3 fractional orders frequency were obtained. In 1994, Saito [9] studied the synchronous vibration of a rotor system with oil-film-bearing supports through the

Y. Xu
School of Aerospace Engineering, Xi'an Jiaotong University, Xi'an, P. R. China

A. C. J. Luo (✉)
Department of Mechanical and Mechatronics Engineering, Southern Illinois University Edwardsville, Edwardsville, IL, USA
e-mail: aluo@siue.edu

J. Zhang (ed.), *Dynamics and Fault Diagnosis of Nonlinear Rotors and Impellers*, Nonlinear Systems and Complexity 34, https://doi.org/10.1007/978-3-030-94301-1_1

numerical harmonic balance method. Four categories of the synchronous response were discovered in that study. In 2001, Jiang and Ulbrich [10] studied a Jeffcott rotor with cross-coupling stiffness and annular rub conditions.

For periodic motions in the Jeffcott rotor system, the above mentioned methods based on perturbation analysis are not adequate. In 2012, Luo [11] developed a generalized harmonic balance method for periodic solutions of nonlinear systems. In 2014, Huang and Luo [12] applied the generalized harmonic balance method to a vertically supported nonlinear Jeffcott rotor system. Bifurcation trees of period-1 to period-2 motions were obtained. In 2015, Huang and Luo [13] investigated a buckled nonlinear Jeffcott rotor system through the generalized harmonic balance method. The periodic solutions of period-1 motion were obtained. Quasi-periodic motions and bifurcation trees were predicted via the Hopf bifurcations. The bifurcation trees of period-1 motion to chaos were also presented in Huang and Luo in [14]. However, the generalized harmonic balance method cannot be applied to non-polynomial nonlinear systems.

In 2015, Luo [15] developed an implicit mapping method for periodic motions in nonlinear systems. The analytical predictions of periodic motions in the non-polynomial nonlinear systems can be obtained. In 2018, Guo and Luo [16] obtained the analytical bifurcation trees of period-1 motions to chaos in a periodically excited pendulum with the implicit mapping method. Guo et al. [17] experimentally verified the periodic motions in a Duffing oscillatory system, predicted by the discrete mapping method. Xu and Luo [18, 19] presented a sequence of periodic motions to chaos in van de Pol oscillators. Xing and Luo [20] applied such an implicit mapping method for bifurcation trees of period-1 to period-4 motions in the first-order, time-delay, nonlinear system. Period-1 motion to chaos in a periodically driven, time-delayed, softening Duffing oscillator was also presented in Xing and Luo [21].

In this chapter, periodic motions in a horizontally supported Jeffcott rotor system will be predicted analytically through an implicit mapping method. The corresponding stability and bifurcation of periodic motions are determined through eigenvalue analysis. Numerical simulations will be completed for comparison of the analytical and numerical results.

1.2 Methodology

Consider a nonlinear Jeffcott rotor system as

$$\begin{aligned} M\ddot{X} + C_1\dot{X} + K_1X + K_2X(X^2+Y^2) &= ME\omega^2\cos\omega\tau, \\ M\ddot{Y} + C_2\dot{Y} + K_1Y + K_2Y(X^2+Y^2) &= ME\omega^2\sin\omega\tau - Mg. \end{aligned} \tag{1.1}$$

where X and Y are the displacements in x- and y-directions, $\dot{X} = dX/d\tau$ and $\dot{Y} = dY/d\tau$ are the velocities in x- and y-directions. M is the mass of the disk, C_1 and

C_2 are damping coefficients, K_1 and K_2 are linear and nonlinear stiffness, E is the eccentricity of the disk and ω is rotating speed.The nonlinear rotor system in Eq. (1.1) has equilibriums as $X_e = 0$, $K_1 Y_e + K_2 Y_e^3 = -Mg$. Let $X = \Delta X$, $Y = \Delta Y + Y_e$. Equation (1.1) can be rewritten in non-dimensional form as

$$\begin{aligned} &\ddot{x} + \mu_1 \dot{x} + \omega_1^2 x + \gamma(2xy + xy^2 + x^3) = e\Omega^2 \cos \Omega t, \\ &\ddot{y} + \mu_2 \dot{y} + \omega_2^2 y + \gamma(x^2 + 3y^2 + x^2 y + y^3) = e\Omega^2 \sin \Omega t \end{aligned} \tag{1.2}$$

where

$$\begin{aligned} &x = \Delta X / Y_s,\, y = \Delta Y / Y_s,\, \dot{x} = dx/dt,\, \dot{y} = dy/dt, \\ &t = \tau / \sqrt{M/K_1},\, \mu_i = \frac{C_i}{\sqrt{MK_1}}\ (i = 1, 2),\, \gamma = \frac{K_2 Y_s^2}{K_1}, \\ &\omega_1^2 = 1 + \gamma,\, \omega_2^2 = 1 + 3\gamma,\, e = \frac{E}{Y_s},\, \Omega = \omega \sqrt{\frac{M}{K_1}}. \end{aligned} \tag{1.3}$$

The non-dimensional rotor system of Eq. (1.2) in state space form is

$$\begin{aligned} &\dot{x}_1 = x_2, \\ &\dot{x}_2 = e\Omega^2 \cos \Omega t - \left[\mu_1 \dot{x} + \omega_1^2 x + \gamma\left(2xy + xy^2 + x^3\right)\right]; \\ &\dot{y}_1 = y_2, \\ &\dot{y}_2 = e\Omega^2 \sin \Omega t - \left[\mu_2 \dot{y} + \omega_2^2 y + \gamma\left(x^2 + 3y^2 + x^2 y + y^3\right)\right]. \end{aligned} \tag{1.4}$$

For the implicit mapping method, consider a small time interval $t \in [t_{k-1}, t_k]$, the above equations can be discretized for a period-m motion ($m = 1, 2, 3, \cdots$) as

$$\begin{aligned} x_{1,k}^{(m)} &= x_{1,k-1}^{(m)} + h x_{2,kc}^{(m)}, \\ x_{2,k}^{(m)} &= x_{2,k-1}^{(m)} - \{h\mu_1 x_{2,kc}^{(m)} + \omega_1^2 x_{1,kc}^{(m)} + [\gamma 2 x_{1,kc}^{(m)} y_{1,kc}^{(m)} + x_{1,kc}^{(m)} (y_{1,kc}^{(m)})^2 + (x_{1,kc}^{(m)})^3] \\ &\quad - e\Omega^2 \cos \Omega (t_{k-1}^{(m)} + \tfrac{1}{2} h)\}; \\ y_{1,k}^{(m)} &= y_{1,k-1}^{(m)} + h y_{2,kc}^{(m)}, \\ y_{2,k}^{(m)} &= y_{2,k-1}^{(m)} - \{\mu_2 y_{2,kc}^{(m)} + \omega_2^2 y_{1,kc}^{(m)} + \gamma[(x_{1,kc}^{(m)})^2 + 3(y_{1,kc}^{(m)})^2 \\ &\quad + (x_{1,kc}^{(m)})^2 y_{1,kc}^{(l)} + (y_{1,kc}^{(m)})^3] - e\Omega^2 \sin \Omega (t_{k-1}^{(m)} + \tfrac{1}{2} h)\}; \end{aligned} \tag{1.5}$$

where

$$\begin{aligned} x_{1,kc}^{(m)} &= \frac{1}{2}\left(x_{1,k}^{(m)} + x_{1,k-1}^{(m)}, y_{1,kc}^{(m)}\right) = \frac{1}{2}\left(y_{1,k}^{(m)} + y_{1,k-1}^{(m)}\right), \\ x_{2,kc}^{(m)} &= \frac{1}{2}\left(x_{2,k}^{(m)} + x_{2,k-1}^{(m)}, y_{2,kc}^{(m)}\right) = \frac{1}{2}\left(y_{2,k}^{(m)} + y_{2,k-1}^{(m)}\right); \\ t_k^{(m)} &= t_0 + 2\pi k / (\Omega N),\, k = 1, 2, \cdots, mN,\, m = 1, 2, \cdots \end{aligned} \tag{1.6}$$

For a point $\mathbf{x}_k^{(m)} = (x_{1,k}^{(m)}, x_{2,k}^{(m)}, y_{1,k}^{(m)}, y_{2,k}^{(m)})^{\mathrm{T}}$, define a mapping $P_k^{(m)} : \mathbf{x}_{k-1}^{(m)} \to \mathbf{x}_k^{(m)}$ with $\mathbf{x}_k^{(m)} = P_k^{(m)} \mathbf{x}_{k-1}^{(m)}$. A relation between two adjoining node vector $\mathbf{x}_{k-1}^{(m)}$ and $\mathbf{x}_k^{(m)}$ is determined by Eq. (1.5) in the nonlinear rotor system.

The mappings for the period-m motions of the rotor system can be expressed as

$$\begin{aligned}
&P_1^{(m)} : \mathbf{x}_0^{(m)} \to \mathbf{x}_1^{(m)} && \Rightarrow \mathbf{x}_1^{(m)} = P_1^{(m)} \mathbf{x}_0^{(m)},\\
&P_2^{(m)} : \mathbf{x}_1^{(m)} \to \mathbf{x}_2^{(m)} && \Rightarrow \mathbf{x}_2^{(m)} = P_2^{(m)} \mathbf{x}_1^{(m)} = P_2^{(m)} \circ P_1^{(m)} \mathbf{x}_0^{(m)},\\
&\vdots\\
&P_{mN}^{(m)} : \mathbf{x}_{mN-1}^{(m)} \to \mathbf{x}_{mN}^{(m)} && \Rightarrow \mathbf{x}_{mN}^{(m)} = P_{mN}^{(m)} \mathbf{x}_{mN-1}^{(m)} = P_{mN}^{(m)} \circ \cdots \circ P_2^{(m)} \circ P_1^{(m)} \mathbf{x}_0^{(m)}
\end{aligned} \tag{1.7}$$

where "$\circ$" represents the action between mappings. The corresponding algebraic equations of the general mapping $P_k^{(m)}$ can be given as

$$\begin{aligned}
&\mathbf{g}_k^{(m)}(\mathbf{x}_{k-1}, \mathbf{x}_k) = \mathbf{0}\\
&(k = 1, 2, \cdots, N,\ \ m = 1, 2, \cdots),
\end{aligned} \tag{1.8}$$

where $\mathbf{g}_k^{(m)} = (g_{1k}, g_{2k}, g_{3k}, g_{4k})^{\mathrm{T}}$ with

$$\begin{aligned}
g_{1k}^{(m)} &= x_{1,k}^{(m)} - (x_{1,k-1}^{(m)} + h x_{2,kc}^{(m)}),\\
g_{2k}^{(m)} &= x_{2,k}^{(m)} - x_{2,k-1}^{(m)} + h\{\mu_1 x_{2,kc}^{(m)} + \omega_1^2 x_{1,kc}^{(m)}\\
&\quad + \gamma[2x_{1,kc}^{(m)} y_{1,kc}^{(m)} + x_{1,kc}^{(m)} (y_{1,kc}^{(m)})^2 + (x_{1,kc}^{(m)})^3] - e\Omega^2 \cos\Omega(t_{k-1}^{(m)} + \tfrac{1}{2}h)\};\\
g_{3k}^{(m)} &= y_{1,k}^{(m)} - (y_{1,k-1}^{(m)} + h y_{2,kc}^{(m)}),\\
g_{4k}^{(m)} &= y_{2,k}^{(m)} - y_{2,k-1}^{(m)} + \{\mu_2 y_{2,kc}^{(m)} + \omega_2^2 y_{1,kc}^{(m)}\\
&\quad + \gamma[(x_{1,kc}^{(m)})^2 + 3(y_{1,kc}^{(m)})^2 + (x_{1,kc}^{(m)})^2 y_{1,kc}^{(l)} + (y_{1,kc}^{(m)})^3] - e\Omega^2 \sin\Omega(t_{k-1}^{(m)} + \tfrac{1}{2}h)\};
\end{aligned} \tag{1.9}$$

For period-m motions of the nonlinear rotor system, the periodicity condition is

$$\mathbf{x}_N^{(m)} = \mathbf{x}_0^{(m)} \quad \Rightarrow (x_{1,N}^{(m)}, x_{2,N}^{(m)}, y_{1,N}^{(m)}, y_{2,N}^{(m)})^{\mathrm{T}} = (x_{1,0}^{(m)}, x_{2,0}^{(m)}, y_{1,0}^{(m)}, y_{2,0}^{(m)})^{\mathrm{T}}. \tag{1.10}$$

There are $4(mN + 1)$ discretized algebraic equations in Eqs. (1.8)–(1.10) for period-m motions of the nonlinear rotor system. Solving these $4(mN + 1)$ discretized algebraic equations will give the periodic solutions of period-m motions. Once the periodic solutions are obtained, the stability and bifurcations of the period-m motions in the nonlinear rotor system can be carried out by eigenvalue analysis.

Consider a small perturbation $\Delta\mathbf{x}_k^{(m)}$ at $\mathbf{x}_k^{(m)*}$, that's $\mathbf{x}_k^{(m)} = \mathbf{x}_k^{(m)*} + \Delta\mathbf{x}_k^{(m)}$. The linearized equation of Eq. (1.8) can be given as

$$\left.\frac{\partial \mathbf{g}_k^{(m)}}{\partial \mathbf{x}_{k-1}^{(m)}}\right|_{\left(\mathbf{x}_{k-1}^{(m)*},\mathbf{x}_k^{(m)*}\right)} \Delta \mathbf{x}_{k-1}^{(m)} + \left.\frac{\partial \mathbf{g}_k^{(m)}}{\partial \mathbf{x}_k^{(m)}}\right|_{\left(\mathbf{x}_{k-1}^{(m)*},\mathbf{x}_k^{(m)*}\right)} \Delta \mathbf{x}_k^{(m)} = \mathbf{0} \tag{1.11}$$

Rearrange Eq. (1.11) as

$$\Delta \mathbf{x}_k^{(m)} = -\left[\frac{\partial \mathbf{g}_k^{(m)}}{\partial \mathbf{x}_k^{(m)}}\right]^{-1} \left[\frac{\partial \mathbf{g}_k^{(m)}}{\partial \mathbf{x}_{k-1}^{(m)}}\right]_{\left(\mathbf{x}_{k-1}^{(m)*},\mathbf{x}_{k-1}^{(m)*}\right)} \Delta \mathbf{x}_{k-1}^{(m)} \equiv DP_k^{(m)} \Delta \mathbf{x}_{k-1}^{(m)} \tag{1.12}$$

where DP_k is the Jacobian matrix of the mapping $P_k^{(m)}$ as

$$DP_k^{(m)} = \left[\frac{\partial \mathbf{x}_k^{(m)}}{\partial \mathbf{x}_{k-1}^{(m)}}\right]_{(\mathbf{x}_{k-1}^*,\mathbf{x}_k^*)}$$

$$\text{for } k = 1, 2, \cdots, mN, \quad m = 1, 2, \cdots \tag{1.13}$$

where the components are listed in Appendix. For period-m motions, the final variation $\Delta \mathbf{x}_{mN}$ can be obtained as

$$\Delta \mathbf{x}_{mN}^{(m)} = DP^{(m)} \Delta \mathbf{x}_0^{(m)} = \underbrace{DP_{mN}^{(m)} \cdot DP_{mN-1}^{(m)} \cdot \cdots \cdot DP_2^{(m)} \cdot DP_1^{(m)}}_{mN-\text{multiplication}} \Delta \mathbf{x}_0^{(m)} \tag{1.14}$$

where

$$DP^{(m)} = \left[\frac{\partial \mathbf{x}_{mN}^{(m)}}{\partial \mathbf{x}_0^{(m)}}\right]_{\left(\mathbf{x}_{mN}^{(m)*},\mathbf{x}_{mN-1}^{(m)*},\cdots,\mathbf{x}_0^{(m)*}\right)} = \prod_{k=mN}^{1} \left[\frac{\partial \mathbf{x}_k^{(m)}}{\partial \mathbf{x}_{k-1}^{(m)}}\right]_{\left(\mathbf{x}_{k-1}^{(m)*},\mathbf{x}_k^{(m)*}\right)}. \tag{1.15}$$

$DP^{(m)}$ is the resultant Jacobian matrix, corresponding to the mappings of the period-m motions of the nonlinear rotor system. The stability and bifurcations of the period-m motions are determined by the eigenvalues of such a Jacobian matrix

$$\left|DP^{(m)} - \lambda \mathbf{I}_{4\times4}\right| = 0 \tag{1.16}$$

The stability and bifurcation conditions for period-m motions of the nonlinear rotor system are defined as follows (Luo [15])

1. If all the magnitudes of all the eigenvalues are less than one (i.e., $|\lambda_i| < 1$, $i = 1$, 2, 3, 4), the period-m motion is stable.

2. If at least one magnitude of the eigenvalues is bigger than one (i.e., $|\lambda_i| > 1$, $i \in \{1,2,3,4\}$), the period-m motion is unstable.
3. If there are magnitudes of the eigenvalues equal to one (i.e., $|\lambda_i| = 1$, $i \in \{1,2,3,4\}$), the stable or unstable period-m motion will has the stability switching or bifurcation conditions with higher order singularity.

The bifurcation conditions are presented as follows.

4. If $\lambda_i = 1$ with $|\lambda_j| < 1$ ($i, j \in \{1,2,3,4\}$, $i \neq j$), a saddle-node bifurcation (SN) occurs.
5. If $\lambda_i = -1$ with $|\lambda_j| < 1$ ($i, j \in \{1,2,3,4\}$, $i \neq j$), a period-doubling bifurcation (PD) occurs.
6. If $|\lambda_{i,j}| = 1$ ($i, j \in \{1, 2, 3, 4\}$, $\lambda_i = \bar{\lambda}_j$), a Neimark bifurcation (NB) occurs.

1.3 Bifurcation Trees

In this section, the semi-analytical solutions of period-m motions varying with excitation frequency in the nonlinear rotor system will be presented through the periodic nodes $(x_{1,k}, x_{2,k}, y_{1,k}, y_{2,k})^{\mathrm{T}}$ with $\mathrm{mod}(k, mN) = 0$ ($k = 0, 1, \cdots, mN$; $m = 1, 2, 4$). Consider the parameters for a bifurcation scenario of the nonlinear rotor system as

$$\begin{aligned} &\mu_1 = \mu_2 = 0.1118, \omega_1^2 = 62.4883, \omega_2^2 = 185.4650, \\ &\gamma = 61.4883, e = -1.0202. \end{aligned} \tag{1.17}$$

In Fig. 1.1, solid and dashed curves represent stable and unstable periodic solutions, respectively. The acronyms “SN”, “NB” and “PD” represent the saddle-node, Neimark, and period-doubling bifurcations, respectively.

The bifurcation trees of period-1 to period-4 motions varying with excitation frequency are presented for $\Omega \in (3.6, 4.0)$ in Fig. 1.1. The periodic node displacement $x_{1,k}$ in the x-direction is presented in Fig. 1.1(a). The stable period-1 motion is for $\Omega \in (3.6, 3.8411)$. A period-doubling bifurcation occurs at $\Omega \approx 3.8411$. The period-1 motion is unstable for $\Omega \in (3.8411, 3.9211)$ and another period-doubling bifurcation occurs at $\Omega \approx 3.9211$. The two period-doubling bifurcations is a pair for a branch of period-2 motion. The period-1 motion is stable from $\Omega \approx 3.9211$ to 3.9503 and unstable from $\Omega \approx 3.9503$ to 3.8347. Another pair of period-doubling bifurcation is at $\Omega \approx 3.9503$ and $\Omega \approx 3.8347$ for the second branch of period-2 motion. The period-1 motion is stable for $\Omega \in (3.8347, 4.0)$. The first branch of the period-2 motion is stable from $\Omega \approx 3.8411$ to 3.7284. A period-doubling bifurcation of the period-2 motion happens at $\Omega \approx 3.7284$ for a period-4 motion. The period-2 motion is unstable between $\Omega \approx 3.7284$ and 3.7634. Two Neimark bifurcations happen at $\Omega \approx 3.7634$ and $\Omega \approx 3.9040$ where a qusai-periodic motion relative to period-2 motion occurs. The period-2 motion is stable for $\Omega \in (3.9040, 3.9211)$ and merges back to the stable period-1 motion. The second branch of period-2

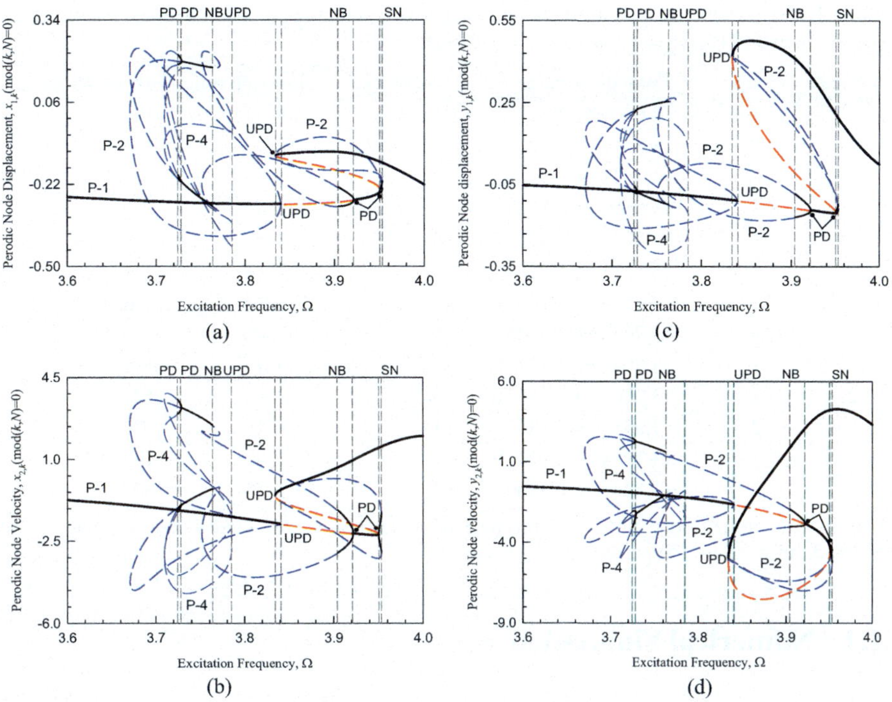

Fig. 1.1 Bifurcation trees for periodic motions varying with excitation frequency $\Omega \in (3.6, 4.0)$. (**a**) displacement x_1, (**b**) velocity x_2, (**c**) displacement y_1, (**d**) velocity y_2, ($\mu_1 = \mu_2 = 0.1118$, $\omega_1^2 = 62.4883, \omega_2^2 = 185.4650, \gamma = 61.4883, e = -1.0202$)

motion exists from $\Omega \approx 3.9503$ to 3.8347. There is a saddle-node bifurcation at $\Omega \approx 3.9532$, and the second period-2 motion is stable for $\Omega \in (3.9503, 3.9532)$ and unstable from $\Omega \approx 3.9532$ to 3.8347. The period-4 motion is stable for $\Omega \approx 3.7284$ to 3.7247, and unstable from $\Omega \approx 3.7247$ to 3.7850. A period-doubling bifurcation of period-4 motion occurs at $\Omega \approx 3.7247$ for period-8 motion. The detailed stability and bifurcations are presented in Table 1.1. The periodic node velocity $x_{2,k}$ in the x-direction is presented in Fig. 1.1(b). The same frequency ranges of the stable and unstable periodic motions are same as stated before, and bifurcation frequency points are the same. The periodic node displacement $y_{1,k}$ and velocity $y_{2,k}$ in the y-direction are presented in Fig. 1.1(c) and (d). Complicated routes from period-1 to period-4 motion are obtained. Due to the period-doubling bifurcations, the bifurcation trees of period-1 motion to chaos will continue to develop. Further, the route from period-1 motion to chaos can be obtained in the nonlinear rotor system.

Table 1.1 The stability and bifurcations of periodic motions in the nonlinear rotor system ($\mu_1 = \mu_2 = 0.1118, \omega_1^2 = 62.4883, \omega_2^2 = 185.4650, \gamma = 61.4883, e = -1.0202$)

Period	No.	Frequency range	Stability	Frequency	Bifurcations
P-1	1	(3.600 ➔ 3.8411)	S	3.8411	PD
	2	(3.8411 ➔ 3.9211)	U	3.9211	PD
	3	(3.9211 ➔ 3.9503)	S	3.9503	PD
	4	(3.9503 ➔ 3.8347)	U	3.8347	PD
	5	(3.8347 ➔ 4.000)	S	–	–
P-2	1	(3.8411 ➔ 3.7284)	U	3.7284	PD
	2	(3.7284 ➔ 3.7634)	S	3.7634	NB
	3	(3.7634 ➔ 3.9040)	U	3.9040	NB
	4	(3.9040 ➔ 3.9211)	S	3.9211	SN
P-2	1	(3.9503 ➔ 3.9532)	S	3.9532	SN
	2	(3.9532 ➔ 3.8347)	U	3.8347	SN
P-4	1	(3.7284 ➔ 3.7247)	S	3.7247	PD
	2	(3.7247 ➔ 3.7850)	U	3.7850	SN

Notice: *S* stable periodic motion; *U* unstable periodic motion. *SN* saddle node bifurcation; *NB* Neimark bifurcation; *PD* period doubling bifurcation

1.4 Numerical Simulation

In this section, numerical simulations of periodic motions in the nonlinear rotor system will be carried out for the stable period-1, 2 and 4 motions. A midpoint integration scheme is employed for numerical simulations. The initial conditions are from the semi-analytical solutions of periodic motions. The harmonic spectrums and harmonic phases are computed via discrete Fourier series. In all plots, numerical and analytical solutions are plotted by solid curves and circular symbols, respectively. The 'I.C.' is for initial conditions and the '1T', '2T' and '3T' represent points at the first, second and third periods of periodic motions.

In Fig. 1.2, the numerical simulation of a stable period-1 motion at $\Omega = 3.835$ in the nonlinear rotor system is presented. The initial condition is $\mathbf{x}_0 \approx (-0.1199, -0.4698, 0.4283, -4.7571)^{\mathrm{T}}$. The phase trajectories (x_1, x_2) and (y_1, y_2) in the x and y-directions are presented in Fig. 1.2(a) and (b). The slow-varying motion is observed in the x-direction compared to the y-direction. The displacement and velocity orbits are presented in Fig. 1.2(c) and (d). Such two plots can be experimentally observed. The harmonic amplitudes in both x- and y-directions for the stable period-1 motion are plotted in Fig. 1.2(e). For displacement x_1, the constant term is $A_{1,0} = -a_{1,0} \approx 0.017401$. The main harmonic amplitudes of displacement x_1 are $A_{1,1} \approx 0.438116$, $A_{1,2} \approx 0.472294$, $A_{1,3} \approx 0.150837$, $A_{1,4} \approx 0.025829$, $A_{1,5} \approx 0.015877$. The other harmonic amplitudes are $A_{1,j} \in (1 \times 10^{-3}, 1 \times 10^{-12})$ $(j = 6, 7, \cdots, 40)$ with $A_{1,40} = 2.44 \times 10^{-12}$. For displacement y_1, the constant term is $A_{2,0} = -a_{2,0} \approx 0.240339$. The main harmonic amplitudes of displacement y_1 are $A_{2,1} \approx 0.137948$, $A_{2,2} \approx 0.136004$, $A_{2,3} \approx 0.564790$, $A_{2,4} \approx 0.145320$, $A_{2,5} \approx 0.037449$, $A_{2,6} \approx 0.045360$, $A_{2,7} \approx 0.025960$, $A_{2,8} \approx 0.010614$. The other

Fig. 1.2 Numerical simulation for stable period-1 motion for $\Omega = 3.835$. (**a**) phase trajectory (x_1, x_2), (**b**) phase trajectory (y_1, y_2), (**c**) displacement orbit (x_1, y_1), (**d**) velocity plane (x_2, y_2), (**e**) Harmonic Amplitude $A_{i,j}$, (**f**) Harmonic phase $\varphi_{i,j}$ $(i = 1, 2; j = 1, 2, \cdots, 40)$. $(\mu_1 = 0.1118, \mu_2 = 0.1118, \omega_1^2 = 62.4883, \omega_2^2 = 185.4650, \gamma = 61.4883, e = -1.0202$

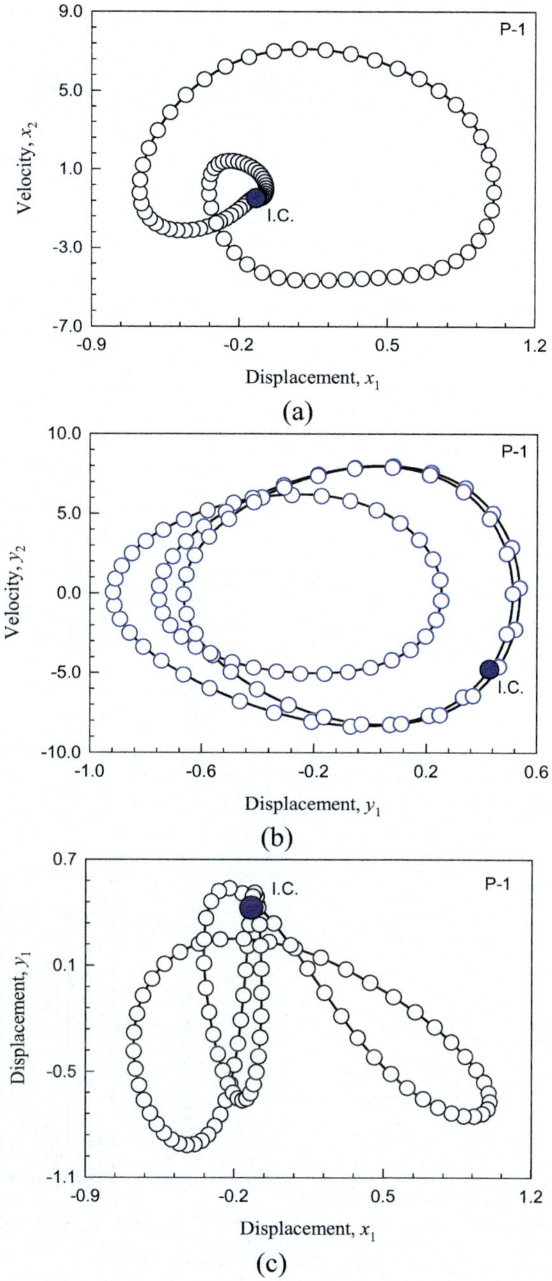

Fig. 1.2 (continued)

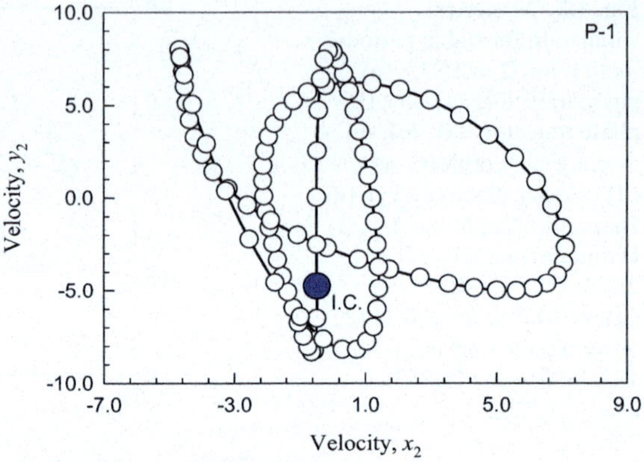

(d)

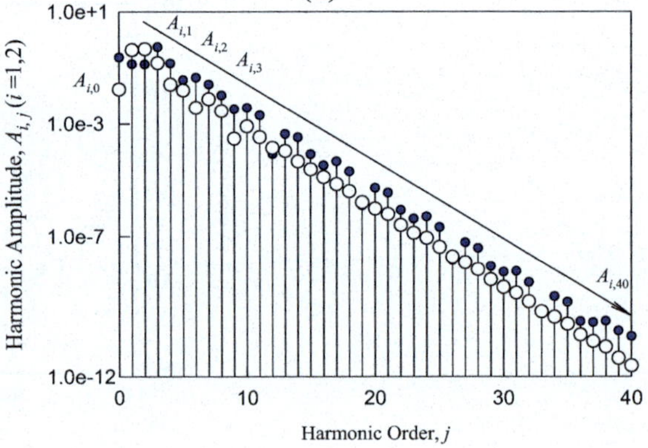

(e)

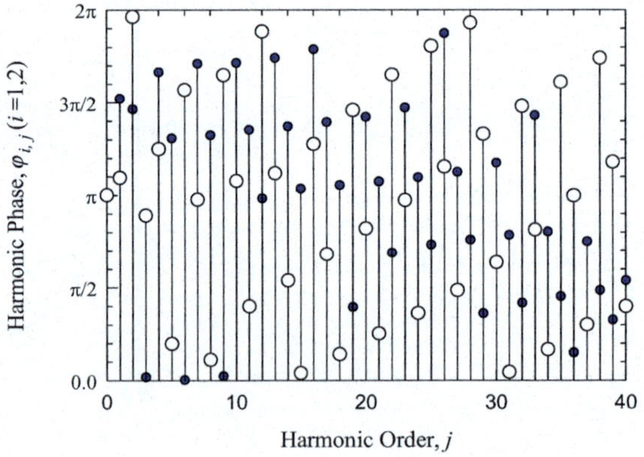

(f)

harmonic amplitudes are $A_{2,j} \in (10^{-3}, 10^{-11})$ $(j = 8, 10, \cdots, 40)$ with $A_{2,40} = 2.67 \times 10^{-11}$. The harmonic amplitude spectrum provides the quantity levels for each harmonic terms in the nonlinear rotor system. Most harmonic amplitudes in the x-direction are less than the y-direction. The harmonic phases for the displacements x_1 and y_1 are plotted in Fig. 1.2(f) for $\varphi_{i,j} \in (0, 2\pi)$ $(i = 1, 2; j = 1, 2, \cdots, 40)$. For such a stable period-1 motion, 40 harmonic terms are needed to keep the accuracy of 1.0×10^{-11} in both x- and y-directions.

In Fig. 1.3, the numerical simulation for a stable period-2 motion is presented for $\Omega = 3.73$ with the initial conditions of $\mathbf{x}_0 \approx (0.199031, -0.919596, -0.079434, 2.247463)^{\mathrm{T}}$. Two phase trajectories (x_1, x_2) and (y_1, y_2) are presented in Fig. 1.3(a) and (b). The trajectory in the y-direction is more complex than the x-direction. The displacement orbit and velocity plane are plotted in Fig. 1.3(c) and (d). The complex motion in the nonlinear rotor system is observed. The harmonic amplitudes in x- and y-directions for displacements are plotted in Fig. 1.3(e). For displacement x_1, the constant term is $A_{1,0/2} = -a_{1,0}^{(2)} \approx 5.962828\text{e-}3$. The main harmonic amplitudes of displacement x_1 for a period-2 motion are $A_{1,1/2} \approx 0.032936$, $A_{1,1} \approx 0.226585$, $A_{1,3/2} \approx 0.179999$, $A_{1,2} \approx 0.268119$, $A_{1,5/2} \approx 0.182072$, $A_{1,3} \approx 0.015802$, $A_{1,7/2} \approx 0.024006$, $A_{1,4} \approx 0.003398$, $A_{1,9/2} \approx 0.018599$, $A_{1,5} \approx 0.010368$, $A_{1,11/2} \approx 0.011399$. The other harmonic amplitudes are $A_{1,j/2} \in (10^{-3}, 10^{-11})$ $(j = 12, 13, \cdots, 64)$ with $A_{1,32} = 1.71 \times 10^{-11}$. For displacement y_1, the constant term is $A_{2,0/2} = -a_{2,0}^{(2)} \approx 0.082557$. The main harmonic amplitudes of displacement y_1 for a period-2 motion are $A_{2,1/2} \approx 0.035543$, $A_{2,1} \approx 0.085015$, $A_{2,3/2} \approx 0.022988$, $A_{2,2} \approx 0.017443$, $A_{2,5/2} \approx 0.045355$, $A_{2,3} \approx 0.111092$, $A_{2,7/2} \approx 0.276854$, $A_{2,4} \approx 0.036636$, $A_{2,9/2} \approx 0.055871$, $A_{2,5} \approx 0.012439$. The other harmonic amplitudes are $A_{1,j/2} \in (1 \times 10^{-3}, 1 \times 10^{-12})$ $(j = 11, 12, \cdots, 64)$ with $A_{1,32} = 7.72 \times 10^{-11}$. The corresponding harmonic phases are plotted in Fig. 1.3(f) for $\varphi_{i,j/2} \in (0, 2\pi)$ $(i = 1, 2; j = 1, 2, \cdots, 64)$. At least 64 harmonic terms are needed in the stable period-2 motion for keeping the accuracy of 1.0×10^{-11} in the x- and y-directions.

The numerical simulation for a stable period-4 motion is shown in Fig. 1.4 with the excitation frequency of $\Omega = 3.727$. The initial condition is $\mathbf{x}_0 \approx (0.191174, -1.099559, -0.081135, 2.399290)^{\mathrm{T}}$. The phase trajectories (x_1, x_2) and (y_1, y_2) are shown in Fig. 1.4(a) and (b). The period-doubling phenomenon can be observed by comparing to phase trajectories of period-2 motion. The period-4 motions become much complex. The displacement orbit (x_1, y_1) and velocity plane (y_1, y_2) are plotted in Fig. 1.4(c) and (d). The knotted movements become more chaotic. The harmonic amplitudes for displacements in the x- and y-directions are shown in Fig. 1.4(e). For displacement x_1 of the stable period-4 motion, the constant term is $A_{1,0/4} = -a_{1,0}^{(4)} \approx 6.06821 \times 10^{-3}$. The main harmonic amplitudes of displacement x_1 are $A_{1,1/4} \approx 0.000591$, $A_{1,1/2} \approx 0.032833$, $A_{1,3/4} \approx 0.001481$, $A_{1,1} \approx 0.225636$, $A_{1,5/4} \approx 0.004185$, $A_{1,3/2} \approx 0.180554$, $A_{1,7/4} \approx 0.010799$, $A_{1,2} \approx 0.268854$, $A_{1,9/4} \approx 0.018515$, $A_{1,5/2} \approx 0.185633$, $A_{1,11/4} \approx 0.004375$, $A_{1,3} \approx 0.015469$, $A_{1,13/4} \approx 0.001054$, $A_{1,7/2} \approx 0.024153$, $A_{1,15/4} \approx 0.000164$, $A_{1,4} \approx 0.003333$, $A_{1,17/4} \approx 0.000835$, $A_{1,9/2} \approx 0.018773$,

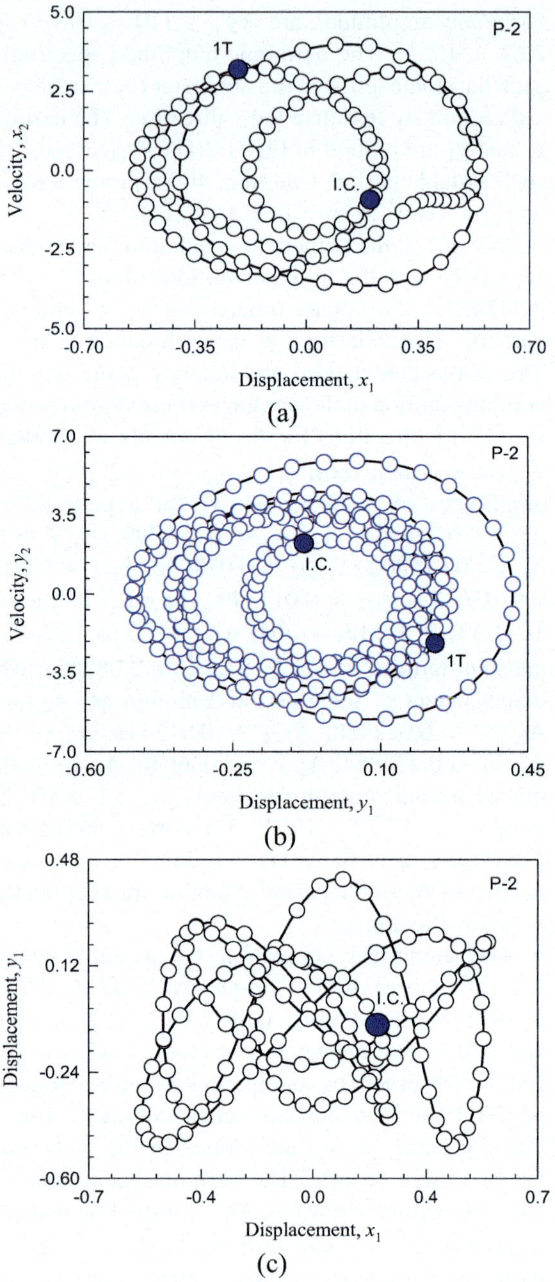

Fig. 1.3 Numerical simulation for stable period-2 motion for $\Omega = 3.73$. (**a**) phase trajectory (x_1, x_2), (**b**) phase trajectory (y_1, y_2), (**c**) displacement orbit (x_1, y_1), (**d**) velocity plane (x_2, y_2), (**e**) harmonic amplitude $A_{i,\,j/2}$, (**f**) harmonic phase $\varphi_{i,\,j/2}$ $(i = 1, 2; j = 1, 2, \cdots, 64)$. $(\mu_1 = 0.1118$, $\mu_2 = 0.1118$, $\omega_1^2 = 62.4883, \omega_2^2 = 185.4650, \gamma = 61.4883$, $e = -1.0202$

Fig. 1.3 (continued)

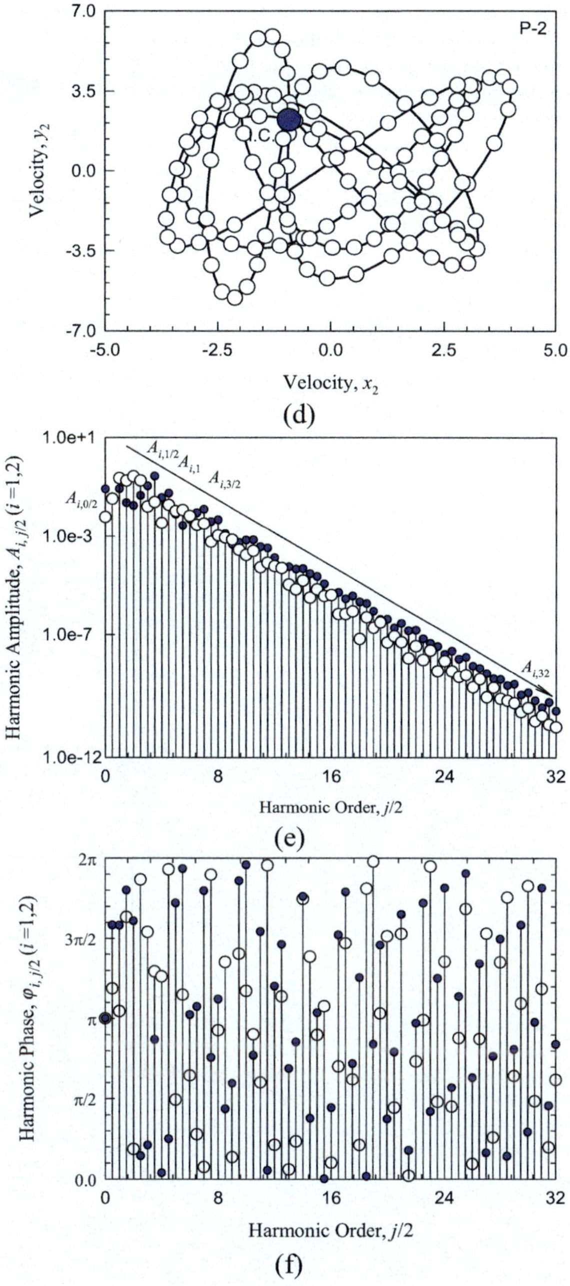

Fig. 1.4 Numerical simulation for stable period-4 motion for $\Omega = 3.727$. (**a**) phase trajectory (x_1, x_2), (**b**) phase trajectory (y_1, y_2), (**c**) displacement orbit (x_1, y_1), (**d**) velocity plane (x_2, y_2), (**e**) Harmonic Amplitude $A_{i,\,j/4}$, (**f**) Harmonic Phase $\varphi_{i,\,j/4}$ $(i = 1, 2; j = 1, 2, \cdots, 132)$. $(\mu_1 = 0.1118, \mu_2 = 0.1118, \omega_1^2 = 62.4883, \omega_2^2 = 185.4650, \gamma = 61.4883, e = -1.0202$

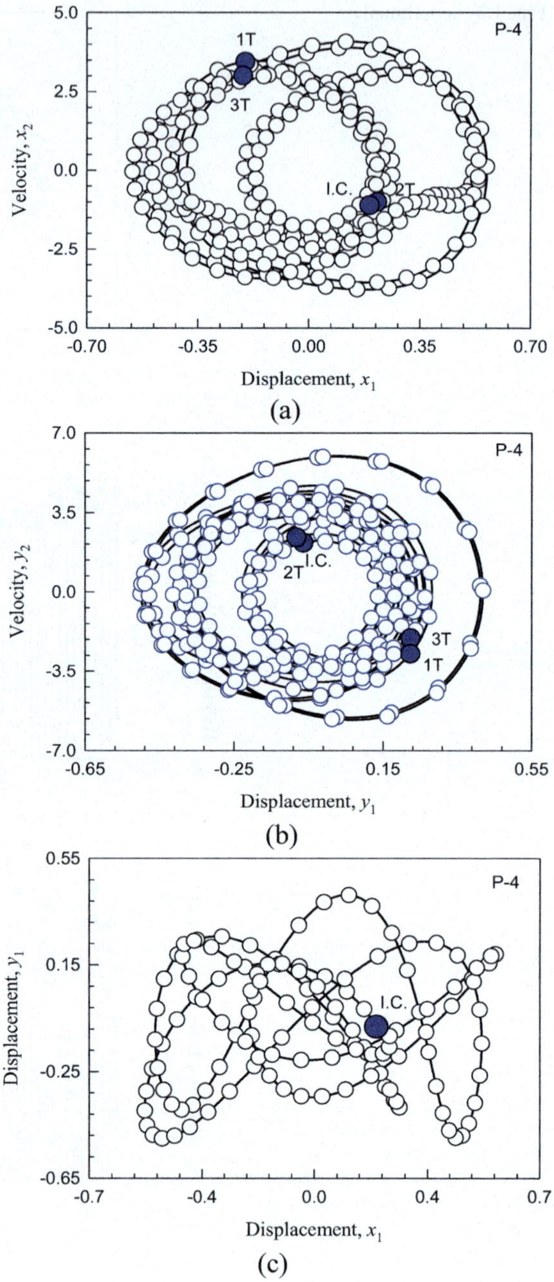

Fig. 1.4 (continued)

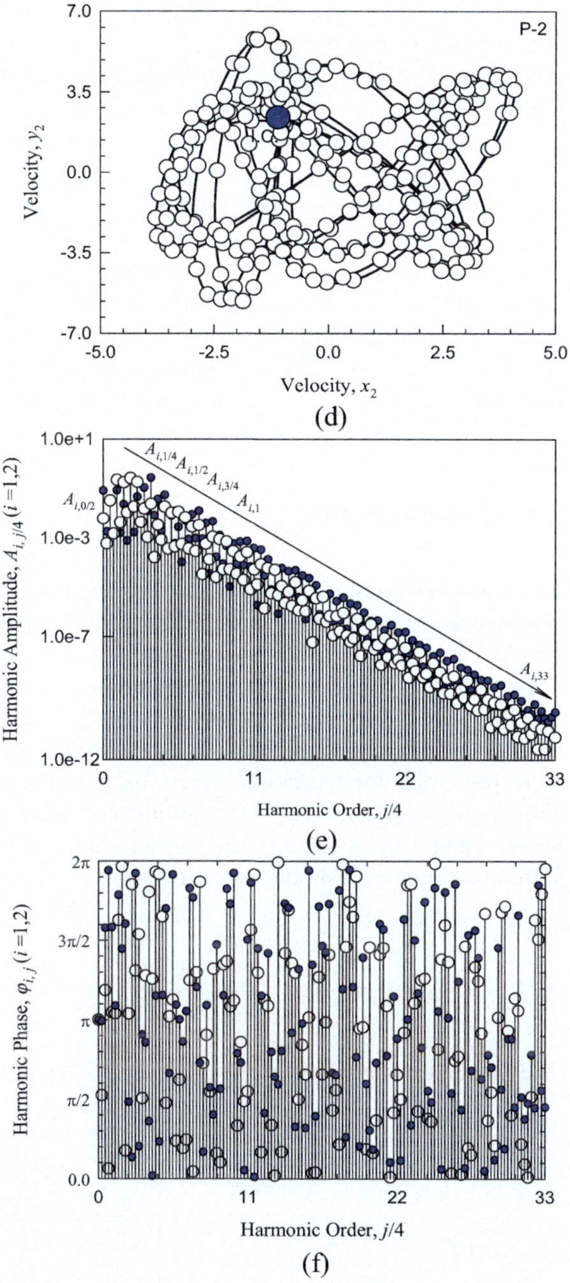

$A_{1,\,19/4} \approx 0.000362$, $A_{1,\,5} \approx 0.010428$, $A_{1,\,21/4} \approx 0.000444$, $A_{1,\,11/2} \approx 0.011466$. The other harmonic amplitudes are $A_{1,\,j/4} \in (1 \times 10^{-3}, 1 \times 10^{-12})$ $(j = 23, 24, \cdots, 132)$ with $A_{1,\,33} = 8.32 \times 10^{-12}$. For displacement y_1 of a stable period-4 motion, the constant term is $A_{2,0/4} = -a_{2,0}^{(4)} \approx 0.083820$. The main harmonic amplitudes are $A_{2,\,1/4} \approx 0.001799$, $A_{2,\,1/2} \approx 0.035631$, $A_{2,\,3/4} \approx 0.001431$, $A_{2,\,1} \approx 0.085958$, $A_{2,\,5/4} \approx 0.001555$, $A_{2,\,3/2} \approx 0.023505$, $A_{2,\,7/4} \approx 0.000775$, $A_{2,\,2} \approx 0.017614$, $A_{2,\,9/4} \approx 0.001666$, $A_{2,\,5/2} \approx 0.046873$, $A_{2,\,11/4} \approx 0.004952$, $A_{2,\,3} \approx 0.110777$, $A_{2,\,13/4} \approx 0.012642$, $A_{2,\,7/2} \approx 0.279169$, $A_{2,\,15/4} \approx 0.008325$, $A_{2,\,4} \approx 0.037357$, $A_{2,\,17/4} \approx 0.056246$, $A_{2,\,9/2} \approx 0.012677$. The other harmonic amplitudes are $A_{2,\,j/4} \in (1.0 \times 10^{-3}, 1.0 \times 10^{-11})$ $(j = 19, 20, \cdots, 132)$ with $A_{2,\,33} = 8.55 \times 10^{-11}$. The corresponding harmonic phases are shown in Fig. 1.4(f) for $\varphi_{i,\,j/4} \in (0, 2\pi)$ $(i = 1, 2; j = 1, 2, \cdots, 132)$. For keeping the accuracy of 1.0×10^{-11} for the stable period-4 motions, at least 132 harmonic terms should be considered.

1.5 Conclusions

The period-m motions (m = 1, 2, 4) in the nonlinear Jeffcott rotor system were obtained semi-analytically through the implicit mapping method. Two branches of period-1 to period-2 motions were presented. One branch of period-1 to period-4 motions was obtained. The corresponding stability and bifurcations of period-m motions were carried out by eigenvalues analysis. Numerical simulations of periodic motions were completed. Harmonic amplitudes and phases for period-m motions were presented for harmonic effects on periodic motions. Indeed, the nonlinear Jeffcott rotor system has very complicated dynamics, and the nonlinear modal shapes cannot be observed. The further studies on the dynamics of the nonlinear Jeffcott rotor system should be completed.

A.1 Appendix

The Jacobian matrix DP_k of the mapping $P_k^{(m)}$ is

$$DP_k^{(m)} = \left[\frac{\partial \mathbf{x}_k^{(m)}}{\partial \mathbf{x}_{k-1}^{(m)}}\right]_{(\mathbf{x}_{k-1}^*, \mathbf{x}_k^*)} = \begin{bmatrix} \frac{\partial x_{1,k}^{(m)}}{\partial x_{1,k-1}^{(m)}} & \frac{\partial x_{1,k}^{(m)}}{\partial x_{2,k-1}^{(m)}} & \frac{\partial x_{1,k}^{(m)}}{\partial y_{1,k-1}^{(m)}} & \frac{\partial x_{1,k}^{(m)}}{\partial y_{2,k-1}^{(m)}} \\ \frac{\partial x_{2,k}^{(m)}}{\partial x_{1,k-1}^{(m)}} & \frac{\partial x_{2,k}^{(m)}}{\partial x_{2,k-1}^{(m)}} & \frac{\partial x_{2,k}^{(m)}}{\partial y_{1,k-1}^{(m)}} & \frac{\partial x_{2,k}^{(m)}}{\partial y_{2,k-1}^{(m)}} \\ \frac{\partial y_{1,k}^{(m)}}{\partial x_{1,k-1}^{(m)}} & \frac{\partial y_{1,k}^{(m)}}{\partial x_{2,k-1}^{(m)}} & \frac{\partial y_{1,k}^{(m)}}{\partial y_{1,k-1}^{(m)}} & \frac{\partial y_{1,k}^{(m)}}{\partial y_{2,k-1}^{(m)}} \\ \frac{\partial y_{2,k}^{(m)}}{\partial x_{1,k-1}^{(m)}} & \frac{\partial y_{2,k}^{(m)}}{\partial x_{2,k-1}^{(m)}} & \frac{\partial y_{2,k}^{(m)}}{\partial y_{1,k-1}^{(m)}} & \frac{\partial y_{2,k}^{(m)}}{\partial y_{2,k-1}^{(m)}} \end{bmatrix}_{(\mathbf{x}_{k-1}^*, \mathbf{x}_k^*)}$$

$$\text{for } k = 1, 2, \cdots, mN, \quad m = 1, 2, \cdots \tag{1.18}$$

with

$$\begin{aligned}
\frac{\partial x_{1,k}}{\partial x_{1,k-1}} &= \Delta^{-1}\left(\Delta_{02}\Delta_{21} - \Delta_{11}^2\right), & \frac{\partial x_{1,k}}{\partial x_{2,k-1}} &= \Delta^{-1}\Delta_{02}\Delta_{12},\\
\frac{\partial x_{1,k}}{\partial y_{1,k-1}} &= \Delta^{-1}\left(\Delta_{02} - \Delta_{22}\right)\Delta_{11}, & \frac{\partial x_{1,k}}{\partial y_{2,k-1}} &= -\Delta_{11}\Delta_{12}\Delta^{-1},\\
\frac{\partial y_{1,k}}{\partial x_{1,k-1}} &= \Delta^{-1}\left(\Delta_{01} - \Delta_{21}\right)\Delta_{11}, & \frac{\partial y_{1,k}}{\partial x_{2,k-1}} &= -\Delta_{11}\Delta_{12}\Delta^{-1},\\
\frac{\partial y_{1,k}}{\partial y_{1,k-1}} &= \Delta^{-1}\left(\Delta_{01}\Delta_{22} - \Delta_{11}^2\right), & \frac{\partial y_{1,k}}{\partial y_{2,k-1}} &= \Delta^{-1}\Delta_{01}\Delta_{12};
\end{aligned} \tag{1.19}$$

$$\begin{aligned}
\frac{\partial x_{2,k}}{\partial x_{1,k-1}} &= \Delta^{-1}\left(\Delta_{01}\Delta_{02} + \Delta_{02}\Delta_{21} - 2\Delta_{11}^2\right)\Delta_{12},\\
\frac{\partial x_{2,k}}{\partial x_{2,k-1}} &= \Delta^{-1}\left(\Delta_{01}\Delta_{02} + \Delta_{02}\Delta_{12}^2 - \Delta_{11}^2\right),\\
\frac{\partial x_{2,k}}{\partial y_{1,k-1}} &= \Delta^{-1}\left(\Delta_{02} - \Delta_{22}\right)\Delta_{11}\Delta_{12}, \quad \frac{\partial x_{2,k}}{\partial y_{2,k-1}} = -\Delta_{11}\Delta_{12}^2\Delta^{-1},\\
\frac{\partial y_{2,k}}{\partial x_{1,k-1}} &= \Delta^{-1}\left(\Delta_{01} - \Delta_{21}\right)\Delta_{11}\Delta_{12}, \quad \frac{\partial y_{2,k}}{\partial x_{2,k-1}} = -\Delta_{11}\Delta_{12}^2\Delta^{-1},\\
\frac{\partial y_{2,k}}{\partial y_{1,k-1}} &= \Delta^{-1}\left(\Delta_{01}\Delta_{02} + \Delta_{01}\Delta_{22} - 2\Delta_{11}^2\right)\Delta_{12},\\
\frac{\partial y_{2,k}}{\partial y_{2,k-1}} &= \Delta^{-1}\left(\Delta_{01}\Delta_{02} + \Delta_{01}\Delta_{12}^2 - \Delta_{11}^2\right);
\end{aligned} \tag{1.20}$$

where

$$\begin{aligned}
&\Delta = \Delta_{11}^2 - \Delta_{01}\Delta_{02},\\
&\Delta_{01} = \gamma\left[y_{1,kc} + x_{1,kc}^2 + \tfrac{1}{2}\left(x_{1,kc}^2 + y_{1,kc}^2\right)\right] + \tfrac{1}{2}\omega_1^2 + \tfrac{1}{h^2}\left(2 + u_1 h\right),\\
&\Delta_{02} = \gamma\left[3y_{1,kc} + y_{1,kc}^2 + \tfrac{1}{2}\left(x_{1,kc}^2 + y_{1,kc}^2\right)\right] + \tfrac{1}{2}\omega_2^2 + \tfrac{1}{h^2}\left(2 + u_2 h\right),\\
&\Delta_{11} = \gamma x_{1,kc}\left(1 + y_{1,kc}\right), \quad \Delta_{12} = \tfrac{2}{h},\\
&\Delta_{21} = \gamma\left(y_{1,kc} + x_{1,kc}^2\right) + \tfrac{1}{2}\omega_1^2 - \tfrac{1}{h^2}\left(2 + u_1 h\right),\\
&\Delta_{22} = \gamma\left[3y_{1,kc} + y_{1,kc}^2 + \tfrac{1}{2}\left(x_{1,kc}^2 + y_{1,kc}^2\right)\right] + \tfrac{1}{2}\omega_2^2 - \tfrac{1}{h^2}\left(2 + u_2 h\right).
\end{aligned} \tag{1.21}$$

References

1. Jeffcott, H.H. 1919. The lateral vibration of loaded shafts in the neighbourhood of a whirling speed. The effect of want of balance. *The London, Edinburgh, and Dublin Philosophical Magazine and Journal of Science* 37 (219): 304–314.
2. Smith, D.M. 1933. The motion of a rotor carried by a flexible shaft in flexible bearings. *Proceedings of the Royal Society of London, Series A, Containing Papers of a Mathematical and Physical Character* 142 (846): 92–118.
3. Yamamoto, T. 1955. On the vibration of a shaft supported by bearing having radial clearance. *Transactions of the Japanese Society of Mechanical Engineering* 21 (103): 182–192.

4. Yamamoto, T. 1955. On the critical speed of a shaft of sub-harmonic oscillation, Transactions of the Japanese Society of. *Mechanical Engineering* 21 (111): 853–858.
5. Ehrich, F.F. 1964. Shaft whirl induced by rotor internal damping. *Journal of Applied Mechanics* 31 (2): 279–282.
6. Gunter, E.J. 1970. Influence of flexibly mounted rolling element bearings on rotor response, Part 1-Linear analysis. *Journal of Lubrication Technology* 92 (1): 59–69.
7. Barrett, L.E., E.J. Gunter, and P.E. Allaire. 1978. Optimum bearing and support damping for unbalance response and stability of rotating machinery. *ASME Journal of Engineering for Power* 100 (1): 89–94.
8. Childs, D.W. 1982. Fractional-frequency rotor motion due to nonsymmetric clearance effects. *ASME Journal of Engineering for Power* 104 (3): 533–541.
9. Saito, S. 1994. Nonlinear vibration of horizontal Jeffcott rotors supported by oil film damper bearings having no centering springs: calculation of nonlinear imbalance response and forced damped natural frequency. *JSME International Journal. Ser. C, Dynamics, Control, Robotics, Design and Manufacturing* 37 (1): 55–60.
10. Jiang, J., and H. Ulbrich. 2001. Stability analysis of sliding whirl in a nonlinear Jeffcott rotor with cross-coupling stiffness coefficients. *Nonlinear Dynamics* 24 (3): 269–283.
11. Luo, A.C.J. 2012. *Continuous dynamical systems*. Glen Carbon/Beijing: L & H Scientific and Higher Education Press Limited.
12. Huang, J., and A.C.J. Luo. 2014. Analytical periodic motions and bifurcations in a nonlinear rotor system. *International Journal of Dynamics and Control* 2 (3): 425–459.
13. ———. 2015. Periodic motions and bifurcation trees in a buckled, nonlinear Jeffcott rotor system. *International Journal of Bifurcation and Chaos* 25 (1): 1550002.
14. ———. 2015. Analytical solutions of period-1 motions in a buckled, nonlinear Jeffcott rotor system. *International Journal of Dynamics and Control* 4 (4): 376–383.
15. Luo, A.C.J. 2015. Periodic flows to chaos based on discrete implicit mappings of continuous nonlinear systems. *International Journal of Bifurcation and Chaos* 25 (3): 1550044.
16. Guo, Y., and A.C.J. Luo. 2016. Periodic motions to chaos in pendulum. *International Journal of Bifurcation and Chaos* 26 (9): 1650159.
17. Guo, Y., A.C.J. Luo, Z. Reyes, and A. Reyes. 2018. On experimental periodic motions in a Duffing oscillatory circuit. *Journal of Vibration Testing and Systems Dynamics* 3 (1): 55–70.
18. Xu, Y., and A.C.J. Luo. 2018. A series of symmetric period-1 motions to chaos in a two-degree-of-freedom van der Pol-Duffing oscillator. *Journal of Vibration Testing and System Dynamics* 2 (2): 119–153.
19. ———. 2018. Sequent period-(2m-1) motions to chaos in the van der Pol oscillator. *International Journal of Dynamics and Control*, 1–13. https://doi.org/10.1007/s40435-018-0468-1.
20. Xing, S., and A.C.J. Luo. 2020. On period-1 motions to chaos in a 1-dimensional, time-delay, nonlinear system. *International Journal of Dynamics and Control*. 8: 44–50
21. Xing, S., and A.C. Luo. 2019. Bifurcation trees of period-1 motions in a periodically excited, softening Duffing oscillator with time-delay. *International Journal of Dynamics and Control*. 7: 842–855

Chapter 2
Fault Diagnosis of Gear Rotor System Based on Collaborative Filtering Recommendation Method

Guangbin Wang, Qingkai Han, and Tengqiang Wang

2.1 Introduction

The gear rotor system is the core component of a large number of energy power equipment such as compressors, wind generators and other speed increase box devices. Normally, it runs under heavy loads, high speeds and harsh environments, and various failures are inevitable. Among them, the pitting of gears, cracks, fractures, and combined failures often occur, as shown in Fig. 2.1. If the gear failure is not diagnosed and maintained in time, it will cause excessive vibration and temperature rise of the parts, and in severe cases, the gear box and the whole machine will fail.

There are many reasons for the failure of gears, such as insufficient precision of the gear itself, errors in the assembly process, or different heat treatment processes and operating conditions, which can lead to failures. When a faulty gear is running, its vibration signal exhibits very strong nonlinear and non-stationary dynamic characteristics. In fact, once the gear is damaged, it must be stopped immediately. The result is that the fault state data is much smaller than the normal state data, which ultimately leads to a decrease in the accuracy of fault state recognition based on this unbalanced state data.

G. Wang
School of Mechanical Engineering, Lignnan Norman University, Zhanjiang, Guangdong, P. R. China

Q. Han (✉)
School of Mechanical Engineering and Automation, Northeastern University, Shenyang, Liaoning, P. R. China
e-mail: qhan@mail.neu.edu.cn

T. Wang
School of Mechanical engineering, Hunan University of Science and Technology, Xiangtan, P. R. China

J. Zhang (ed.), *Dynamics and Fault Diagnosis of Nonlinear Rotors and Impellers*, Nonlinear Systems and Complexity 34, https://doi.org/10.1007/978-3-030-94301-1_2

(a) Gear broken tooth (b) Pitting on the tooth surface

Fig. 2.1 Fault profile of gear broken teeth and pitting corrosion

Unbalanced data usually exhibits unbalanced categories, uneven sample distribution, and overlapping sample classes. Balancing the data has become a key link in training classification models and data mining. Data balancing generally uses oversampling or under-sampling techniques. Oversampling is to reconstruct a balanced data set by increasing the number of samples of the minority class, and retaining the original category information in the process of reconstructing the data. The main algorithms include SMOTE [1], LN-SMOTE [2], SMOTE-RSB algorithm [3] etc. The basic idea of Minority Oversampling Technology (SMOTE) is to artificially synthesize new samples based on the characteristics of minority samples, and add them to the original data set, that is, to achieve balance by inserting new samples into minority samples. Its essence is random oversampling. However, using this simple copy strategy to balance the data set will lead to overfitting of the model, which will make the information learned by the model not general enough, and sample aliasing will occur in the process of artificial synthesis, and eventually lead to a decrease in the accuracy of fault state recognition.

The collaborative filtering recommendation method [4–7] is an effective and intelligent method to find suitable conclusions in a large amount of complicated information. It can construct an evaluation model based on the user's interest in the information score, and predict the user's high-value preference for certain information. The advantage of collaborative filtering recommendation is that it does not require too much knowledge in a specific field, and a good recommendation effect can be obtained through statistical-based machine learning algorithms [8–10]. In recent years, scholars have gradually dealt with the information overload problem of monitoring data of mechanical equipment based on collaborative filtering recommendation technology. Xue Peng uses Pearson method and vector cosine method to calculate the degree of similarity of faults in collaborative filtering theory, and applies collaborative filtering theory to the field of civil aircraft fault diagnosis [11]. Guo Mingliang uses a collaborative filtering recommendation algorithm based on real-time operating status data to recommend possible faults in online EMUs [12]. In these two papers, the collaborative filtering recommendation algorithm is applied to mechanical fault diagnosis, which is essentially based on the method of sorting the similarity of the fault signals to recommend the fault type. The matrix factorization collaborative filtering algorithm based on data association features is widely used in web page information recommendation and interest product recommendation algorithms due to its fast convergence speed and high efficiency in filtering redundant information, but it cannot compare users' preferences for web information and interest products to construct a score. Matrix to construct a gear failure scoring matrix. Hence, a gear fault feature-state joint scoring matrix is innovatively designed, and a new collaborative filtering recommendation model is further constructed based on this, achieving good results in the fault diagnosis of the gear rotor system.

2.2 Adaptive Oversampling Technology and Multi-scale Entropy Analysis Method

2.2.1 Adaptive Oversampling Technology

Aiming at the problem that there are too many vibration signal samples in the normal state and few failure samples as the gear is running, an improved self-adaptive synthesis minority sample oversampling method is proposed to detect the pitting, pitting, and pitting of a small number of gears in the collected large amount of gear running data The broken tooth fault data is adaptively synthesized.

The specific algorithm flow is as follows,

1. Calculate the degree of imbalance: Assuming that in the gear state data, the amount of fault data is X_s and the amount of normal state data is X_m, then the degree of imbalance is

$$d = \frac{X_s}{X_m}, d \in (0, 1) \tag{2.1}$$

2. Calculate the number of minority samples to be synthesized: From the unbalance in step 1, the number of minority samples to be synthesized is further calculated according to Eq. (2.2),

$$G = (X_m - X_s) \times b, \quad b \in (0, 1) \tag{2.2}$$

   Among them, as b = 1, the number of synthesized new samples is exactly balanced with the number of samples in the majority class.
3. Find the minority sample X from the sample data, and use the Euclidean distance method to find the k nearest neighbors of the minority sample data set $X = \{X1, X2,..., Xn\}$, among these k nearest neighbors, Assuming that Δi is the number of

samples belonging to the majority class in k neighbors, the proportion of samples belonging to the majority class in k neighboring samples is

$$r = \frac{\Delta i}{k}, \quad r \in (0, 1) \tag{2.3}$$

4. For the r_i of each minority sample obtained in step three, normalize r_i as following,

$$r_i = r_i / \sum_{i=1}^{ms} r_i \tag{2.4}$$

5. For each minority gear pitting sample in the original gear experimental sample data, calculate the number of samples that need to synthesize new gear pitting faults according to Eq. (2.5),

$$g_i = r_i \times G \tag{2.5}$$

6. Among the k neighbors of each minority gear pitting failure sample x_i to be synthesized, select 1 shaft electric bearing pitting failure sample x_{zi} to synthesize according to Eq. (2.6).

$$s_i = x_i + (x_{zi} - x_i) \times \lambda \tag{2.6}$$

where $\lambda(0 < \lambda < 1)$ is the scale factor of the minority class of new samples to be synthesized, and the above steps are repeated for adaptive oversampling until the synthesis meets the number required in step 5.

2.2.2 *Multi-scale Permutation Entropy*

Compared with the permutation entropy [13, 14], the multi-scale permutation entropy [15] can be used to calculate and measure the complexity and uncertainty of the sequence from multiple scales, and can accurately reflect the small mutation behaviors of the vibration response of the dynamic system. The algorithm has very good stability and portability [16], and is suitable for the processing and analysis of mechanical equipment vibration response signals. Assuming that the time series $X = \{X(i), i = 1, 2, \ldots, n\}$, first, the multi-scale permutation entropy algorithm performs multi-scale coarse-graining of the time series,

$$y_j^{(x)} = \frac{1}{s}\sum_{i=j-1}^{js} x_i,\ j = 1, 2, \cdots, [N/s] \tag{2.7}$$

Here, s is the scale factor, and $[N/s]$ is the ratio rounding. Reconstruct the sequence $y_j^{(x)}$ as following,

$$Y_l^s = \left\{y_l^s, y_{l+1}^s, \cdots y_{l+(m-1)l}^s\right\} \tag{2.8}$$

Here, $l_1, l_2, ..., l_m$ are the column positions of the elements in the reconstruction component Y_l^s, and Y_l^s are sorted as,

$$y_{l+(l_1-1)t}^s \leq y_{l+(l_2-1)t}^s \leq \cdots \leq y_{l+(l_m-1)t}^s \tag{2.9}$$

A sequence of symbols

$$s(l) = (j_1, j_2 \cdots j_m) \tag{2.10}$$

Among them, $l = 1, 2, ...K$, $K \leq m!$, the permutation entropy of K different symbol sequences of X is

$$H_p(m) = \sum_{j=1}^{m!} P_j \ln P_j \tag{2.11}$$

Here, $0 \leq H_p(m) \leq In(m!)$, as $P_j = 1/m!$, $H_p(m) = -\sum_{j=1}^{m!} P_j In P_j$, there is a maximum value $In(m!)$, and $H_p(m!)$ is normalized, namely,

$$H_p = H_p(m)/\ln(m!) \tag{2.12}$$

H_P is the multi-scale permutation entropy of time series X.

2.2.3 *Wavelet Packet Energy Entropy*

The wavelet packet transform uses multi-scale operations to divide the frequency bands. It decomposes the high frequency bands more finely, and then adaptively selects the corresponding frequency bands for the decomposed signals, and matches them with the signal frequency bands. Compared with the wavelet transform, it has higher time-frequency analysis rate. Wavelet packet energy entropy is a method that combines information entropy and wavelet packet transform, and the concept of information entropy is added to the wavelet transform theory.

Suppose there are vibration signal data of u groups of gears, $S^{(1)},...,S^{(h)}$, $S^{(h+1)},...S^{u}$, and there are v different types of states for these gears, $z^{(1)}$, $z^{(2)},...,$ $z^{(v)}$. Knowing the existence status of the first h groups of training data $S^{(1)},...,$ $S^{(h)}$, the collaborative filtering recommendation system is now used to identify the existence status of the following u-h-1 groups of test data $S^{(h+1)},...,$ $S^{(u)}$.

The i-th group of signal data $S^{(i)}(i = 1,...,h,\ h + 1,...,u)$ is decomposed into a layer using wavelet packets, and its $b = 2^{a}-1$ sub-bands are obtained. The total signal $S^{(i)}$ can be expressed as follows

$$S^{(i)} = S_{a0}^{(i)} + S_{a1}^{(i)} + \cdots + S_{ab}^{(i)} \tag{2.13}$$

Suppose the energy corresponding to $S_{aw}^{(i)}$ $(w = 0, 1, \cdots, b)$ is $E_{aw}^{(i)}$, then

$$E_{aw}^{(i)} = \int \left| S_{aw}^{(i)}(t) \right|^{2} dt \tag{2.14}$$

The total energy of the signal is

$$E^{(i)} = \sum_{w=0}^{b} E_{aw}^{(I)} \tag{2.15}$$

The normalized eigenvector $T^{(i)}$ of constructing $S^{(i)}$ with energy as the element is as follows,

$$T^{(i)} = \left[e_{a0}^{(i)}, e_{a1}^{(i)}, \cdots, e_{ab}^{(i)} \right] \tag{2.16}$$

Among them $e_{aw}^{(i)} = E_{aw}^{(i)}/E^{(i)}$, and $e_{aw}^{(i)} \in (0, 1)$.

The T is the normalized wavelet packet energy entropy characteristic matrix.

2.3 Construction Method of Feature-State Joint Scoring Matrix of Gear Failure

The collaborative filtering recommendation system is built on the corresponding scoring matrix. For the identification of gear status, without a specific scoring rule, the corresponding scoring matrix cannot be established. Based on the multi-scale permutation entropy and wavelet packet energy entropy as eigenvalue elements, a gear fault feature matrix T is constructed. After normalizing the fault features, the feature score matrix $A \in R^{(b + 1) \times u}$ of Eq. (2.17) is obtained. The score can be very high Reflect the degree of "like", 1 means very like, 0 means very dislike. Among them, $e^1, e^2, \ldots, e^h$ represents the training data score value, and $e^{h+1}, e^{h+2}, \ldots, e^u$ represents the test data value.

$$A = \begin{pmatrix} e_{a0}^{(1)} & \cdots & e_{a0}^{(h)} & e_{a0}^{(h+1)} & \cdots & e_{a0}^{(u)} \\ e_{a1}^{(1)} & \cdots & e_{a1}^{(h)} & e_{a1}^{(h+1)} & \cdots & e_{a1}^{(u)} \\ \vdots & \ddots & \vdots & \vdots & \ddots & \vdots \\ e_{ab}^{(1)} & \cdots & e_{ab}^{(h)} & e_{ab}^{(h+1)} & \cdots & e_{ab}^{(u)} \end{pmatrix} \tag{2.17}$$

The gear state type corresponding to the training data is known and can be represented by a matrix of 1, while the gear state corresponding to the test data is an unknown value to be predicted, scores recorded as $P_t^{(i')}$. Normally, we set the initial score to zero, that is $P_t^{(i')} = 0$, and then gear status matrix $B \in R^{v \times u}$ is constructed by Eq. (2.18).

$$B = \begin{pmatrix} 1 & \cdots & \varepsilon & p_1^{(h+1)} & \cdots & p_1^{(u)} \\ \varepsilon & \cdots & \varepsilon & p_2^{(h+1)} & \cdots & p_2^{(u)} \\ \vdots & \ddots & \vdots & \vdots & \ddots & \vdots \\ \varepsilon & \cdots & 1 & p_v^{(h+1)} & \cdots & p_v^{(u)} \end{pmatrix} \tag{2.18}$$

Combine the feature score matrix A and the state score matrix B, the gear fault feature-state joint score matrix C is obtained as following,

$$C = \begin{pmatrix} A \\ B \end{pmatrix} = \begin{pmatrix} e_{a0}^{(1)} & \cdots & e_{a0}^{(h)} & e_{a0}^{(h+1)} & \cdots & e_{a0}^{(u)} \\ e_{a1}^{(1)} & \cdots & e_{a1}^{(h)} & e_{a1}^{(h+1)} & \cdots & e_{a1}^{(u)} \\ \vdots & \ddots & \vdots & \vdots & \ddots & \vdots \\ e_{ab}^{(1)} & \cdots & e_{ab}^{(h)} & e_{ab}^{(h+1)} & \cdots & e_{ab}^{(u)} \\ 1 & \cdots & \varepsilon & p_1^{(h+1)} & \cdots & p_1^{(u)} \\ \varepsilon & \cdots & \varepsilon & p_2^{(h+1)} & \cdots & p_2^{(u)} \\ \vdots & \ddots & \vdots & \vdots & \ddots & \vdots \\ \varepsilon & \cdots & 1 & p_v^{(h+1)} & \cdots & p_v^{(u)} \end{pmatrix} \tag{2.19}$$

Among them $C \in R^{d \times u}$, and $d = b + 1 + v$.

2.4 Recommended Diagnosis Method for Gear Fault Collaborative Filtering

The core idea of collaborative filtering is to find neighbor users similar to the target user in a large number of user groups by analyzing the user's existing behavior, and predict the target user's preferences based on the neighbor user's evaluation of a certain information. The goal of the fault recommendation system can be stated as: For the fault features that have not been scored in the state i, through the recommendation system, the prediction score of the fault feature i is given, and the corresponding recommendation is given to the feature i accordingly. The idea of the collaborative filtering algorithm based on matrix decomposition is to decompose the higher-dimensional score matrix into two lower-dimensional matrices. Let the high-dimensional matrix be R_{mn}, the two low-dimensional matrices are the state latent factor matrix P_{km} and the characteristic latent factor matrix Q_{kn}^T, where $k \leftrightarrow min\ (m, n)$ is the number of latent factor features. Specifically, as shown in Eq. (2.20),

$$R_{mn} \approx P_{km}{}^{*} Q_{kn}^{T} \tag{2.20}$$

The collaborative filtering algorithm based on matrix decomposition is used to predict the score of the user's item, based on the learned latent factor matrix and characteristic latent factor matrix, as following,

$$\hat{R} = P^{j*}\left(Q^{i}\right)^{T} \tag{2.21}$$

For the existing n score records, the error square is used to calculate the loss function of each score, as shown in Eq. (2.22)

$$L\,(P, Q, R) = \frac{1}{n}\sum\left(P^{j}, Q^{i}, R_{i}^{(j)}\right) = \frac{1}{n}\sum\left(R_{i}^{(j)} - R_{i}^{(j)}\right) \tag{2.22}$$

To prevent over-fitting, a regularization term is added to $L(P,Q,R)$, namely,

$$L\,(P, Q, R) = \arg\min\left(L\,(P, Q, R) + \lambda\left(\|P\|^{2}\right) + \left(\|Q\|^{2}\right)\right) \tag{2.23}$$

Among them, λ is the regularization coefficient, and further, the gradient descent method is used to deal with the above minimization problem. The core problem of the matrix factorization model is to minimize $L(P,Q,R)$, that is, to minimize the overall loss function of the above formula by finding suitable parameters P and Q.

For gear fault identification, our goal is to decompose the joint scoring matrix C of gear state identification into the product of two feature matrices $\Theta \in R^{u \times k}$ and $X \in R^{d \times k}$, namely $C = X\Theta^{T}$, as shown in Eq. (2.24). According to these two feature matrices, the predicted score $P_t^{(i')}$ of the test data $S^{(i')}$ on the state $Z^{(t)}$ is given.

$$\overset{C}{\begin{pmatrix} e_{a0}^{(1)} & \cdots & e_{a0}^{(h)} & e_{a0}^{(h+1)} & \cdots & e_{a0}^{(u)} \\ e_{a1}^{(1)} & \cdots & e_{a1}^{(h)} & e_{a1}^{(h+1)} & \cdots & e_{a1}^{(u)} \\ \vdots & \ddots & \vdots & \vdots & \ddots & \vdots \\ e_{ab}^{(1)} & \cdots & e_{ab}^{(h)} & e_{ab}^{(h+1)} & \cdots & e_{ab}^{(u)} \\ 1 & \cdots & \varepsilon & p_1^{(h+1)} & \cdots & p_1^{(u)} \\ \varepsilon & \cdots & \varepsilon & p_2^{(h+1)} & \cdots & p_2^{(u)} \\ \vdots & \ddots & \vdots & \vdots & \ddots & \vdots \\ \varepsilon & \cdots & 1 & p_v^{(h+1)} & \cdots & p_v^{(u)} \end{pmatrix}} \simeq \overset{X}{\begin{pmatrix} x_k^{(1)} & \cdots & x_k^{(1)} \\ \vdots & \ddots & \vdots \\ x_k^{(d)} & \cdots & x_k^{(d)} \end{pmatrix}} \cdot \overset{\Theta^T}{\begin{pmatrix} \theta_1^{(1)} & \cdots & \theta_k^{(u)} \\ \vdots & \ddots & \vdots \\ \theta_1^{(u)} & \cdots & \theta_k^{(u)} \end{pmatrix}}^T \tag{2.24}$$

The training of the matrix factorization model can actually be described by the following minimization problem: Find the optimal parameters Θ and X through Eq. (2.25) to minimize the overall loss function $L(\Theta, X, C)$.

$$(\Theta, X) = \arg\min L\left((\Theta, X, C) + \lambda\left(\|\Theta\|^2 + \|X\|^2\right)\right) \tag{2.25}$$

Among them, λ is the regularization coefficient.

Finally, the gradient descent method and the alternating least square method are used to optimize the parameters Θ and X, and then the prediction score $P_t^{(i')}$ of the test data $S^{(i')}$ to the state $Z^{(t)}$ is obtained as,

$$p_t^{(i')} = \theta^{(i')}\left(X^{(t)}\right)^T \tag{2.26}$$

Then the state $Z^{(t)}$ corresponding to the highest score $P_t^{(i')}$ is the state of the recognized test data $S^{(i')}$.

The gear collaborative filtering fault identification based on matrix decomposition has strong scalability, high calculation efficiency for massive gear monitoring data, and good processing ability for sparse joint scoring matrix.

2.5 Collaborative Filtering Recommended Gear Fault Diagnosis Model Fused with Multi-scale and Multi-domain Entropy

Multi-scale permutation entropy can effectively characterize the complexity of time series signals, and has a good characterization of the dynamic mutation and randomness of the sequence, but it is more difficult to identify different types of faults, especially compound faults, under a unified scale. The entropy value of the fault sequence has a certain difference, but there are also times when the entropy value also overlaps as time changes. Wavelet energy entropy has the ability to recognize the time-frequency distribution of non-stationary time-varying signals, and it is also difficult to distinguish early weak faults. After performing adaptive over-sampling and balancing of different types of sampled data, the time domain and time-frequency domain characteristics of the balanced data are merged to construct a gear fault feature-state joint scoring matrix, establish a collaborative filtering recommended fault diagnosis model, and optimize model parameters. Following this idea, the steps of the gear fault diagnosis algorithm fused with multi-scale and multi-domain entropy can be obtained as follows.

1. Sampling the vibration data of the gear's various fault states. Suppose the sampling time sequence is $X = \{X_1,X_2,...,X_s\}$, which is the type of fault state, then $X_s = (x_{s1},x_{s2},...,x_{si})$ is the data of each state, and i is the number of data. In engineering, the number of samples in the normal state is the largest, which is much larger than the data in the fault state, which is $sn_{normal} \approx sn_{fault}$.
2. Calculate the unbalanced d of different fault state data to normal state data and the number b of small fault data to be synthesized. Use the adaptive oversampling technique to obtain relatively balanced fault samples. Suppose it is n here, Then the status data of each type after balance is $X = \{X_1,X_2,...,X_s\}$, of which $X_s = (xi)$ and $i = 1,2,...,n$.
3. Find the normalized multi-scale permutation entropy $H_p = H_p(m)/In(m!)$ of the vibration signal of each fault state of the gear, and use the method in Sect. 2.2.3 to find the multi-scale wavelet energy entropy $T(i) = [e_{a0}{}^{(i)},e_{a1}{}^{(i)},...,e_{ab}{}^{(i)}]$ of the gear state signal.
4. Using the method in Sect. 2.3, find the feature joint scoring matrix C of gear fault, using 1 to represent the fault state and 0 to represent the non-fault state, and construct the fault state-feature scoring matrix R of the gear fault category.
5. Eigen value decomposition of matrix C, establish a collaborative filtering recommendation model, use gradient descent and alternating least squares to optimize parameters Θ and X, and determine the optimal recommendation model.
6. Based on the optimal recommendation model, learn the state data of the test gear to obtain the predicted score $P_t{}^{(i')}$ of the data to be tested for state $Z^{(t)}$, and the state $Z^{(t)}$ corresponding to the highest score $P_t{}^{(i')}$ is the state of the identified test data $S^{(i')}$.

2.6 Experiment and Analysis of Gear Fault Diagnosis Cases

2.6.1 Gear Fault Diagnosis Experiment

The gear failure experiment was carried out on a failure simulation test bench. Figure 2.2(a) shows the overall structure, and Fig. 2.2(b) shows the internal structure of the gearbox. The test rig is driven by a motor and is transmitted to the input shaft of the gearbox through couplings, transmission shafts, and pulleys. The gear on the input shaft drives the intermediate shaft to mesh with the gears, and the other gear on the intermediate shaft transmits power to the output shaft through the output shaft meshing gear.

Vibration acceleration sensors are used to collect vibration signals at multiple points on the gearbox. Figure 2.3 shows a sensor layout diagram, with a total of 2 in the horizontal direction and 6 in the vertical direction, a total of 8 vibration acceleration sensors. In the experiment, the current of the electromagnetic brake is set to 55 mA, that is, the braking torque is 1790 N.mm, the sampling frequency of the acquisition system is 16KHz, and the transmission ratio of the pulley is 3.56.

In the experiment, 4 gears with different degrees of pitting failure (Fig. 2.4) and 5 gears with different degrees of broken tooth failure (Fig. 2.5) were made on the meshing gears at the input shaft end of the gearbox. The vibration of the sensor is most obvious in the vertical direction of the input shaft, so take the three vibration acceleration signals at 1, 2, and 6 here as the analysis signal. The experiment simulates four working conditions of normal gears and pinion gears with different degrees of pitting failures, large gears with different degrees of broken teeth, and large gears with broken teeth and simultaneous pinion pitting.

In the input meshing gear, the number of teeth of the small gear is 36, and the number of teeth of the large gear is 101. Then the meshing frequency is 303.37 Hz at the motor speed of 30 Hz. Because the motor drives the input shaft to rotate through

Fig. 2.2 Gear failure simulation test bench (**a**) and internal structure diagram of the gearbox (**b**)

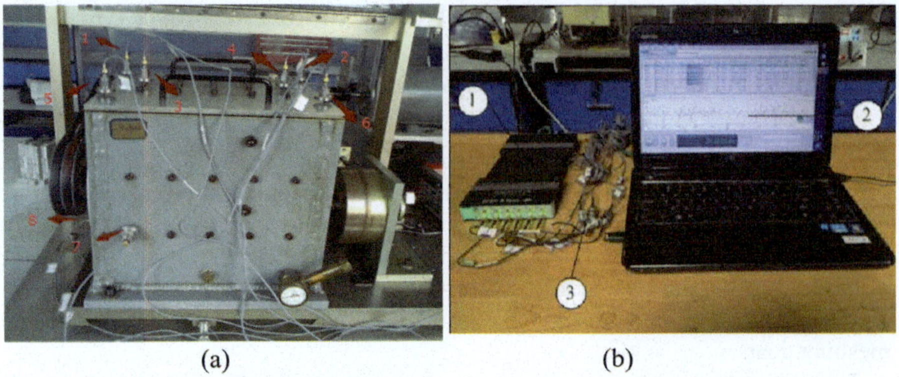

Fig. 2.3 Sensor measuring point layout (**a**) and BK's signal acquisition system (**b**)

Fig. 2.4 Four gears with different degrees of pitting corrosion failure (the tooth surface contains 1, 2, 3, 4 pits respectively)

Fig. 2.5 Five kinds of gears with different degrees of broken teeth (the width of the broken teeth is 1/5, 2/5, 3/5, 4/5, 1 of the tooth width, respectively)

the pulley, the transmission ratio has certain fluctuations, so the meshing frequency is also It fluctuates around 303 Hz. The time domain and frequency domain diagrams of the vibration signals in the four fault states are shown in Fig. 2.6. From the frequency spectrum, it can be seen that the characteristic frequency of the gear in the normal state is the meshing frequency and its multiplier. Pitting is similar to wear, and the characteristic frequency is expressed frequency and its division frequency. The frequency and its sub-multiplication, the characteristic frequency of the broken tooth fault is the meshing frequency and its multiplication, and sidebands

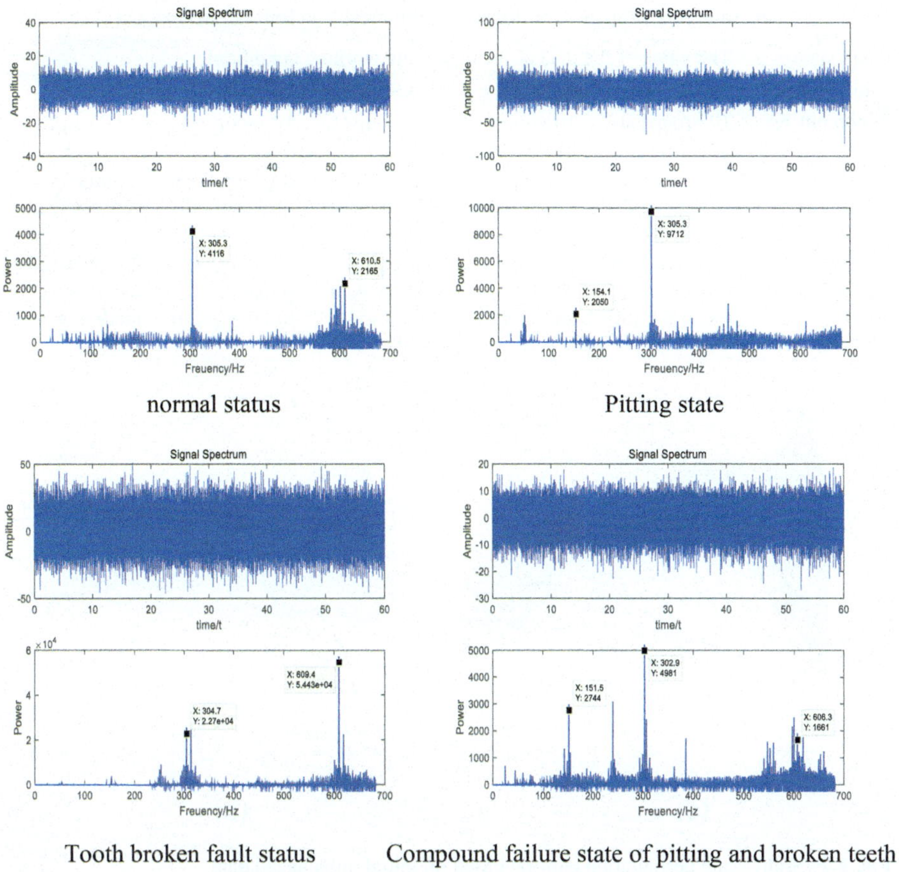

Fig. 2.6 Time domain and frequency domain diagrams of vibration signals in four fault states

appear near this frequency. Compound faults have both sub-multiplications and multiplications of the meshing frequency, as well as sidebands.

2.6.2 Algorithm Verification and Analysis

Using the data acquisition strategy listed in Table 2.1, 240 sets of data are obtained under normal operating conditions, of which 120 sets are training data, 60 sets are cross-validation data, and 60 sets are test data. There are 40 groups of each type of failure data, of which 20 groups are training data, 10 groups are cross-validation data, and 10 groups are test data. The sampling frequency is 16 kHz. At this time, the normal data is much larger than the failure data of each kind, so it belongs to the unbalanced distribution data.

Table 2.1 Data sampling strategies for different failure levels in four failure states

Fault state	The value of the data according to the working conditions				
Normal gear	240 groups of data(3Sensor signal×80 sampling periods)				
Pinon pitting	10 groups (1 pit)	10 groups (2 pits)	10 groups (3 pits)	10 groups (4 pits)	
Big gear broken tooth	8 groups (1/5 broken tooth)	8 groups (2/5 broken tooth)	8 groups (3/5 broken tooth)	8 groups (4/5 broken tooth)	8 groups (broken teeth)
Pinion pitting+broken tooth	8 groups (1/5 broken tooth + 1 pit)	8 groups (2/5 broken tooth 2 pits)	8 groups (3/5 broken tooth + 3 pits)	8 groups (4/5 broken tooth + pits)	8 groups (all the broken teeth + 4 pits)

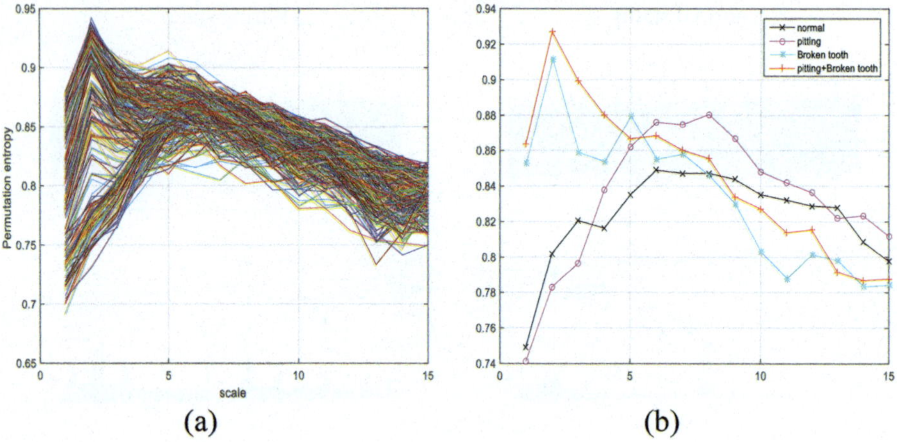

Fig. 2.7 Multi-scale permutation entropy based on initial unbalanced data

1. **Collaborative filtering recommendation diagnosis based on multi-scale permutation entropy of initial unbalanced data**

The multi-scale permutation entropy spectrum of the gear vibration signal is shown in Fig. 2.7, Fig. 2.7(a) is the multi-scale permutation entropy of all data, Fig. 2.7(b) is the multi-scale data of each fault in the Fig. 2.7(a). It can be seen from Fig. 2.7, as the scale increases, the permutation entropy is a process that first increases and then gradually decreases. Under a uniform scale, the permutation entropy of different types of faults is still different.

For the cross-validation set and test data set, the fault recognition accuracy rate is shown in Table 2.2. It can be seen that the overall recognition rate of collaborative filtering recommendation diagnosis for unbalanced data is not high, and the cross-validation set is less than 75%. Under the same parameters, the recognition rate of the test set data is lower, most of which are less than 70%. For this kind of fault, it is almost impossible to identify the compound fault of broken tooth and pitting.

Table 2.2 Correct fault recognition rate of multi-scale permutation entropy under different regular coefficients λ and the number of eigenvalues k

		Total recognition rate(%)	Total recognition rate(%)	Recognition rate of each fault state (%)			
Regular coefficient λ	Number of eigenvalues k	Cross-validation data set	Test data set	Normal	Pitting	Broken tooth	Compound fault
0.001	10	72.22	66.67	100	0	0	0
0.0015	11	73.33	70	95	60	0	0
0.002	8	73.33	66.67	99	100	0	0
0.0025	6	75.56	65.56	98.33	100	0	0
0.003	11	73.33	67.78	100	10	0	100
0.0035	13	67.78	66.89	100	20	0	0
0.004	6	74.44	68.89	98.33	30	0	0

Table 2.3 Correct fault recognition rate of wavelet packet energy entropy under different regular coefficients λ and the number of eigenvalues k

Regular coefficient λ	Number of eigenvalues k	Total recognition rate(%) Cross-validation data set	Total recognition rate(%) Test data set	Recognition rate of each fault state (%) Normal	Pitting	Broken tooth	Compound fault
0.001	10	74.44	70	100	30	0	0
0.0015	7	77.78	74.44	98.33	80	0	0
0.002	5	77.78	77.77	98.33	60	50	0
0.0025	6	71.11	67.78	95	40	0	0
0.003	5	72.22	60	90	0	0	0
0.0035	9	76.67%	66.67%	100	0	0	0
0.004	4	78.88	74.44	98.33	30	10	40

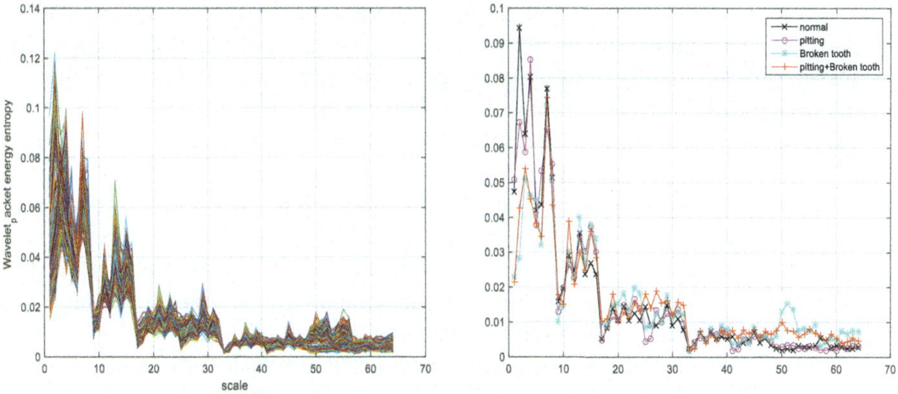

(a) **(b)**

Fig. 2.8 Wavelet packet energy entropy based on initial unbalanced data

2. **Collaborative filtering recommendation diagnosis based on wavelet packet energy entropy based on initial unbalanced data**

   The 5-layer wavelet packet energy entropy of the gear vibration signal is shown in Fig. 2.8. Figure 2.8(a) shows the wavelet packet energy entropy of 64 frequency bands of all data, Fig. 2.8(b) shows the value of each fault from Fig. 2.8(a). It can be seen from Fig. 2.8, as a whole, wavelet energy is mainly concentrated in the low frequency band, and the energy entropy of different faults cannot be distinguished.

   For the cross-validation set and the test data set, the fault recognition accuracy rate is shown in Table 2.3. It can be seen that the overall recognition rate of collaborative filtering recommendation diagnosis for unbalanced data is not high, about 75% for the cross-validation set. Under the same parameters, the recognition rate of the test set data is lower, between 65% and 75%. For a single type of fault, the composite fault of broken tooth and pitting + broken tooth cannot be identified.

3. **Multi-domain mixed entropy based collaborative filtering recommendation diagnosis based on initial unbalanced data**

   Multi-scale permutation entropy is the expression of information entropy of time series, and wavelet packet energy entropy is time-frequency domain information. Multi-scale entropy and wavelet packet energy entropy are merged into feature sets to construct collaborative filtering features, and the scoring matrix is used for gear fault diagnosis experiments. The results are shown in Table 2.4.

   It can be seen from Table 2.5 that the fault recognition rate of the cross-validation set can reach more than 95%, and for the test data set, the fault recognition accuracy rate is also around 90%. It can well identify normal and pitting fault conditions, and has a good ability to identify broken tooth faults, and the recognition effect of pitting + broken tooth composite faults is poor.

From the perspective of collaborative filtering diagnosis of initial unbalanced data, the recognition effect of separate multi-scale permutation entropy and wavelet packet energy entropy is not good, but the fusion diagnosis of the two features greatly improves the effect, as shown in Fig. 2.9.

4. **Collaborative filtering recommendation diagnosis based on multi-scale permutation entropy of balanced data**

   Since there are 240 sets of normal state data and only 40 sets of each type of fault state data, the unbalance degree is 40/240 = 1/6. For each type of fault, adaptive over-sampling technology is adopted, taking b = 1, the number of over-sampling is 200 groups, after over-sampling, each type of fault state data is also 240 groups, which is completely balanced with the normal state.

   Calculating the permutation entropy under different scales, the result is similar to Fig. 2.7. As the scale increases, the permutation entropy also increases first and then gradually decreases. Under a uniform scale, the permutation entropy of different types of faults is still different. Its fault recognition accuracy rate is shown in Table 2.5. For the cross-validation set, the fault recognition rate is mostly above 90%, and the recognition rate of the test set data is mostly about 90%. As far as a single type of fault is concerned, most of the faults can be better identified.

5. **Collaborative filtering recommendation diagnosis based on wavelet packet energy entropy based on balanced data**

   Calculate the wavelet energy entropy in different frequency bands. The result is similar to Fig. 2.8. On the whole, the wavelet energy is mainly concentrated in the low frequency band, and the wavelet packet energy entropy of different types of faults cannot be distinguished. Its fault recognition accuracy rate is given in Table 2.6. For the cross-validation set, the fault recognition rate is mostly above 90%, and the recognition rate of the test set data is mostly about 90%. As far as a single type of fault is concerned, most of the faults can be better identified.

6. **Multi-domain mixed entropy based collaborative filtering recommendation diagnosis based on balanced data**

   Multi-scale entropy and wavelet packet energy entropy are combined to form a hybrid feature set, a collaborative filtering feature score matrix is constructed, and a gear fault diagnosis experiment is performed. The experimental results are given in Table 2.7. It can be seen that the fault recognition rate of the cross-validation set is above 98%, and the fault recognition accuracy rate of the test data set is also above 95%. The recognition rates of three single faults of normal, pitting, and broken teeth are all above 96%, and the recognition rates of pitting + broken tooth composite faults are all above 80%.

From the perspective of collaborative filtering diagnosis of balanced data, the recognition effect of single multi-scale permutation entropy and wavelet packet energy entropy has reached more than 90%, and the diagnosis effect of the fusion diagnosis of the two features has reached more than 95%. It can be seen that the effect of fault diagnosis is the best based on the balanced data set of multiple fault

Table 2.4 Correct recognition rate of faults of multi-domain mixed entropy under different regular coefficients λ and the number of eigenvalues k

Regular coefficient λ	Number of eigenvalues k	Total recognition rate(%)	Total recognition rate(%)	Recognition rate of each fault state (%)			
		Cross-validation data set	Test data set	normal	Pitting	Broken tooth	Compound fault
0.001	8	95.56	87.78	100	90	100	0
0.0015	6	97.77	88.89	100	100	0	100
0.002	9	96.67	94.44	100	100	50	100
0.0025	12	97.78	94.44	100	90	70	90
0.003	13	98.90	91.11	100	100	100	20
0.0035	12	95.56	94.44	100	100	100	50
0.004	13	93.33	93.33	100	100	100	40

Table 2.5 Correct recognition rate of faults of multi-scale permutation entropy under different regular coefficients λ and the number of eigenvalues k

		Total recognition rate(%)	Total recognition rate(%)	Recognition rate of each fault state (%)			
Regular coefficient λ	Number of eigenvalues *k*	Cross-validation data set	Test data set	normal	Pitting	Broken tooth	Compound fault
0.001	6	79.17	71.67	76.67	78.33	65	66.67
0.0015	6	91.67	88.33	91.67	78.33	90	93.33
0.002	8	95.41	70.42	83.33	93.33	68.33	36.67
0.0025	9	95.41	92.08	80	100	91.67	96.67
0.003	12	96.25	91.67	96.67	91.67	96.67	81.67
0.0035	9	93.33	85.42	93.33	100	98.33	50
0.004	12	95.83	95.42	100	95	90	96.67

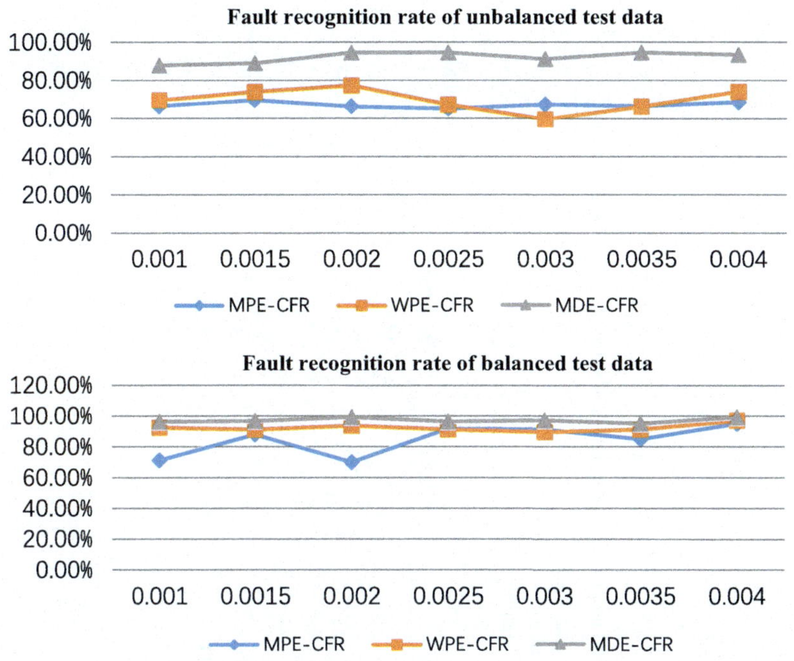

Fig. 2.9 The fault recognition rate of the three algorithms on the test data (λ is the abscissa)

types, and the multi-domain hybrid feature information that integrates multi-scale permutation entropy and wavelet packet energy entropy.

2.6.3 Discussion of Algorithm Effect

Three algorithms are proposed, namely, the collaborative filtering recommendation based on multi-scale permutation entropy (MPE-CFR) algorithm, the collaborative filtering recommendation based on wavelet packet energy entropy (WPE-CFR) algorithm, and the collaborative filtering recommendation based on multi-domain hybrid entropy (MDE-CFR) algorithm (Fig. 2.10).

2.7 Conclusion

The measured gearbox vibration signal has complex nonlinear and non-stationary characteristics, as pitting, cracking or fracture of the gear occurs in the gear rotor system. In the actual use of the gear, the fault data that can be collected is generally far less than the normal state data, leading to the gear state data to be unbalanced. In

Table 2.6 Correct recognition rate of faults under different regularity coefficient λ and number of eigenvalues k of wavelet packet energy entropy

Regular coefficient λ	Number of eigenvalues k	Total recognition rate(%)	Total recognition rate(%)	Recognition rate of each fault state (%)			
		Cross-validation data set	Test data set	normal	Pitting	Broken tooth	Compound fault
0.001	9	97.5	92.92	95	100	80	96.67
0.0015	12	96.67	91.67	95	91.67	96.67	83.33
0.002	10	95	94.17	86.67	100	93.33	96.67
0.0025	7	95.83	91.67	96.67	100	73.33	96.67
0.003	12	95	90	91.67	86.67	86.67	95
0.0035	8	95.41	91.67	71.67	100	95	100
0.004	8	96.25	97.5	98.33	100	98.33	93.33

Table 2.7 Correct recognition rate of faults under different regularity coefficients λ and the number of eigenvalues k of multi-domain mixed entropy

Regular coefficient λ	Number of eigenvalues k	Total recognition rate(%)	Total recognition rate(%)	Recognition rate of each fault state (%)			
		Cross-validation data set	Test data set	normal	Pitting	Broken tooth	Compound fault
0.001	11	99.58	96.17	100	100	100	88.33
0.0015	13	100	96.67	96.67	100	100	90
0.002	12	99.58	99.17	100	100	96.67	100
0.0025	11	100	96.25	100	100	100	85
0.003	9	98.75	97.08	98	100	96.67	93.33
0.0035	13	99.58	95	100	100	98.33	81.67
0.004	10	100	99.17	100	100	98.33	98.33

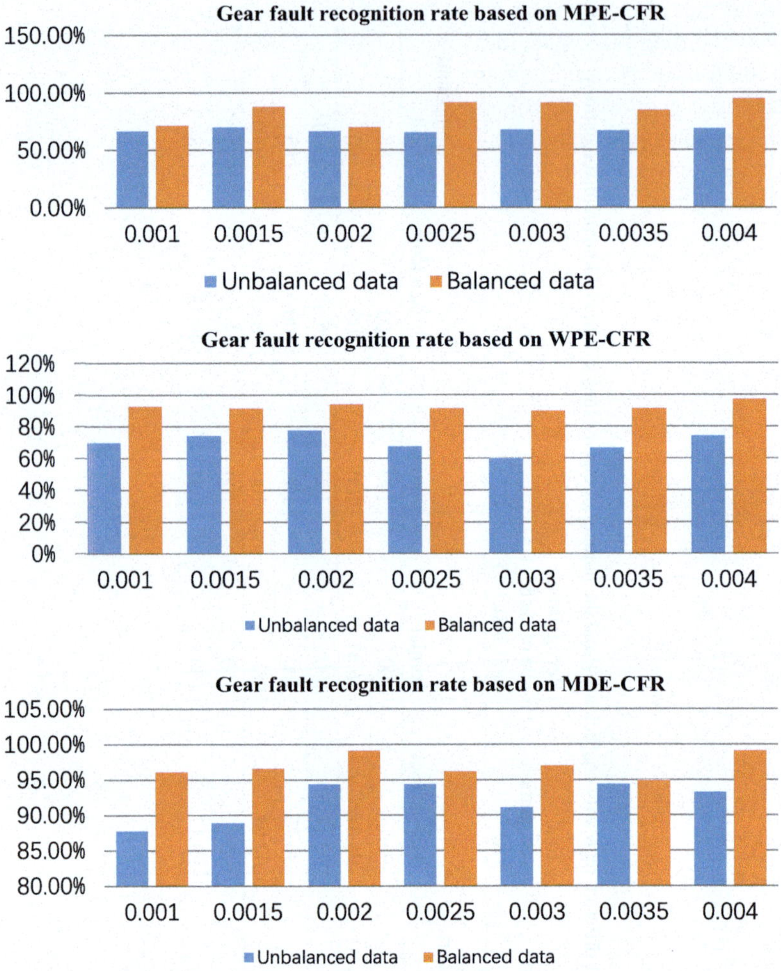

Fig. 2.10 Fault recognition rate of the three algorithms for balanced and unbalanced data sets (λ is the abscissa)

order to solve the problem of fault identification of the complex nonlinear vibration signal of the gear under the condition of data imbalance, in-depth theoretical analysis and experimental study are carried out, and the conclusions can be drawn as follows,

1. Faced with the problem of unbalanced gear multi-state data, an adaptive minority sample over-sampling data balancing method is proposed, which can determine the increased number of samples required to balance the data and adaptively create new data to solve gear failures The data is far less than the normal data, which causes the problem of reduced fault recognition effect; experiments have

proved that after adopting adaptive minority oversampling to balance the data types, the diagnostic accuracy of collaborative filtering recommendation based on multi-scale permutation entropy has increased from about 70% to 90% The diagnostic accuracy of collaborative filtering recommendation based on multi-layer wavelet energy entropy has increased from about 75% to more than 90%.

2. Both an innovative design of the gear fault feature-state joint scoring matrix and an innovative design of the gear fault feature-state joint scoring matrix are designed, in order to solve the problem that the collaborative filtering recommendation algorithm cannot construct a scoring matrix for judging users' preferences to specific information as used in fault diagnosis. The entropy values in the time domain and time-frequency domain are merged to construct a synergy Filter recommended gear fault diagnosis model, use gradient descent algorithm and alternating least square method to optimize model parameters, and propose a collaborative filtering recommended gear diagnosis method based on multi-domain hybrid entropy fusing multi-scale permutation entropy and wavelet packet energy entropy; the experiment proves the fusion Multi-scale permutation entropy and multi-layer wavelet energy entropy are used as feature quantities for gear fault diagnosis. The diagnosis and recognition rate for unbalanced data can reach more than 90%, and the diagnosis and recognition rate for balanced data can reach more than 95%.
3. Multi-domain hybrid entropy, which combines multi-scale permutation entropy and wavelet packet energy entropy as the characteristic value, has a strong fault recognition ability, which can reach more than 90%. Therefore, extracting multi-domain mixed entropy features and normalizing to construct a collaborative filtering score matrix for fault recommendation diagnosis is also an effective method to solve the problem of data imbalance classification.

A.1 Appendix 1

Gear Fault Diagnosis Flowchart based on mix of Multi-scale and Multi-domain Entrop

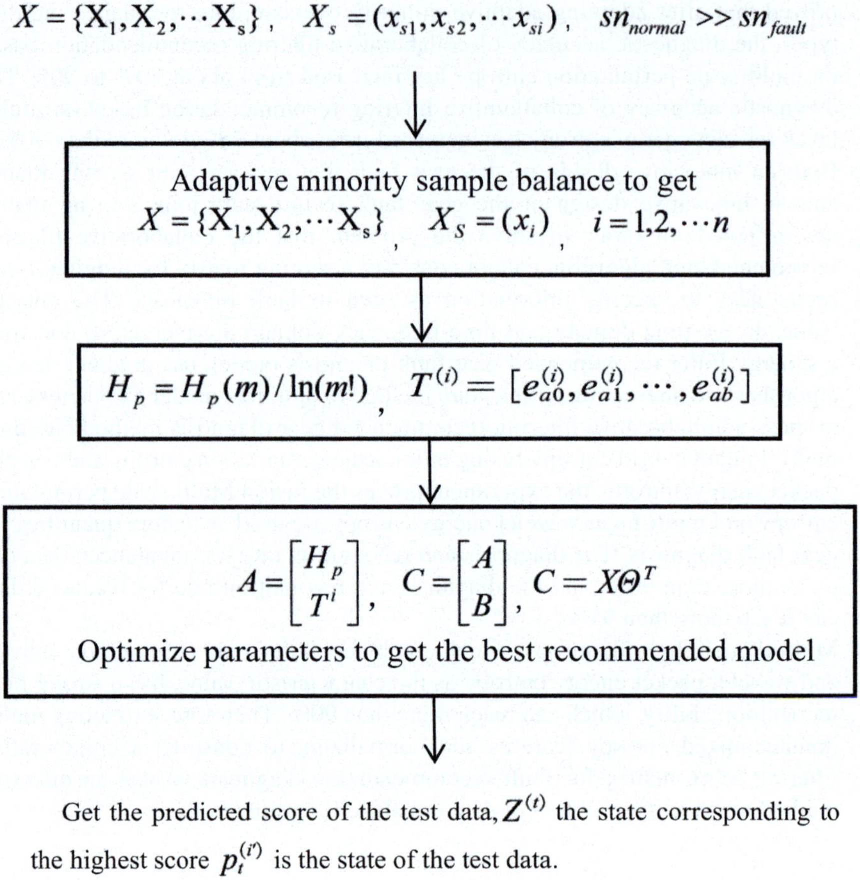
$X=\{X_1,X_2,\cdots X_s\}$, $X_s=(x_{s1},x_{s2},\cdots x_{si})$, $sn_{normal} >> sn_{fault}$
Adaptive minority sample balance to get
$X=\{X_1,X_2,\cdots X_s\}$, $X_S=(x_i)$, $i=1,2,\cdots n$
$H_p=H_p(m)/\ln(m!)$, $T^{(i)}=[e_{a0}^{(i)},e_{a1}^{(i)},\cdots,e_{ab}^{(i)}]$
$A=\begin{bmatrix}H_p\\T^i\end{bmatrix}$, $C=\begin{bmatrix}A\\B\end{bmatrix}$, $C=X\Theta^T$
Optimize parameters to get the best recommended model
Get the predicted score of the test data, $Z^{(t)}$ the state corresponding to the highest score $p_t^{(i')}$ is the state of the test data.

B.1 Appendix 2

Examples of main algorithm codes

```
%1.Gear Fault Diagnosis algorithm based on mix of Multi-scale and Multi-
domain Entropy
load train.mat  %Characteristics of training samples
load test.mat   %Characteristics of test-validation samples
Load canshu.mat  %Algorithm parameters
[m,n]=size(train);
[m_2,n_2]=size(test);
fault_begin=canshu(1)+1;  %canshu(1):feature number
fault_end=canshu(1)+1+canshu(2);   %canshu(2):Number of fault types
% Get R matrix for X_train
R=(train~=0);
cv_unknow=[cv(1:canshu(1),:);
test_unknow=[test(1:canshu(1),:);
%% Use X_cv to choose the best parameters
num_features=[6 7 8 9 10 11 12 13];
lambda=0.001.*ones(1,length(num_feature));
 train2 = [cv_unknow train];
 R2 = [(cv_unknow ~= 0) R];
 num_data = size(train2, 2);
 num_characteristic = size(train2, 1);
% Try every parameter
params=[num_features;lambda];
params_numbers=size(params,2);
Accuracy_matrix=zeros(params_numbers,1);
for i=1:params_numbers
    num_features=params(1,i);
    lambda=params(2,i);
    X = 0.1*randn(num_characteristic, num_features);
    Theta =0.1*randn(num_data, num_features);
    initial_parameters = [X(:); Theta(:)];
    options = optimset('GradObj', 'on', 'MaxIter',1500);
    theta = fmincg (@(t)(cofiCostFunc(t,train2, R2, num_data,
num_characteristic, num_features, lambda)), initial_parameters, options);
    % Unfold the returned theta back into U and W
    X = reshape(theta(1:num_characteristic*num_features), num_characteristic,
num_features);
    Theta = reshape(theta(num_characteristic*num_features+1:end), ...
            num_data, num_features);
    fprintf('Recommender system learning completed.\n');
```

```
      p = X * Theta';
      p1=p(:,1:n_2);)
    [~,I_1]=max(test(fault_begin:fault_end,:));
    [~,I_2]=max(p1(fault_begin:fault_end,:));
    a=I_1-I_2;
    Accuracy=((sum(a==0))/size(a,2))*100;
    Accuracy_matrix(i,1)=Accuracy
  end

  % 2.Oversampling of minority samples
  function [final_features final_mark] = SMOTE(original_features, origi-
nal_mark)
  ind = find(original_mark == 1);
  P = original_features(ind,:);
  T = P';
  X = T;
  I = nearestneighbour(T, X, 'NumberOfNeighbours', 3);
  I = I';
  [r c] = size(I);
  S = [];
  th=0.3;
  for i=1:r
     for j=2:c
        index = I(i,j);
          new_P=P(i,:)+((P(index,:)-P(i,:))*rand);
        S = [S;new_P];
     end
  end
  original_features = [original_features;S];
  [r c] = size(S);
  mark = ones(r,1);
  original_mark = [original_mark;mark];
  train_incl = ones(length(original_mark), 1);
  I = nearestneighbour(original_features', original_features', 'NumberOfNeigh-
bours', 3);
  I = I';
  for j = 1:length(original_mark)
     neighbors = I(j, 2:3);
    len = length(find(original_mark(neighbors) ~= original_mark(j,1)));
     if(len >= 2)
       if(original_mark(j,1) ==1)
          train_incl(neighbors(original_mark(neighbors) ~= original_mark(j,1)),1)
= 0;
```

```
else
train_incl(j,1) = 0;
end
end
end
final_features = original_features(train_incl == 1, :);
final_mark = original_mark(train_incl == 1, :);
 end
% 3.Multi-scale permutation entropy
 X=data;
 mt=15;%largest scale
 m=5;  %embedding dimension
 tau=1;%time delay
 mpe=zeros(1,mt);
 for i=1:mt
    M=fix(length(X)/i);
    y=mean(reshape(X(1:M*i),i,M));
   mpe(i)=pec(y,m,tau);  %i scale Permutation entropy
 end

function [pe ,hist] = pec(y,m,t)
ly = length(y);
permlist = perms(1:m);
[h,l]=size(permlist);  %m!  *m
c(1:length(permlist))=0;
 for j=1:ly-t*(m-1)
     [~,iv]=sort(y(j:t:j+t*(m-1)));
     for jj=1:h
         if (abs(permlist(jj,:)-iv))==0
             c(jj) = c(jj) + 1 ;
         end
     end
 end
hist = c;
c=c(find(c~=0));
p = c/sum(c);
pe = -sum(p.*log(p));
pe=pe/log(factorial(m));
 %4.Wavelet packet energy entropy
 function [X_E]=Wavelet_packet(x)
  [p,q]=size(x);
  windowSize = 5;
```

```
b = (1/windowSize)*ones(1,windowSize);
a = 1;
x1=filter(b,a,x);
n=6;
E=zeros(2^n,q);
for i=1:q
    x1(:,i)= x1(:,i)-mean(x1(:,i));
    x1(:,i)=detrend(x1(:,i));
    wpt1=wpdec(x1(:,i),n,'db1','shannon');
    for j=1:2^n
        E(j,i)=norm(wpcoef(wpt1,[n,j-1]),2);
    end
end
end
 [X_E] = Normalize_column(E)
```

References

1. Chawla N V, Bowyer K W, Hall L O, et al. SMOTE: synthetic minority over-sampling technique [J]. Journal of Artificial Intelligence Research, 2002, 16(1): 321–357.
2. Maciejewski, T., and J. Stefanowski. 2011. Local neighbourhood extension of SMOTE for mining imbalanced data [C]. In Computational Intelligence and Data Mining. IEEE, pp. 104–111.
3. Ramentol, E., Y. Caballero, R. Bello, et al. 2012. SMOTE-RSB: A hybrid pre-processing approach based on oversampling and under sampling for high imbalanced data sets using SMOTE and rough sets theory [J]. *Knowledge & Information Systems* 33 (2): 245–265.
4. Adomavicius, G., and A. Tuzhilin. 2005. Toward the next generation of recommender systems: a survey of the state-of-the-art and possible extensions [J]. *IEEE Transactions on Knowledge and Data Engineering.* 17 (6): 734–749.
5. Cai, Y., H.F. Leung, Q. Li, et al. 2014. Typicality based collaborative filtering recommendation [J]. *IEEE Transactions on Knowledge & Data Engineering* 26 (3): 766–779.
6. Adomavicius, G., and Y.O. Kwon. 2012. Improving aggregate recommendation diversity using Ranking based techniques [J]. *IEEE Transactions on Knowledge & Data Engineering* 24 (5): 896–911.
7. Sarwat, M., J.J. Levandoski, A. Eldawy, et al. 2014. LARS: An efficient and scalable location aware recommender system [J]. *IEEE Transactions on Knowledge & Data Engineering* 26 (6): 1–1.
8. Zenebe, A., and A.F. Norcio. 2009. Representation, similarity measures and aggregation methods using fuzzy sets for content-based recommender systems [J]. *Fuzzy Sets and Systems* 160 (1): 76–94.
9. Chen, H., L.I. Ren-Fa, Y.F. Liu, et al. 2010. Algorithms recommend research on personalized search engine [J]. *Application Research of Computers* 27 (1): 47–48.
10. Cui, Laizhong, Ou Peng, Fu Xianghua, Zhenkun Wen, and Lu. Nan. 2016. A novel multi objective evolutionary algorithm for recommendation systems [J]. *Journal of Parallel and Distributed Computing.* 10: 69–73.

11. Peng, Xue. 2014. Intelligent fault diagnosis method of civil aircraft based on collaborative filtering theory [J]. *Journal of Civil Aviation University of China* 32 (4): 23–26.
12. Mingliang, Guo. 2015. *Research and implementation of decision recommendation technology for EMU operation and maintenance based on knowledge base[D]*. Beijing: Beijing Jiaotong University.
13. Bandt, C., and B. Pompe. 2002. Permutation entropy: a natural complexity measure for time series [J]. *Physical Review Letters, The American Physiological Society* 88 (17): 174102. (1-4).
14. Yan, R.Q., Y.B. Liu, and R.X. Gao. 2012. Permutation entropy: A nonlinear statistical measure for status characterization of rotary machines [J]. *Mechanical Systems and Signal Processing* 29: 474–484.
15. Ersoy, Ö., and E.A. Öykum. 2019. Classification of hepatitis viruses from sequencing hromatograms using multiscale permutation entropy and support vector machines [J]. *Entropy* 21 (12): 1–17.
16. Dong, Z.L., J.D. Zheng, H.Y. Pan, et al. 2020. Rolling bearing fault diagnosis method based on compound multi-scale permutation entropy and FO-SVM [J]. *Noise and Vibration Control* 40 (2): 102–108.

Chapter 3
Study on Dynamic Behaviors of Rotor Model with Coupling Faults and Applications of TPOD Method

Kuan Lu, Yongfeng Yang, Jin Chen, Ruijuan Sang, and Yushu Chen

3.1 Introduction

The influences of the faults in the rotating machines on the associated dynamic behavior have been one focus of attention for many researchers. The presence of the faults may lead to a dangerous and catastrophic effects on the dynamic behaviors of rotating structures and lead to serious damage to the rotating machineries. In the normal case, the faults in the rotor system contain the crack fault [1–4], the rub-impact of rotor-to-stator fault [5–8], the pedestal looseness fault [9–12], the ball bearing fault [13–15], the misalignment fault [16–18], etc. In engineering, the coupling faults (the looseness and rub-impact, the crack and rub-impact, etc.) usually occur in the actual rotor systems. The amplitude varies when the cracked rotor is operating, which will lead to rub-impact with other units of the rotating machine. However, if the rub-impact fault occurs, the crack of shaft or impeller may occur on account of the harsh operating conditions and the huge impact. Few researchers studied the rotor systems with the coupling faults, since they are more complicated than the systems with only single fault. Hence, the dynamical behaviors of the rotor systems with the complex coupling faults should be investigated further.

The rotor systems with faults are high-dimensional and nonlinear, they are difficult to study the qualitative behaviors and the calculations are extremely

K. Lu
Institute of Vibration Engineering, Northwestern Polytechnical University, Xi'an, P. R. China

College of Mechanical Engineering, Beijing University of Technology, Beijing, P. R. China

Y. Yang · J. Chen · R. Sang
College of Mechanical Engineering, Beijing University of Technology, Beijing, P. R. China
e-mail: lukuan@nwpu.edu.cn

Y. Chen
School of Astronautics, Harbin Institute of Technology, Harbin, P. R. China

J. Zhang (ed.), *Dynamics and Fault Diagnosis of Nonlinear Rotors and Impellers*, Nonlinear Systems and Complexity 34, https://doi.org/10.1007/978-3-030-94301-1_3

expensive, so the low-dimensional models should be provided to represent the high-dimensional ones. The common order reduction methods include the center manifold method [19–21], the inertial manifold method [22–24], the Galerkin method [25–27], the Lyapunov-Schmidt (L-S) method [28, 29], the POD method [30–35], and other order reduction methods [36–38], were summarized by Rega and Steindl in their applied studies of nonlinear dynamics [39, 40].

POD method is an effective and powerful method for data analysis, aimed at obtaining low-order modes of the original system. The results show that POD method can not only reduce the DOFs of a system greatly but also improve the computational efficiency [41]. Many researchers applied POD method in the rotor systems, to study the nonlinear dynamical behaviors. Lu, etc. [42] proposed the TPOD method based on the inertial manifold method and applied it in the 23-DOFs rotor system, the efficiency of the order reduction method was verified by the comparison of the bifurcation diagram and the relative error. More, POD method was applied in the rotor-bearing system to discuss the singularity and the prime vibration of the reduce system [43]. Lu, etc. [44] applied the TPOD method to reduce a 7-DOFs rotor system supported by a pair of ball bearings with pedestal looseness to a 2-DOFs one. Further, the comparisons of topological structures in bifurcation verify the efficiency of the order reduction method, and the stability analysis of the reduced model is studied. The TPOD method was also applied in the rotor-bearing systems with looseness at one end and both ends, respectively. The POM energy method was used to confirm the dimension of the reduced systems, and the energy also gave an expression to the physical significance of the TPOD method [45].

The aim of this Chapter is to apply TPOD method for order reduction to the rotor-bearing model, so that to study the dynamical behaviors of the rotor model with the coupling faults. A 24-DOFs rotor model with both crack and rub-impacted faults is established, it is compared with the cracked and the rub-impact rotor model to demonstrate the complexity of the coupling fault. Then the POM energy method is used to confirm the dimension of the reduced model. Finally, the efficiency of TPOD method is presented by comparing with the bifurcation diagrams.

3.2 Rotor Model with the Coupling Faults

As shown in Fig. 3.1, the high-pressure rotor system supported by a pair of sliding bearings with both crack and rub-impact faults is modeled as a 24-DOFs system.

Assume that the axial and torsional vibrations of the system and the gyroscopic moment are neglectful. $O_i(i = 1...12)$ are the geometric centers of the discs. $m_i(i = 1...12)$ are the equivalent lumped masses. $c_i(i = 1...12)$ are the equivalent damping coefficients at the position of the lumped masses. $k_i(i = 1...11)$ are the equivalent stiffness of the corresponding discs. Further, assume that the crack occurs on the shaft between the first and second disc. The rotor model is supported by a pair of liquid-film bearings on both ends.

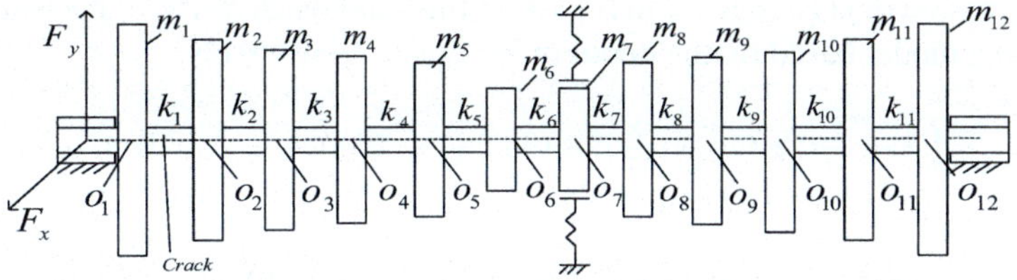

Fig. 3.1 Rotor model with coupling faults (crack and rub-impact)

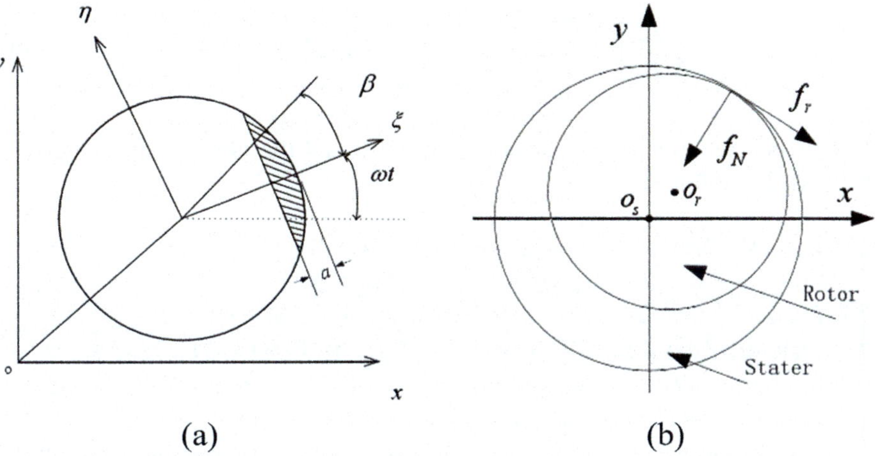

Fig. 3.2 Schematic diagram of the cracked section and rubbing model. (**a**) The cracked model. (**b**) The rub-impact model

The schematic diagram of the cracked section is shown in Fig. 3.2(a). α is the depth of crack, β the intersection angle between the unbalance and the crack normal vector. Law of open and close describes the effects of crack intersection angle to the shaft stiffness. As usual, the crack model contains three kinds of models: the first one is the cosine model; the second one is the square wave model and the third one is the model integrates cosine and square wave.

In Fig. 3.2(a), the initial intersection angle of the cracked normal vector and x axis is 0. We consider the case of gravity dominance, the open and close function of crack $f(\varphi)$ can be expressed as the function of rotating angle ωt. k is the cracked stiffness of rotating shaft. Δk_ξ and Δk_η are the stiffness variation of cracked normal and tangential vector, respectively. The stiffness of rotating shaft with crack is expressed as following,

$$\begin{bmatrix} k_x & k_{xy} \\ k_{yx} & k_y \end{bmatrix} = \begin{bmatrix} k & 0 \\ 0 & k \end{bmatrix} - f(\varphi) \begin{bmatrix} \Delta k_\xi \cos^2(\omega t) + \Delta k_\eta \sin^2(\omega t) & (\Delta k_\xi - \Delta k_\eta)\sin(\omega t)\cos(\omega t) \\ (\Delta k_\xi - \Delta k_\eta)\sin(\omega t)\cos(\omega t) & \Delta k_\xi \sin^2(\omega t) + \Delta k_\eta \cos^2(\omega t) \end{bmatrix} \tag{3.1}$$

The expression of $f(\varphi)$ is given in Eq. (3.2), this model can describe the square wave and cosine model based on the depth of the crack, respectively.

$$f(\varphi) = \begin{cases} \frac{1}{2} + \frac{2}{\pi}\cos(\varphi) - \frac{2}{3\pi}\cos(3\varphi) + \frac{2}{5\pi}(5\varphi) - \ldots & \mathrm{a/R} \le 0.5 \\ (1 + \cos\varphi)/2 & \mathrm{a/R} > 0.5 \end{cases} \tag{3.2}$$

The rub-impact model is shown in Fig. 3.2(b), it is combined with linear contact force and Coulomb friction, and the formula is expressed as Eq. (3.3)

$$\begin{bmatrix} P_x \\ P_y \end{bmatrix} = \begin{cases} 0 & \mathrm{r} \le \mathrm{r}_0 \\ k_b\left(1 - \frac{r_0}{r}\right)\begin{bmatrix} x - \mu_0 y \\ \mu_0 x + y \end{bmatrix} & \mathrm{r} > r_0 \end{cases} \tag{3.3}$$

The dynamical equation is shown in Eq. (3.4) and the dimensionless process is expressed as follows:

$$\begin{cases} m_1\ddot{X}_1 + c_1\dot{X}_1 + k_1(X_1 - X_2) - f(\varphi)\left(\Delta k_\xi \cos^2(\omega t) + \Delta k_\eta \sin^2(\omega t)\right)(X_1 - X_2) \\ \quad - f(\varphi)\left(\Delta k_\xi - \Delta k_\eta\right)\sin(\omega t)\cos(\omega t)(Y_1 - Y_2) = F_x\left(X_1, Y_1, \dot{X}_1, \dot{Y}_1\right) + m_1 E_1\omega^2\cos(\omega t + \beta_1) \\ m_1\ddot{Y}_1 + c_1\dot{Y}_1 + k_1(Y_1 - Y_2) - f(\varphi)\left(\Delta k_\xi - \Delta k_\eta\right)\sin(\omega t)\cos(\omega t)(X_1 - X_2) \\ \quad - f(\varphi)\left(\Delta k_\xi \sin^2(\omega t) + \Delta k_\eta \cos^2(\omega t)\right)(Y_1 - Y_2) = F_y\left(X_1, Y_1, \dot{X}_1, \dot{Y}_1\right) + m_1 E_1\omega^2\cos(\omega t + \beta_1) - m_1 g \\ m_2\ddot{X}_2 + c_2\dot{x}_2 + k_1(X_2 - X_1) + k_2(X_2 - X_3) - f(\varphi)\left(\Delta k_\xi \cos^2(\omega t) + \Delta k_\eta \sin^2(\omega t)\right)(X_2 - X_1) \\ \quad - f(\varphi)\left(\Delta k_\xi - \Delta k_\eta\right)\sin(\omega t)\cos(\omega t)(Y_2 - Y_1) = m_2 E_2\omega^2\cos(\omega t + \beta_2) \\ m_2\ddot{Y}_2 + c_2\dot{Y}_2 + k_1(Y_2 - Y_1) + k_2(Y_2 - Y_3) - f(\varphi)\left(\Delta k_\xi - \Delta k_\eta\right)\sin(\omega t)\cos(\omega t)(X_2 - X_1) \\ \quad - f(\varphi)\left(\Delta k_\xi \sin^2(\omega t) + \Delta k_\eta \cos^2(\omega t)\right)(Y_2 - Y_1) = m_2 E_2\omega^2\sin(\omega t + \beta_2) - m_2 g \\ m_3\ddot{X}_3 + c_3\dot{x}_3 + k_2(X_3 - X_2) + k_3(X_3 - X_4) = m_3 E_3\omega^2\cos(\omega t + \beta_3) \\ m_3\ddot{Y}_3 + c_3\dot{Y}_3 + k_2(Y_3 - Y_2) + k_3(Y_3 - Y_4) = m_3 E_3\omega^2\sin(\omega t + \beta_3) - m_3 g \\ \vdots \\ m_6\ddot{X}_6 + c_6\dot{x}_6 + k_5(X_6 - X_5) + k_6(X_6 - X_7) = m_6 E_6\omega^2\cos(\omega t + \beta_6) \\ m_6\ddot{Y}_6 + c_6\dot{Y}_6 + k_5(Y_6 - Y_5) + k_6(Y_6 - Y_7) = m_6 E_6\omega^2\sin(\omega t + \beta_6) - m_6 g \\ m_7\ddot{X}_7 + c_7\dot{x}_7 + k_6(X_7 - X_6) + k_7(X_7 - X_8) = m_7 E_7\omega^2\cos(\omega t + \beta_7) + P_x \\ m_7\ddot{Y}_7 + c_7\dot{Y}_7 + k_6(Y_7 - Y_6) + k_7(Y_7 - Y_8) = m_7 E_7\omega^2\sin(\omega t + \beta_7) - m_7 g + P_y \\ m_8\ddot{X}_8 + c_8\dot{x}_8 + k_7(X_8 - X_7) + k_8(X_8 - X_9) = m_8 E_8\omega^2\cos(\omega t + \beta_8) \\ m_8\ddot{Y}_8 + c_8\dot{Y}_8 + k_7(Y_8 - Y_7) + k_8(Y_8 - Y_9) = m_8 E_8\omega^2\sin(\omega t + \beta_8) - m_8 g \\ \vdots \\ m_{11}\ddot{X}_{11} + c_{11}\dot{x}_{11} + k_{10}(X_{11} - X_{10}) + k_{11}(X_{11} - X_{12}) = m_{11} E_{11}\omega^2\cos(\omega t + \beta_{11}) \\ m_{11}\ddot{Y}_{11} + c_{11}\dot{Y}_{11} + k_{10}(Y_{11} - Y_{10}) + k_{11}(Y_{11} - Y_{12}) = m_{11} E_{11}\omega^2\sin(\omega t + \beta_{11}) - m_{11} g \\ m_{12}\ddot{X}_{12} + c_{12}\dot{x}_{12} + k_{11}(X_{12} - X_{11}) = F_x(x_{12}, y_{12}, \dot{x}_{12}, \dot{y}_{12}) + m_{12} E_{12}\omega^2\cos(\omega t + \beta_{12}) \\ m_{12}\ddot{Y}_{12} + c_{12}\dot{Y}_{12} + k_{11}(Y_{12} - Y_{11}) = F_y(x_{12}, y_{12}, \dot{x}_{12}, \dot{y}_{12}) + m_{12} E_{12}\omega^2\sin(\omega t + \beta_{12}) - m_{12} g \end{cases} \tag{3.4}$$

$$\tau = \omega t,\, x_i = \frac{X_i}{c},\, y_i = \frac{Y_i}{c},\, \dot{x}_i = \frac{dx_i}{d\tau},\, \dot{y}_i = \frac{dy_i}{d\tau},$$
$$\ddot{x}_i = \frac{d\dot{x}_i}{d\tau},\, \ddot{y}_i = \frac{d\dot{y}_i}{d\tau},\quad M_1 = \frac{m_1 c\omega^2}{sP},\, M_{12} = \frac{m_{12} c\omega^2}{sP}$$

The dimensionless governing equation is expressed as Eq. (3.5). f_x and f_y are the model of dimensionless nonlinear oil-film force, $f_x = \frac{F_x}{sP}$, $f_y = \frac{F_y}{sP}$. $G(x, y, \alpha) = \frac{2}{(1-x^2-y^2)^{1/2}}\left[\frac{\pi}{2} + \arctan\frac{y\cos\alpha - x\sin\alpha}{(1-x^2-y^2)^{1/2}}\right]$ and $V(x, y, \alpha) = \frac{2+(y\cos\alpha - x\sin\alpha)G(x,y,\alpha)}{1-x^2-y^2}$ are the $S(x, y, \alpha) = \frac{x\cos\alpha + y\sin\alpha}{1-(x\cos\alpha + y\sin\alpha)^2}$, α directional components of bearing nonlinear oil-film force. $\alpha = \arctan\frac{y+2\dot{x}}{x-2\dot{y}} - \frac{\pi}{2}\left(\frac{(y+2\dot{x})(x-2\dot{y})}{|y+2\dot{x}||x-2\dot{y}|}\right) - \frac{\pi}{2}\left(\frac{y+2\dot{x}}{|y+2\dot{x}|}\right)$ is the bearing clearance, $s = \frac{\mu\omega RL}{P}\left(\frac{R}{c}\right)^2\left(\frac{L}{2R}\right)^2$ is the Sommerfeld number. μ is the lubricating oil viscosity, L the bearing length, R the radius of the bearing, ω the external excitation, P the loading, and τ the dimensionless time. The detailed values of the corresponding parameters of the rotor system are listed as in the Appendix 1. The nonlinear oil-film force [46] of x and y directions can be found in Appendix 2. Also, the formulas of the corresponding parameters in Appendix 2 are given.

$$\begin{cases}
\ddot{x}_1 + \frac{c_1}{m_1\omega}\dot{x}_1 + \frac{k_1}{m_1\omega^2}(x_1 - x_2) - \frac{f(\varphi)(\Delta k_\xi\cos^2(\tau) + \Delta k_\eta\sin^2(\tau))}{m_1\omega^2}(x_1 - x_2) \\
\quad - \frac{f(\varphi)(\Delta k_\xi - \Delta k_\eta)\sin(\tau)\cos(\tau)}{m_1\omega^2}(y_1 - y_2) = \frac{1}{M_1} f_x(x_1, y_1, \dot{x}_1, \dot{y}_1) + \frac{E_1}{c}\cos(\tau + \beta_1) \\
\ddot{y}_1 + \frac{c_1}{m_1\omega}\dot{y}_1 + \frac{k_1}{m_1\omega^2}(y_1 - y_2) - \frac{f(\varphi)(\Delta k_\xi - \Delta k_\eta)\sin(\tau)\cos(\tau)}{m_1\omega^2}(x_1 - x_2) \\
\quad - \frac{f(\varphi)(\Delta k_\xi\sin^2(\tau) + \Delta k_\eta\cos^2(\tau))}{m_1\omega^2}(y_1 - y_2) = \frac{1}{M_1} f_y(x_1, y_1, \dot{x}_1, \dot{y}_1) + \frac{E_1}{c}\sin(\tau + \beta_1) - \frac{g}{\omega^2 c} \\
\ddot{x}_2 + \frac{c_2}{m_2\omega}\dot{x}_2 + \frac{k_1}{m_2\omega^2}(x_2 - x_1) + \frac{k_2}{m_2\omega^2}(x_2 - x_3) - \frac{f(\varphi)(\Delta k_\xi\cos^2(\tau) + \Delta k_\eta\sin^2(\tau))}{m_2\omega^2}(x_2 - x_1) \\
\quad - \frac{f(\varphi)(\Delta k_\xi - \Delta k_\eta)\sin(\tau)\cos(\tau)}{m_2\omega^2}(y_2 - y_1) = \frac{E_2}{c}\cos(\tau + \beta_2) \\
\ddot{y}_2 + \frac{c_2}{m_2\omega}\dot{y}_2 + \frac{k_1}{m_2\omega^2}(y_2 - y_1) + \frac{k_2}{m_2\omega^2}(y_2 - y_3) - \frac{f(\varphi)(\Delta k_\xi - \Delta k_\eta)\sin(\tau)\cos(\tau)}{m_2\omega^2}(x_2 - x_1) \\
\quad - \frac{f(\varphi)(\Delta k_\xi\sin^2(\tau) + \Delta k_\eta\cos^2(\tau))}{m_2\omega^2}(y_2 - y_1) = \frac{E_2}{c}\sin(\tau + \beta_2) - \frac{g}{\omega^2 c} \\
\vdots \\
\ddot{x}_7 + \frac{c_7}{m_7\omega}\dot{x}_7 + \frac{k_6}{m_7\omega^2}(x_7 - x_6) + \frac{k_7}{m_7\omega^2}(x_7 - x_8) = \frac{E_7}{c}\cos(\tau + \beta_7) + p_x \\
\ddot{y}_7 + \frac{c_7}{m_7\omega}\dot{y}_7 + \frac{k_6}{m_7\omega^2}(y_7 - y_6) + \frac{k_7}{m_7\omega^2}(y_7 - y_8) = \frac{E_7}{c}\sin(\tau + \beta_7) - \frac{g}{\omega^2 c} + p_y \\
\vdots \\
\vdots \\
\ddot{x}_{11} + \frac{c_{11}}{m_{11}\omega}\dot{x}_{11} + \frac{k_{10}}{m_{11}\omega^2}(x_{11} - x_{10}) + \frac{k_{11}}{m_{11}\omega^2}(x_{11} - x_{12}) = \frac{E_{11}}{c}\cos(\tau + \beta_{11}) \\
\ddot{y}_{11} + \frac{c_{11}}{m_{11}\omega}\dot{y}_{11} + \frac{k_{10}}{m_{11}\omega^2}(y_{11} - y_{10}) + \frac{k_{11}}{m_{11}\omega^2}(y_{11} - y_{12}) = \frac{E_{11}}{c}\sin(\tau + \beta_{11}) - \frac{g}{\omega^2 c} \\
\ddot{x}_{12} + \frac{c_{12}}{m_{12}\omega}\dot{x}_{12} + \frac{k_{11}}{m_{12}\omega^2}(x_{12} - x_{11}) = \frac{1}{M_{12}} f_x(x_{12}, y_{12}, \dot{x}_{12}, \dot{y}_{12}) + \frac{E_{12}}{c}\cos(\tau + \beta_{12}) \\
\ddot{y}_{12} + \frac{c_{12}}{m_{12}\omega}\dot{y}_{12} + \frac{k_{11}}{m_{12}\omega^2}(y_{12} - y_{11}) = \frac{1}{M_{12}} f_y(x_{12}, y_{12}, \dot{x}_{12}, \dot{y}_{12}) + \frac{E_{12}}{c}\sin(\tau + \beta_{12}) - \frac{g}{\omega^2 c}
\end{cases} \quad (3.5)$$

The dimensionless equation can be rewritten as Eq. (3.6), in which $\mathbf{z} = (x_1, y_1, \cdots, x_{12}, y_{12})^T$, the expression of $\bar{\mathbf{c}}$, $\bar{\mathbf{k}}$, $\bar{\mathbf{f}}$ are shown as above,

$$\ddot{\mathbf{z}} = -\bar{\mathbf{c}}\dot{\mathbf{z}} - \bar{\mathbf{k}}\mathbf{z} + \bar{\mathbf{f}} \quad (3.6)$$

To facilitate the theory analysis, we implement the Taylor series expansion of the oil-film force, thus α can be rewritten as,

$$\alpha = \arctan\frac{y+2\dot{x}}{x-2\dot{y}} - \frac{\pi}{2}\left(\frac{(y+2\dot{x})\,(x-2\dot{y})}{|y+2\dot{x}|\,|x-2\dot{y}|}\right) - \frac{\pi}{2}\left(\frac{y+2\dot{x}}{|y+2\dot{x}|}\right) \tag{3.7}$$

For the convenience of calculation, Eq. (3.6) is written as Eq. (3.8) briefly,

$$\ddot{\mathbf{Z}} = -\mathbf{C}\dot{\mathbf{Z}} - \mathbf{K}\mathbf{Z} + \mathbf{F} \tag{3.8}$$

In Eq. (3.8), **C** is the damping matrix, **K** the stiffness matrix, **F** the force vector, which includes oil-film and external excitation etc. $\mathbf{Z} = [z_1\ z_2 \ldots z_{24}]^T$ denotes $[x_1\ y_1 \ldots x_{12}\ y_{12}]^T$ in Eq. (3.6).

3.3 Discussions on Dynamical Behaviors

In this section, the nonlinear dynamical behaviors of the rotor, with the coupling faults by comparing the coupling faults with the single fault (crack or rub-impact), will be studied in detail. Then we will discuss the effects of the systematic parameters to the coupling fault. Here we define the initial values as $z_i = 0 (i = 1, 2, \ldots, 24)$, $\dot{z}_5 = 0.8$, $\dot{z}_6 = 0.5$, $\dot{z}_i = 0.001$ $(i = 1, 2, \ldots, 24, i \neq 5, 6)$, and the integral step is $\pi/256$.

3.3.1 Dynamical Behaviors

Figure 3.3 shows four amplitude-frequency curves of the rotor systems with different faults. That is, the original system without faults; the cracked fault; the rub-impact fault and the coupling fault. 1/2 sub-harmonic vibration occur in the cracked and coupling system at the frequency of $2\omega_c$, ω_c is the natural frequency. From Fig. 3.3, it can be seen that the amplitude of the coupling system is apparently higher than the system with only crack fault, which indicates that rub-impact will increase the response amplitude of the crack. The amplitude of the system with rub-impact bumps up at the frequency of 1000 rad/s (nearby $3\omega_c$), this result illustrates that rub-impact will aggravate the complex motion of the system.

Figure 3.4 shows the bifurcation diagrams of the rotor systems with different faults: (a) the oil-film fault; (b) the crack fault; (c) the rub-impact fault; (d) the coupling fault. From the results, it is clear the bifurcation behaviors vary with different rotor faults. The obvious period-doubling bifurcation occurs at the second order natural frequency of the cracked fault system. Also, the period-doubling bifurcation occurs in the coupling fault system, but the amplitude is much larger. As the rotating speed is 1000 rad/s (nearby $3\omega_c$), the amplitude of the rub-impact

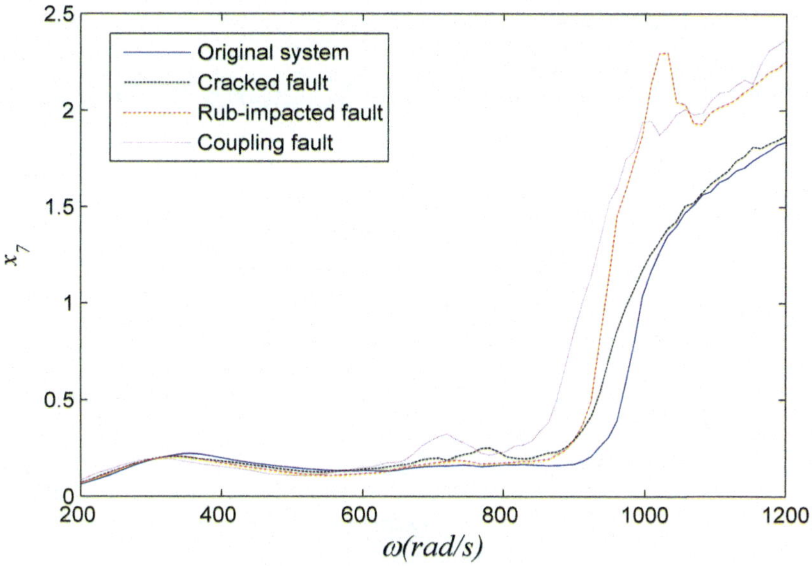

Fig. 3.3 Comparison of amplitude-frequency curves

fault system suddenly increases, meaning that the rub-impact fault can give rise to the complex motion of the system.

As $\omega = 350$ rad/s, near the primary resonance, the time histories of the four systems are given clearly in Fig. 3.5. The curves contain both the transient and steady-state processes. The transient process of the oil-filmed system (Fig. 3.5(a)) vanishes in about two periods, the other three systems (Fig. 3.5(b)–(d)) decayed after three periods. This indicates that the stability of the system with faults can't be kept, in comparison to the normal system. Meanwhile, in Fig. 3.5(d), the steady-state of the coupling fault is not standard sinusoidal, which has a visible difference in the other three systems.

Figure 3.6 shows the trajectories of the orbit of shaft center of the four systems. The motion curves in Fig. 3.6(a) and (c) are all ellipse, and the major axis and minor axis of ellipse of the oil-filmed system is longer than the rub-impact system. The trajectories of the orbit of shaft center of cracked (b) and coupling (d) faults are the closed curves similar to ellipse, but the bottom is a little narrow. The crack fault is main reason to change the ellipse of trajectories of the orbit of shaft center as closed curves similar to ellipse the near the primary reasonance.

The comparisions of the phase portraits (Fig. 3.7) are used to verify the results in Fig. 3.6 based on the same initial values and rotating speed. In Fig. 3.7(a) and (c), it is easy to see that both the phase portraits are closed and similar to circle, and

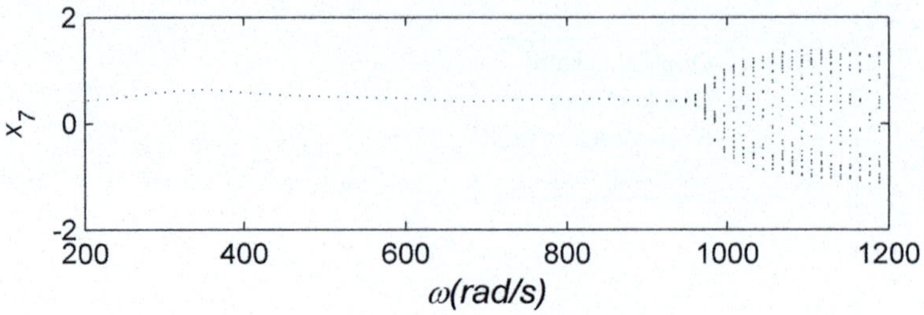

(a) The system without fault

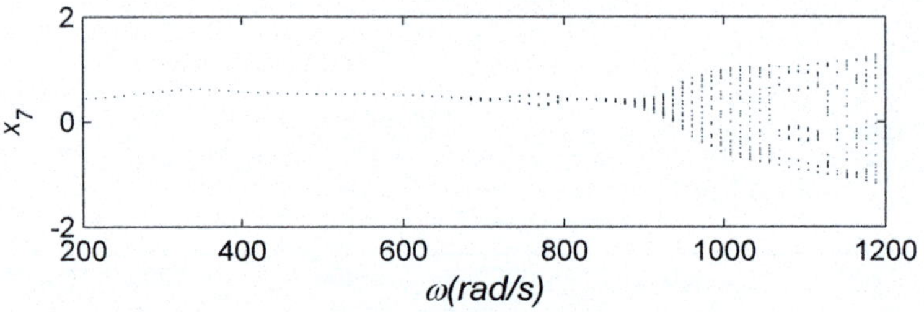

(b) The cracked system

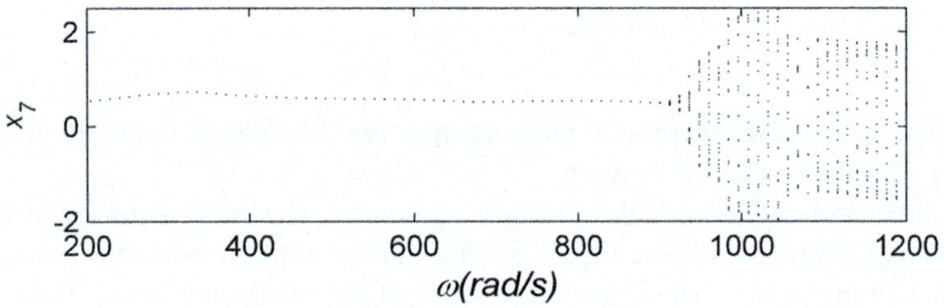

(c) The system with rub-impact

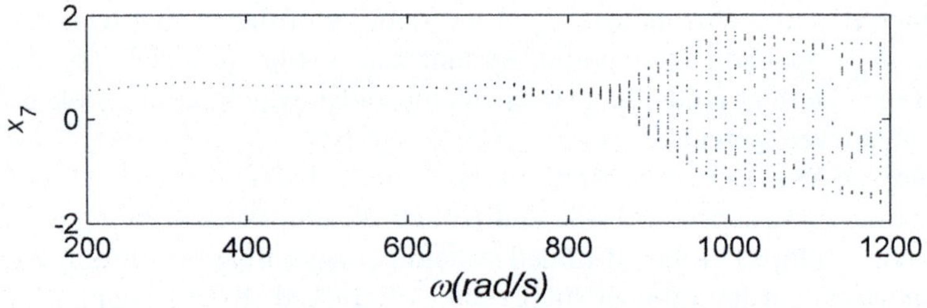

(d) The system with the coupling fault

Fig. 3.4 Bifurcation diagrams

the diameter of the oil-filmed system is larger. In the other two figures, namely, Fig. 3.7(a) and (c), both curves are irregularly closed.

Remark 1 The crack and coupling faults have greater effects on the motion state near the primary resonance. Moreover, the rub-impact fault has more obvious effect on the vibration amplitude at the frequency of primary resonance. The dynamical

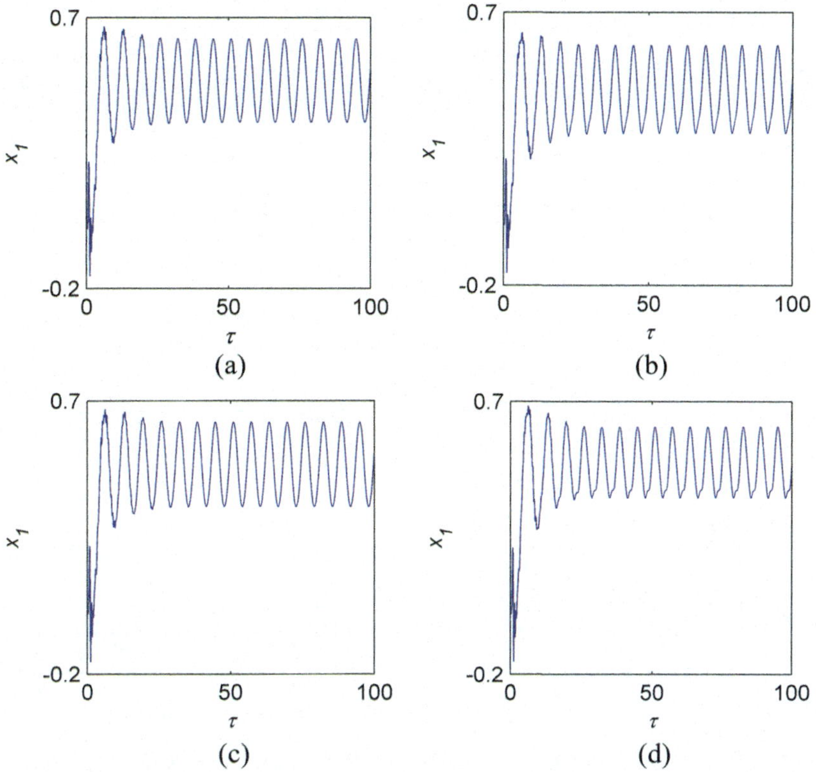

Fig. 3.5 Time histories of x_1 when $\omega = 350$ rad/s. (**a**) The oil-filmed system. (**b**) The cracked system. (**c**) The rub-impacted system. (**d**) The coupled system

behaviors in Figs. 3.5, 3.6, and 3.7 can provide theory guidance to the fault diagnosis of the nonlinear rotor-bearing system.

Similarly, the time histories, trajectories of the orbit of shaft center, phase portraits of the four different systems, at the frequency of 1/2 sub-harmonic vibration (730 rad/s), are studied in Figs. 3.8, 3.9, and 3.10. In Fig. 3.8, the time histories contain the transient process and the stable motion state, the four systems vary from transient to stable state after 15π, so the three different faults have slight effects on the dynamical behaviors of the systems near the frequency $2\omega_c$. The signal is similar to sinusoidal one as the oil-filmed (Fig. 3.8(a)) and cracked system (Fig. 3.8(c)) leads to periodic motion. The signals of the other two systems (Fig. 3.8(b) and (d)) are more complicated relative to oil-filmed and cracked systems.

Figure 3.9 shows the trajectories of the orbit of shaft center of the four different system as $\omega = 730$ rad/s. The trajectories of the oil-filmed and rub-impact systems (Fig. 3.9 (a) and (c)) are similar with ellipse orbit. In comparison with Fig. 3.9(a) and (c), double-loop curves are embodied in the cracked and coupling systems. There is one intersection in the cracked system and two intersections in the coupling system,

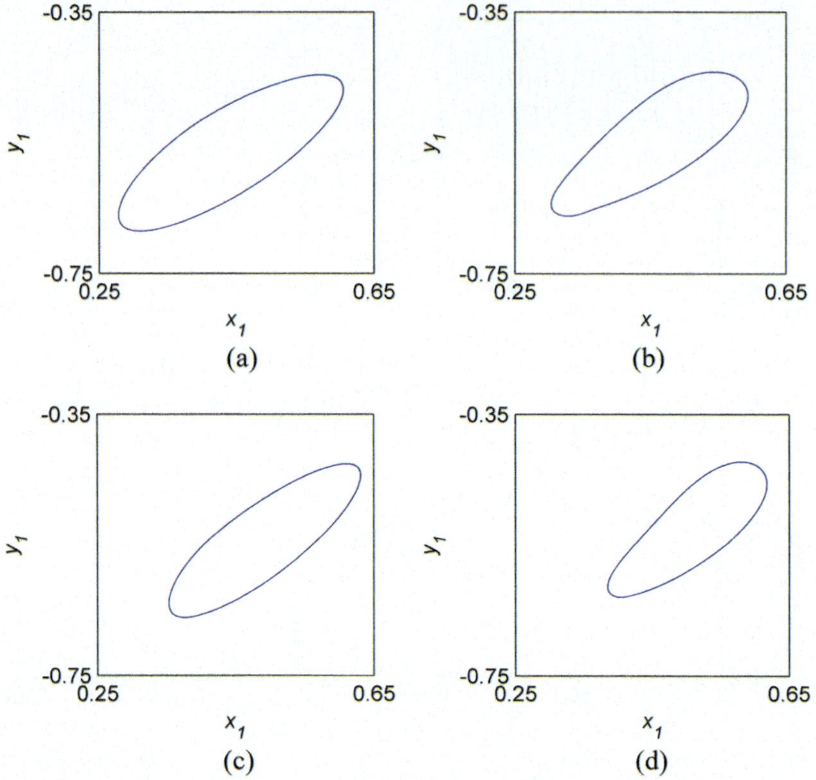

Fig. 3.6 Trajectories of the orbit of shaft center of o_1 when $\omega = 350$ rad/s. (**a**) The oil-filmed system. (**b**) The cracked system. (**c**) The rub-impacted system. (**d**) The coupled system

the results indicate that the crack fault is a key factor to the appearance of double-loop curves at $2\omega_c$, and the rub-impact fault may affect the topological structures of the double-loop curves.

As is clearly seen in Fig. 3.10(a) and (c), the phase portraits of the oil-filmed and rub-impact systems are the orbits are similar with elliptic one. The curves in Fig. 3.10(b) and (d) are both double-loop, and the two loops are closer in the cracked system than in the coupling system. Also, this result verifies that the crack fault can lead to double-loop curves, the rub-impact fault affects the topological structures of phase portraits.

Remark 2 In Figs. 3.8, 3.9, and 3.10, the crack fault has great influence on the rotor system at the frequency of $2\omega_c$. The signal of cracked fault system is more complex, and the double-loop curves occur in both trajectories of the orbit of shaft center and phase portraits. In particular, the period-doubling bifurcation occurs in Fig. 3.4(b). The rub-impact fault will not affect the natural structures of the cracked behaviors.

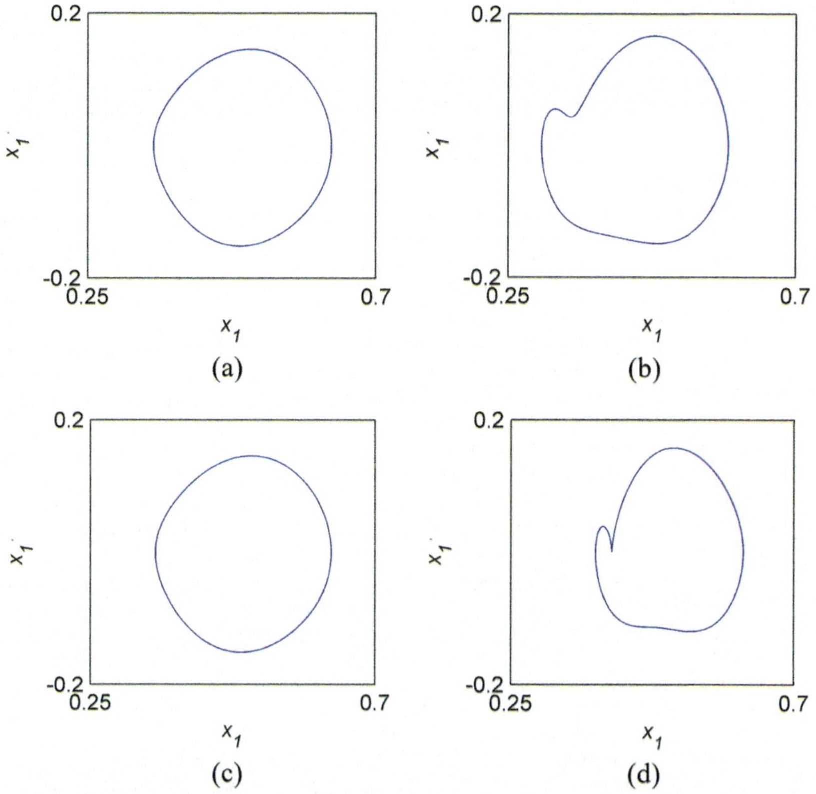

Fig. 3.7 Phase portraits of o_1 when $\omega = 350$ rad/s. (**a**) The oil-filmed system. (**b**) The cracked system. (**c**) The rub-impacted system. (**d**) The coupled system

Hence, the dynamical behaviors near the frequency of $2\omega_c$ can be regarded as the important judgments to verify the crack fault of the rotor-bearing system.

In Fig. 3.11, the four figures (Fig. 3.11(a)–(d)) show the time histories of the rotor-bearing models at the frequency of $3\omega_c$. There exist certain differences between the four faults of the rotor system. The differences of the time histories can provide theoretical guidance to the fault detection of the rotor system.

As is shown in Fig. 3.12, it shows the orbits of shaft center diagrams of the oil-film, crack, rub-impact, coupling fault when $\omega = 1020$ rad/s (the third order natural frequency). Based on the constraint of the displacement of rub-impact fault, the orbit of shaft center is easier than the other three curves. But the displacement is the largest, meaning that the rub-impact fault arouses the larger motion near the third order frequency. The orbit of shaft center of the crack fault is more complex than that of the oil-film fault, and the complexity of the orbit of shaft center of the coupling fault is situated between the oil-film fault and crack fault.

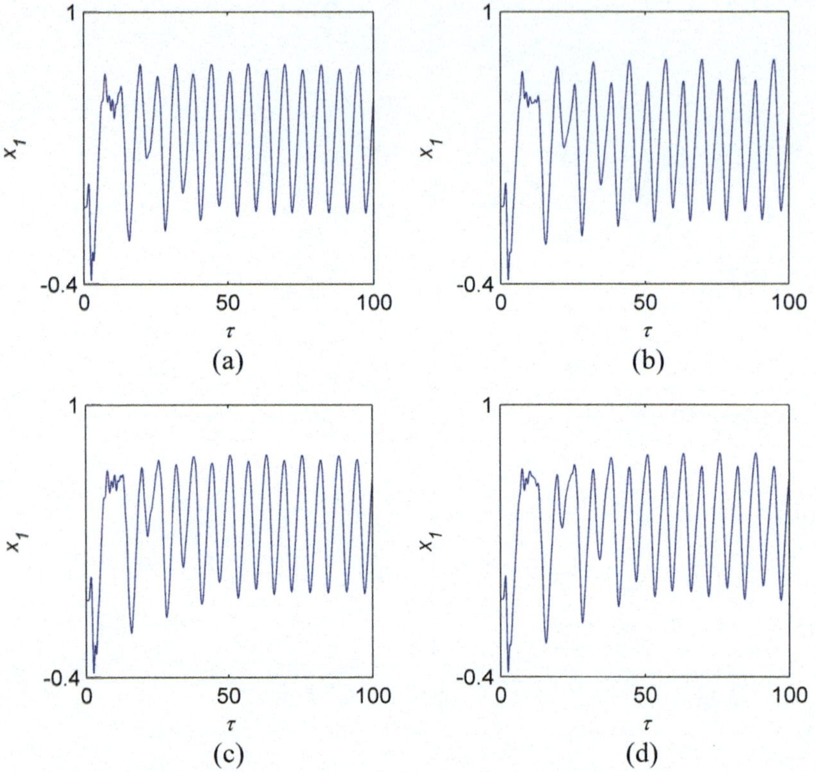

Fig. 3.8 Time history curves of x_1 when $\omega = 730$ rad/s. (**a**) The oil-filmed system. (**b**) The cracked system. (**c**) The rub-impacted system. (**d**) The coupled system

Figure 3.13 shows phase diagrams of oil-film, crack, rub-impact and the coupling systems at the frequency of $\omega = 1020$ rad/s. Oil-film rotor induces complex motion at this moment. The phase diagram of the rub-impact fault is annular narrow band, and it is quasi-periodic attractor. The sort of the four fault cases can be given based on the complexity of the attractor: crack, coupling, oil-film and rub-impact.

Remark 3 By comparing the results in Figs. 3.11, 3.12, and 3.13, we can obtain the conclusions as follows: the crack fault can arouse the 1/2 sub-harmonic vibration, the violent vibration at the frequency of $3\omega_c$ can be aroused by the rub-impact fault. At the frequency of primary resonance, the rub-impact and coupling fault may give rise to the reducing of the first order critical speed, while the crack and coupling fault may induce the structural variation of trajectories of the orbit of shaft center and phase portraits. In addition to, the crack can restrain the vibration amplitude at the frequency of $3\omega_c$ in the coupling system, the crack also arouses the trajectories of the orbit of shaft center and attractor more complex.

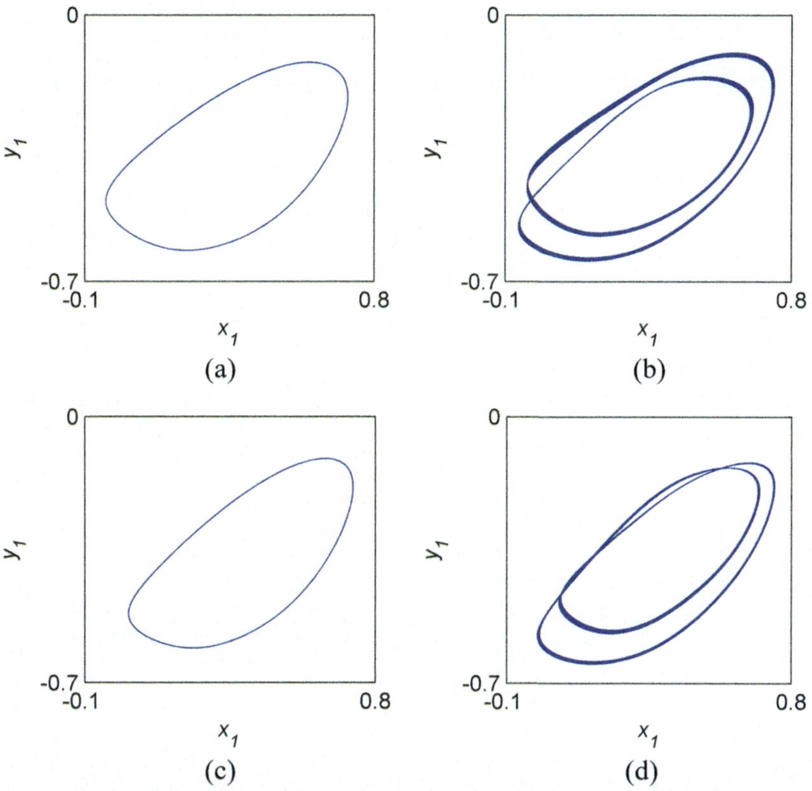

Fig. 3.9 Trajectories of the orbit of shaft center of o_1 when $\omega = 730$ rad/s. (**a**) The oil-filmed system. (**b**) The cracked system. (**c**) The rub-impacted system. (**d**) The coupled system

Remark 4 The shaft stiffness varies with the open and close of crack periodically as the crack fault occurs, which may make the trajectories of the orbit of shaft center and attractor structure more complex. As the full annular rub occurs, the vibration amplitude is restrained. Meanwhile, the frictional force and restriction will reduce the complexity of the attractor.

3.3.2 Effects of Systematic Parameters

The effects of systematic parameters of the rotor system on the dynamical behaviors of the coupling fault will be analyzed in the following. The systematic parameters include cracked depth, clearance between rotor and stator, stator stiffness, eccentricity. The bifurcation and amplitude-frequency behaviors will be provided as the values of the systematic parameters vary.

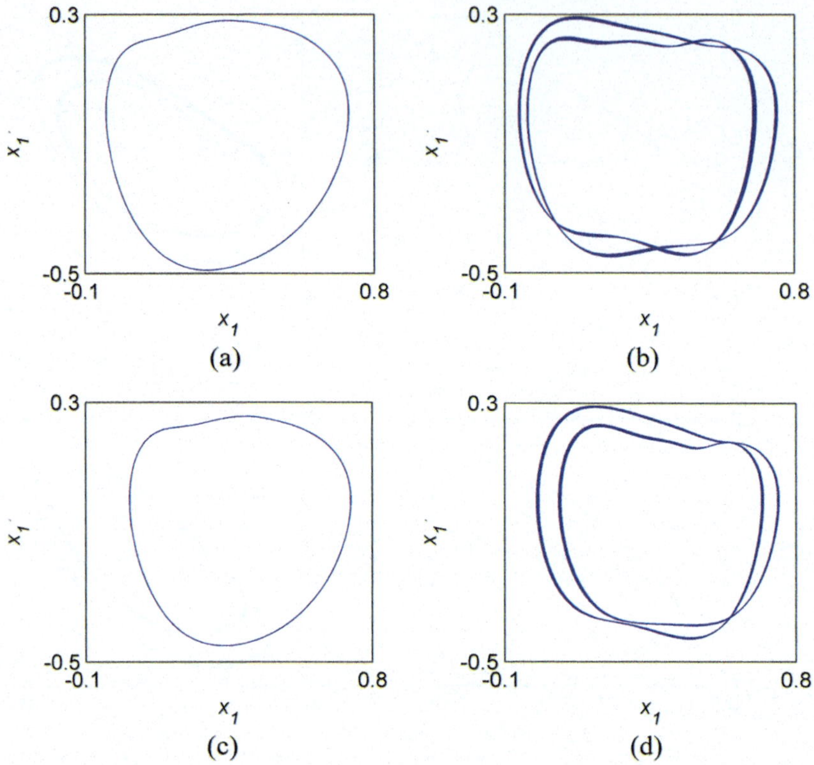

Fig. 3.10 Phase portraits of o_1 when $\omega = 730$ rad/s. (**a**) The oil-filmed system. (**b**) The cracked system. (**c**) The rub-impacted system. (**d**) The coupled system

3.3.2.1 Cracked Depth

The effects of the cracked depth variation on dynamical behaviors of the coupling fault are investigated in this section. As is shown in Fig. 3.14(a) and (b), they could describe the bifurcation diagrams of the coupling fault systems with the cracked depth $a = 0.6R$ and $a = R$. Figure 3.15 shows the comparison of frequency-amplitudes of two cracked depths. As the cracked depth decreases, the variable quantity of stiffness decreases correspondingly, so the first order critical speed of the coupling fault rotor slightly increases. The period-doubling bifurcation vanishes at the frequency of $2\omega_c$, the 1/2 sub-harmonic vibration amplitude of the relevant amplitude-frequency characteristic decreases at the same time (Fig. 3.15). Since the crack fault characteristic does not play the main role and even vanishes, the rub-impact fault emerges, the vibration amplitude suddenly increases at the frequency of $3\omega_c$. This changing demonstrates that the crack is the main factor to generate 1/2 sub-harmonic vibration in the coupling system again, there will be more effects with the increase of the cracked depth. The rub-impact fault will result in the amplitude of

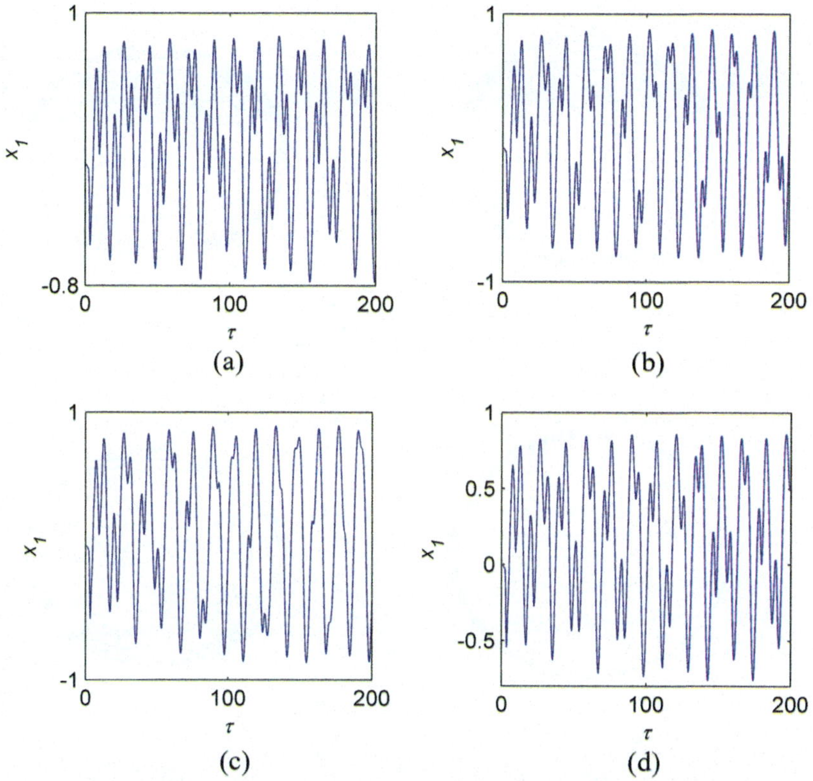

Fig. 3.11 Time histories of x_1 when $\omega = 1020$ rad/s. (**a**) The oil-filmed system. (**b**) The cracked system. (**c**) The rub-impacted system. (**d**) The coupled system

the coupling fault peaking at the frequency of $3\omega_c$, and there is reversible inhibition of the crack to the amplitude aroused by the rub-impact. The restrain will be stronger as the cracked depth increases. As $\omega \in (1100, 1200)$, the cracked depth will have slight impact on the amplitude based on the comparison of the amplitude-frequency curves (Fig. 3.15). In comparison to the bifurcation diagrams (Fig. 3.14), the complex motion of the system changes.

3.3.2.2 Clearances Between Rotor and Stator

This section will study the effects of clearances between rotor and stator to the coupling fault. Similarly, Figs. 3.16 and 3.17 present the bifurcation and amplitude-frenquency as the clearance is 0.12 and 0.18, respectively. As the clearance varies, the clearance has slight impact on the vibration amplitude and critical speed of the coupling system. Near the frequency of $2\omega_c$, the amplitude of 1/2 sub-harmonic resonance decreases as the clearance decreases, the amplitudes of the systems with

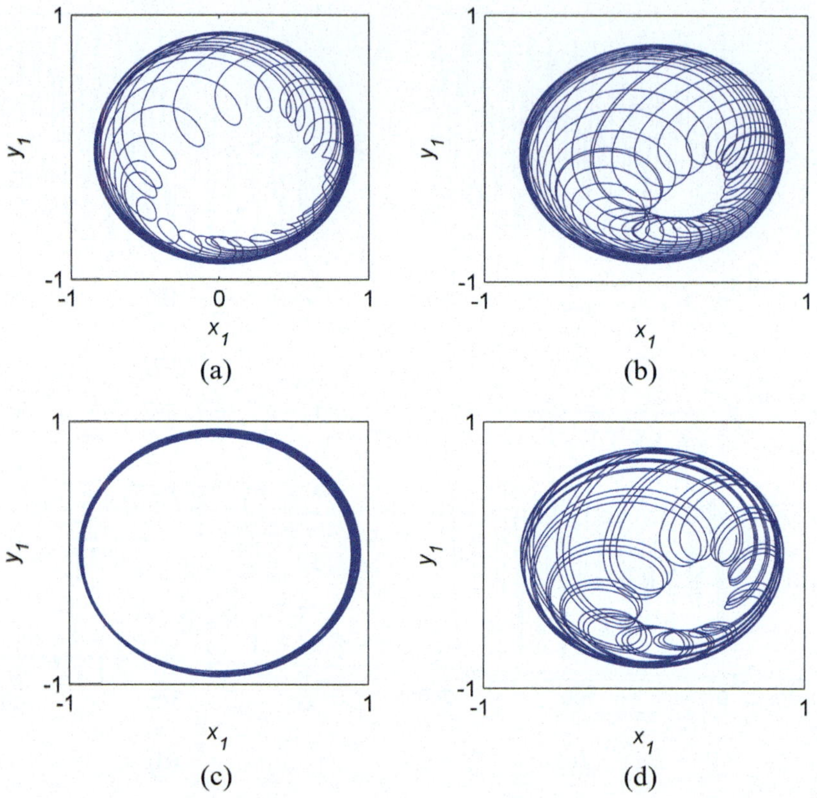

Fig. 3.12 Trajectories of the orbit of shaft center of o_1 when $\omega = 1020$ rad/s. (**a**) The oil-filmed system. (**b**) The cracked system. (**c**) The rub-impacted system. (**d**) The coupled system

different clearance are similar as $\omega \in [2\omega_c, 3\omega_c]$. More, there is a sharp increase of the amplitude as the clearance decreases at the frequency of $3\omega_c$, the rub-impact fault behaviors are enhanced obviously, verifying that the sudden increase of the $3\omega_c$ amplitude of the coupling fault system is primarily caused by the rub-impact fault. At the high frequency (over 1100 rad/s), the amplitude increases slightly as the clearance decreases. Hence, the topological structures of the bifurcation diagram will be more complex in comparison to Fig. 3.16(b).

3.3.2.3 Stator Stiffness

The effects of the stator stiffness variation on the bifurcation behaviors and amplitude-frequency are studied in this section. Two different parameter values are chosen to analyze the bifurcation (Fig. 3.18) and amplitude-frequency (Fig. 3.19) of the coupling fault system: the first is 500(N/m), the second is 1000(N/m). In Fig.

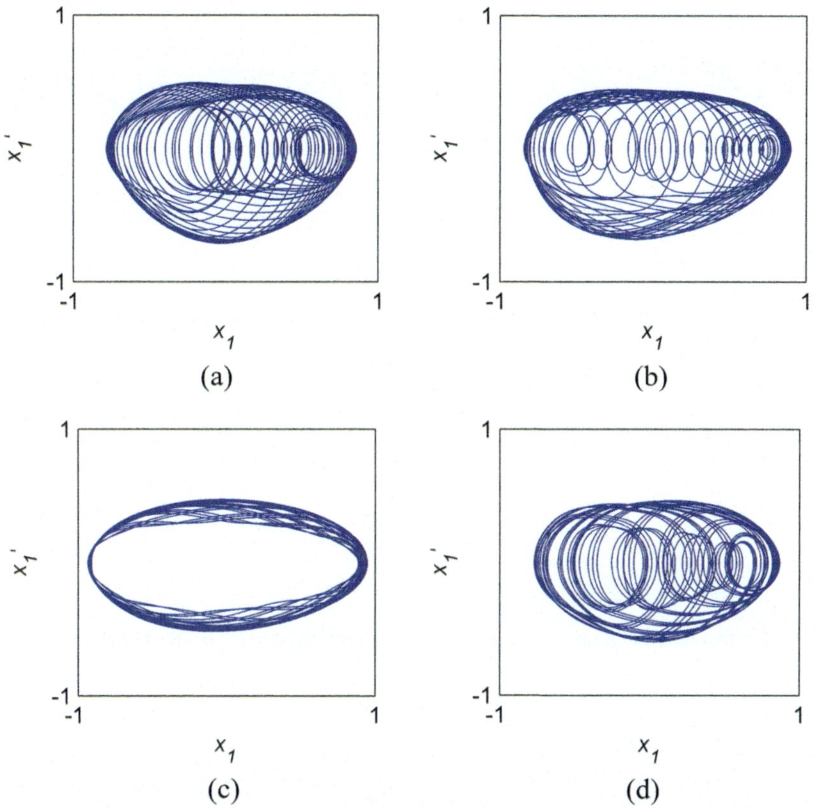

Fig. 3.13 Phase portraits of o_1 when $\omega = 1020$ rad/s. (**a**) The oil-filmed system. (**b**) The cracked system. (**c**) The rub-impacted system. (**d**) The coupled system

3.19, it can see clearly that the variation of the stator stiffness has slight impact on the primary frequency, vibration amplitude and critical speed. 1/2 sub-harmonic resonance amplitude increases as the stiffness decreases. At the frequency of $3\omega_c$, the amplitude changes obviously as stiffness decreases. More, the complex motion region is simpler when the stiffness is $500(N/m)$.

3.3.2.4 Eccentricity

Finally, the effects of the eccentricity variation on the coupling fault system are studied. Figures 3.20 and 3.21 present the comparisons of bifurcation and amplitude-frequency of two different eccentricities (0.00015 mm and 0.00012 mm). As is shown in Fig. 3.21, the amplitude increaces as eccentricity magnifies at the frequency of primary reasonance. 1/2 sub-harmonic resonance is more obvious near the frequency of $2\omega_c$, and it is clear the annulus of the period-doubling bifurcation is

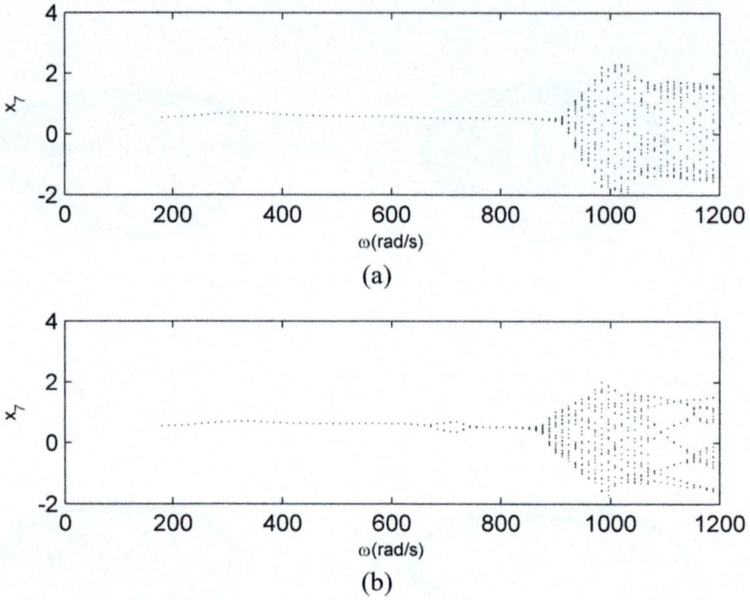

Fig. 3.14 Bifurcation diagrams for coupled fault system. (**a**) $a = 0.6R$. (**b**) $a = R$

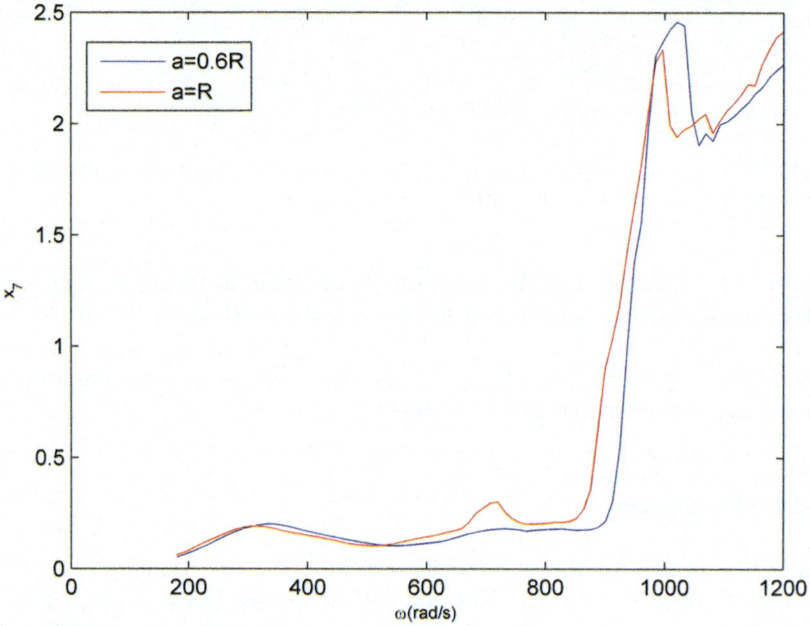

Fig. 3.15 Comparison of the amplitude-frequency curves

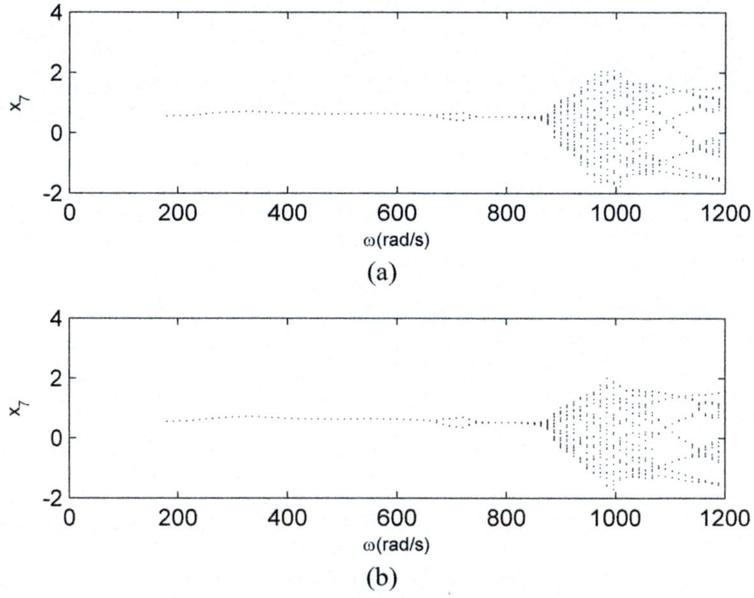

Fig. 3.16 Bifurcation diagrams for coupled fault system. (**a**) $r = 0.12$. (**b**) $r = 0.18$

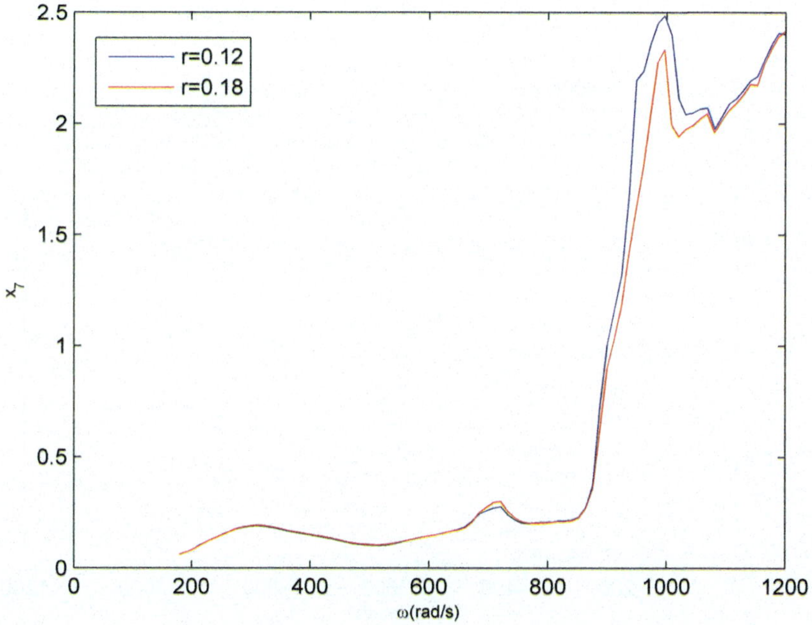

Fig. 3.17 Comparison of the amplitude-frequency curves

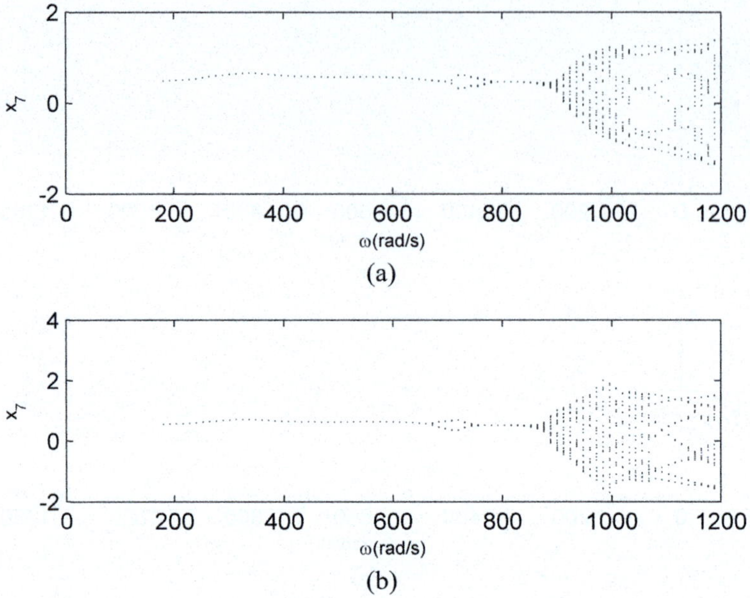

Fig. 3.18 Bifurcation diagrams for coupled fault system. (**a**) $k = 500$. (**b**) $k = 1000$

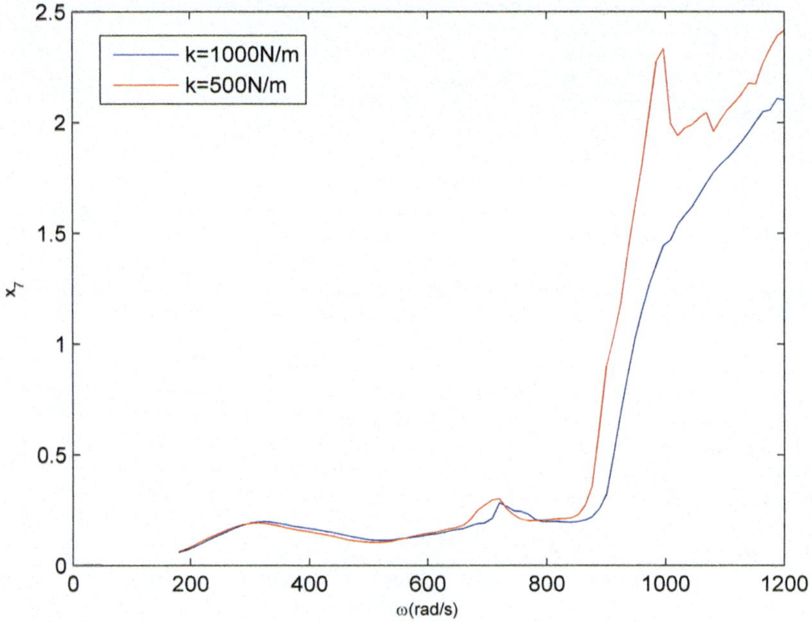

Fig. 3.19 Comparison of the amplitude-versus-frequency

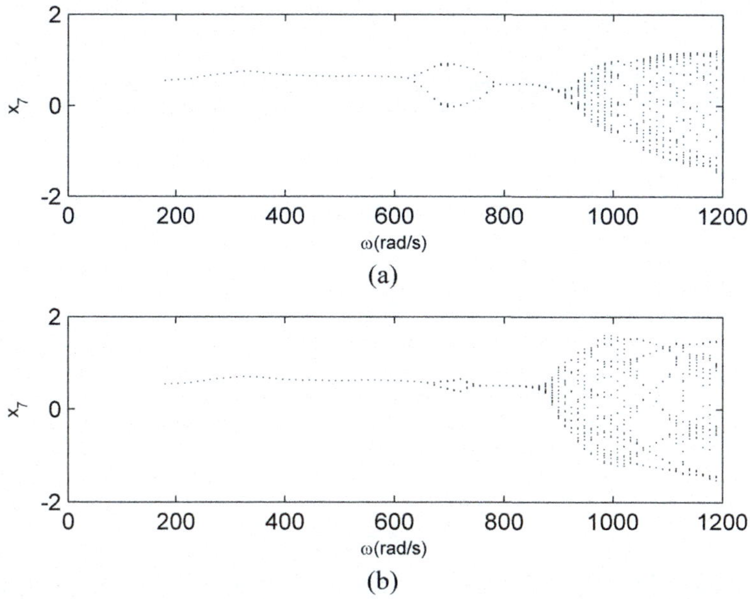

Fig. 3.20 Bifurcation diagrams for coupled fault system. (**a**) $b = 0.00015$. (**b**) $b = 0.00012$

larger. As frequency is over $2\omega_c$, the amplitude decreases as eccentricity increases, this result indicates that increasing the eccentricity appropriately can improve the stability of the system under a certain speed. But with high frequency, the bifurcation behaviors will become more complex as the eccentricity increases. In other words, the nonlinear dynamic behaviors will be more complex.

As is disscussed above, the eccentricity of the system has more impact to dynamical characteristic of the amplitude-frequency and bifurcation at the frequencies of primary resonance, $2\omega_c$, $3\omega_c$, and high frequency. The eccentricity may affect the visualizations of different faults at each frequency range.

Remark 5 The systematic parameters play a key role in the rotor models, and the variations of the parameters will affect the dynamical behaviors of the rotor-bearing system. We make a preliminary study on the effects of cracked depth, clearance between rotor and stator, stator stiffness, eccentricity to the bifurcation diagrams and amplitude-frequency curves. This research is only a beginning to discuss the variations of the dynamical behaviors with the systematic parameters. We can also use the frequency-spectrum analysis to describe the effects of the variations of the parameters.

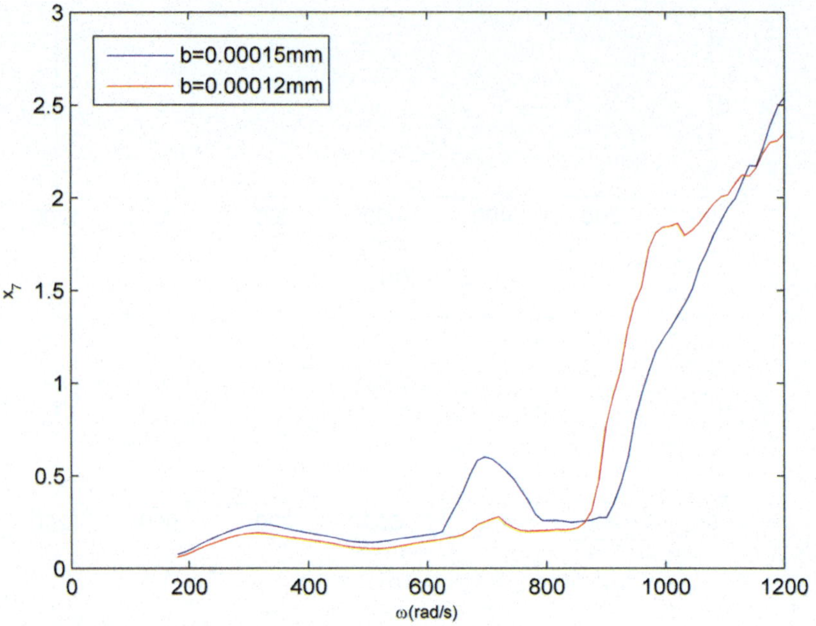

Fig. 3.21 Comparison of the amplitude-frequency curves

3.4 Optimal Reduced Model Based on POM Energy

The first few order POMs pertaining to the rotating speed of the rotor will be studied to confirm the dimension of the reduced model in this section.

In Fig. 3.22, it shows the first three order POM energy curve of the 24-DOFs rotor model with the coupling fault, and it is clear that the three order POMs occupy almost 100% energy as the rotating speed is about 1120 rad/s, so the optimal order reduction model at this speed is obtained based on the POM energy method. The POM energy method has been applied in ref. [44, 45]. Then, the TPOD method [42, 44, 45] is used to reduce the 24-DOFs rotor model to a 3-DOFs one.

The dynamical equation of the order reduction model can be expressed as following,

$$\ddot{\mathbf{P}} = -\mathbf{C}_3\dot{\mathbf{P}} - \mathbf{K}_3\mathbf{P} + \mathbf{F}_3 \tag{3.9}$$

Remark 6 The POM can provide the physical significance of the TPOD method, and POM energy can reveal the amount occupation of the dynamical behaviors of the reduced model relative to the original one. Hence, the POM energy method is applied to confirm the dimension of the order reduction model. Then we use the TPOD method to get the optimal reduced model at the corresponding rotating speed.

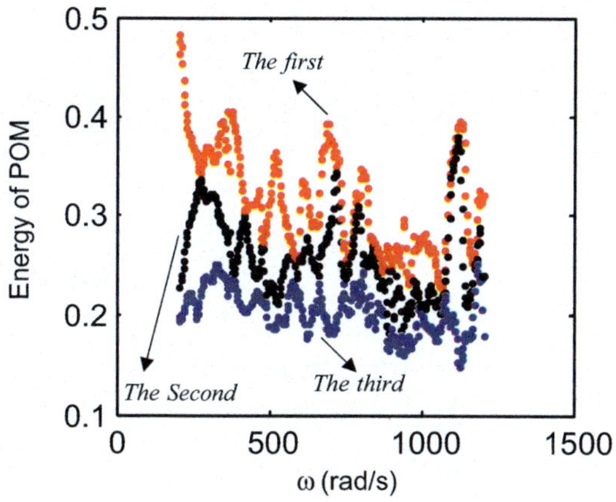

Fig. 3.22 The POM energy of the 24-DOFs model with coupling fault

3.5 Efficiency of the Order Reduction Method

In this section, we will discuss the efficiency of the order reduction method via the bifurcation analysis of the original and the reduced system.

On the basis of POM energy curve (Fig. 3.22), the 3-DOFs reduced model is obtained as the rotating speed is 1120 rad/s. As is shown in Fig. 3.23, we can see clearly that the reduced system (Fig. 3.23(b)) reserves the period-doubling bifurcation, the rotating speed of complex motion and the topological structure of the original one (Fig. 3.23(a)). The comparison of the bifurcation diagrams verifies the efficiency of the TPOD method, we can also use other dynamical behaviors (phase portraits, amplitude-frequency curve, etc.) to analyze the efficiency [44, 45].

Remark 7 The reduced order reduction model obtained by TPOD method reserves the dynamical behaviors of the original one well. TPOD method is proposed by the present authors, it has been applied in the rotor models supported by the sliding bearings or ball bearings with looseness. The efficiency of the order reduction method applied in the rotor system with coupling fault further generalizes this method.

3.6 Conclusions

The nonlinear dynamic behaviors of the rotor system with the coupling faults have been studied in details. A 24-DOFs rotor system supported by a pair of sliding bearings with both crack and rub-impact faults has been established. Then, the coupling faults have been studied via comparing the single fault (crack or rub-impact), and the effects of the systematic parameters (cracked depth, clearance between rotor and stator, stator stiffness, eccentricity) to the coupling faults have also been presented by comparing with the bifurcation and amplitude-frequency

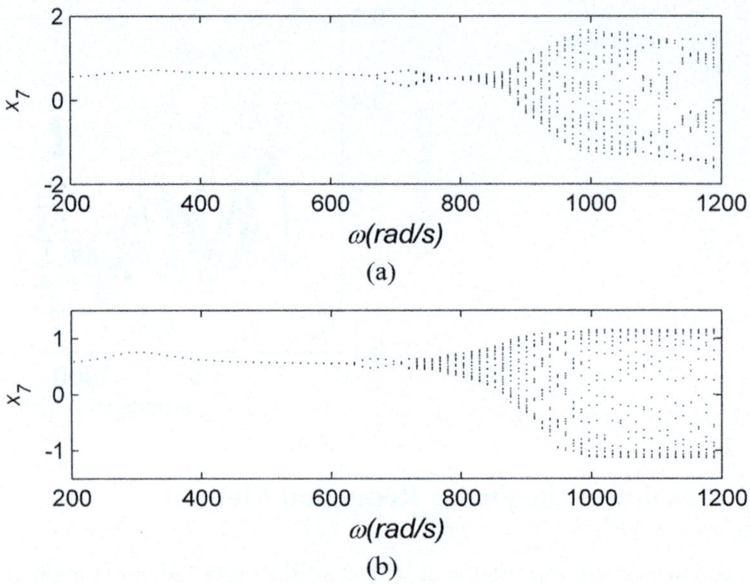

Fig. 3.23 Bifurcation diagrams. (**a**) The original system. (**b**) The reduced system

behaviors. Further, POM energy method has been used to confirm the DOF number of the reduced system, and TPOD has been applied to reduce the original system to a 3-DOFs one, the efficiency of the optimal order reduction model has been verified via the comparison of the bifurcation diagrams. Further studies on this subject will be carried out by the present authors in the two aspects: the first is to study the dynamic behaviors of the rotor system supported by ball bearings with the coupling faults; the second is to study the coupling faults of the rotor system with uncertainties.

Acknowledgements This study was funded by the National Natural Science Foundation of China (Grant No. 12072263, 11972295, 11802235), Natural Science Foundation of Shaanxi Province (Grant No. 2020JQ-129), State Key Laboratory of Mechanical Behavior and System Safety of Traffic Engineering Structures (Grant No. KF2020-26) and China Postdoctoral Science Foundation (Grant No. 2021 M690274, 2021T140033). The authors declare that they have no conflict of interest.

A.1 Appendix 1

$$m_1 = m_{12} = 50kg, m_2 = m_{11} = 47kg, m_3 = m_{10} = 44kg, m_4 = m_9 = 41kg,$$

$$m_5 = m_8 = 38kg, m_6 = m_7 = 35kg, R = 60mm, c_1 = c_{12} = 1800N \cdot s/m,$$

$$c_2 = c_{11} = 1700N \cdot s/m, c_3 = c_{10} = 1600N \cdot s/m, c = 0.11mm, c_4 = c_9 = 1500N \cdot s/m,$$

$$c_5 = c_8 = 1400N \cdot s/m, c_6 = c_7 = 1300N \cdot s/m, k_i = 2.5 \times 10^8 N/m\,(i = 1 \cdots 11),$$

$$L = 30mm, \beta_3 = \frac{\pi}{2}, E_3 = 0.18mm, \mu = 0.018pa \cdot s, \mathrm{E}_i = 0\,(i = 1 \cdots 12, i \neq 3),$$

$$\mu_0 = 0.15, \beta_i = 0\,(i = 1 \cdots 12, i \neq 3), k_b = 1 \times 10^3 N/m, r = 0.18mm$$

B.1 Appendix 2

$$\begin{Bmatrix} f_x \\ f_y \end{Bmatrix} = -\frac{\left[(x-2\dot{y})^2+(y+2\dot{x})^2\right]^{1/2}}{1-x^2-y^2} \times \begin{Bmatrix} 3xV(x, y, \alpha) - \sin\alpha G(x, y, \alpha) - 2\cos\alpha S(x, y, \alpha) \\ 3yV(x, y, \alpha) + \cos\alpha G(x, y, \alpha) - 2\sin\alpha S(x, y, \alpha) \end{Bmatrix}$$

$$\alpha = \arctan\left(\frac{y+2\dot{x}}{x-2\dot{y}}\right) - \frac{\pi}{2}\mathrm{sign}\left(\frac{y+2\dot{x}}{x-2\dot{y}}\right) - \frac{\pi}{2}\mathrm{sign}\,(y+2\dot{x})$$

$$G(x, y, \alpha) = \frac{2}{\left(1-x^2-y^2\right)^{1/2}}\left[\frac{\pi}{2} + \arctan\frac{y\cos\alpha - x\sin\alpha}{\left(1-x^2-y^2\right)^{1/2}}\right]$$

$$V(x, y, \alpha) = \frac{2 + (y\cos\alpha - x\sin\alpha)\,G(x, y, \alpha)}{1 - x^2 - y^2}$$

$$S(x, y, \alpha) = \frac{x\cos\alpha + y\sin\alpha}{1 - (x\cos\alpha + y\sin\alpha)^2}$$

References

1. Sinou, J., and A.W. Lees. 2005. The influences of cracks in rotating shafts. *Journal of Sound and Vibration* 285: 1015–1637.
2. Ren, Z.H., S.H. Zhou, et al. 2015. Crack fault diagnosis of rotor systems using wavelet transform. *Computers and Electrical Engineering* 45: 33–41.
3. Hou, L., Y.S. Chen, et al. 2015. Bifurcation analysis for 2:1 and 3:1 super-harmonic resonances of an aircraft cracked rotor system due to maneuver load. *Nonlinear Dynamics* 81 (1): 531–547.
4. Sinou, J. 2008. Detection of cracks in rotor based on the 2 and 3 super-harmonic frequency components and the crack-unbalance interactions. *Communications in Nonlinear Science and Numerical Simulation* 13 (9): 2024–2040.
5. Ma, H., and T. Yu. 2009. Time-frequency features of two types of coupled rub-impact faults in rotor systems. *Journal of Sound and Vibration* 321: 1109–1128.
6. Jacquet, R.G., M. Torkhani, P. Cartraud, et al. 2013. Rotor to stator contacts in turbomachines. Review and application. *Mechanical Systems and Signal Processing* 40: 401–420.
7. Khanlo, H.M., M. Ghayour, and R.S. Ziaei. 2011. Chaotic vibration analysis of rotating, flexible, continuous shaft-disk system with a rub-impact between the disk and the stator. *Communications in Nonlinear Science and Numerical Simulation* 16: 566–582.
8. Hou, L., Y.S. Chen, and Q.J. Cao. 2014. Nonlinear vibration phenomenon of an aircraft rub-impact rotor system due to hovering flight. *Communications in Nonlinear Science and Numerical Simulation* 19: 286–297.
9. Ren, Z.H., Y.N. Teng, et al. 2009. Pedestal looseness fault analysis of overhanging dual-disc rotor-bearing. *Applied Mechanics and Materials* 16-19: 654–659.
10. Chu, F.L., and Y. Tang. 2001. Stability and nonlinear responses of a rotor-bearing system with pedestal looseness. *Journal of Sound and Vibration* 241 (5): 879–893.
11. Ma, H., X.Y. Zhao, Y.N. Teng, and B.C. Wen. 2011. Analysis of dynamic behaviors for a rotor system with pedestal looseness. *Shock and Vibration* 18: 13–27.
12. Ji, Z., and J.W. Zu. 1998. Method of multiple scales for vibration analysis of rotor-shaft systems with nonlinear bearing pedestal model. *Journal of Sound and Vibration* 218 (2): 293–305.
13. Dolenc, B., P. Boskoski, et al. 2016. Distributed bearing fault diagnosis based on vibration analysis. *Mechanical Systems and Signal Processing* 66-67: 521–532.
14. Sekhar, A.S. 2003a. Crack detection and monitoring in a rotor supported on fluid film bearings: start-up vs run-down. *Mechanical Systems and Signal Processing* 17 (4): 897–901.
15. Sekhar, A. 2003b. Crack detection through wavelet transform for a run-up rotor. *Journal of Sound and Vibration* 259: 461–472.
16. Lu, K., Y.L. Jin, et al. 2021. The applications of POD method in dual rotor-bearing systems with coupling misalignment. *Mechanical Systems and Signal Processing* 150: 107236.
17. Salawu, O. 1997. Detection of structural damage through changes in frequency: a review. *Engineering Structures* 19 (9): 718–723.

18. Adewusi, S., and B. Al-Bedoor. 2002. Experimental study on the vibration of an overhung rotor with a propagating transverse crack. *Shock and Vibration* 9: 91–104.
19. Alho, A., and C. Uggla. 2015. Global dynamics and inflationary center manifold and slow-roll approximants. *Journal of Mathematical Physics* 56: 012502.
20. Valls, C. 2015. Center problem in the center manifold for quadratic and cubic differential systems in R^3. *Applied Mathematics and Computation* 251: 180–191.
21. Verdugo, A., and R. Rand. 2008. Center manifold analysis of a DDE model of gene expression. *Communications in Nonlinear Science and Numerical Simulation* 13: 1112–1120.
22. Kazufumi, I., and K. Karl. 2008. Reduced-order optimal control based on approximate inertial manifolds for nonlinear dynamical systems. *SIAM Journal on Numerical Analysis* 46: 2867–2891.
23. Foial, C., G. Sell, and R. Teman. 1988. Inertial manifolds for nonlinear evolutionary equations. *Journal of Differential Equations* 73: 93–114.
24. Marion, M. 1989. Approximate inertial manifolds for reaction–diffusion equations in high space dimension. *Journal of Dynamics and Differential Equations* 1: 245–267.
25. Kunisch, K., and S. Volkwein. 2001. Galerkin proper orthogonal decomposition methods for parabolic problems. *Numerische Mathematik* 90: 117–148.
26. Marion, M., and R. Temam. 1989. Nonlinear Galerkin methods. *SIAM Journal on Numerical Analysis* 5: 1139–1157.
27. Carlberg, K., C. Bou-Mosleh, and C. Farhat. 2011. Efficient non-linear model reduction via a least-squares Petrove-Galerkin projection and compressive tensor approximations. *International Journal of Numerical Methods in Engineering* 86: 155–181.
28. Guo, S.J., and S.L. Yan. 2016. Hopf bifurcation in a diffusive Lotka-Volterra type system with nonlocal delay effect. *Journal of Differential Equations* 260: 781–817.
29. Nikolic, M., and M. Rajkovic. 2006. Bifurcations in nonlinear models of fluid-conveying pipes supported at both ends. *Journal of Fluids and Structures* 22: 173–195.
30. Stefanescu, R., A. Sandu, and I.M. Navon. 2015. POD/DEIM reduced-order strategies for efficient four dimensional variational data assimilation. *Journal of Computational Physics* 295: 569–595.
31. Lu, K., Y.L. Jin, et al. 2019. Review for order reduction based on proper orthogonal decomposition and outlooks of applications in mechanical systems. *Mechanical Systems and Signal Processing* 123: 264–297.
32. Amabili, M., and C. Touze. 2007. Reduced-order models for nonlinear vibrations of fluid-filled circular cylindrical shells: comparison of POD and asymptotic nonlinear normal modes methods. *Journal of Fluids and Structures* 23: 885–903.
33. Liang, Y.C., H.P. Lee, et al. 2002. Proper orthogonal decomposition and its applications, part I: theory. *Journal of Sound and Vibration* 252: 527–544.
34. Stefanescu, R., and I.M. Navon. 2013. POD/DEIM nonlinear model order reduction of an ADI implicit shallow water equations model. *Computational Physics* 237: 95–114.
35. Stefanescu, R., A. Sandu, and I.M. Navon. 2014. Comparison of POD reduced order strategies for the nonlinear 2D shallow water equations. *International Journal for Numerical Methods in Fluids* 76: 497–521.
36. Yang, H.L., and G. Radons. 2012. Geometry of inertial manifolds probed via a Lyapunov projection method. *Physical Review Letters* 108: 154101.
37. Hamilton, N., M. Tutkun, and R.B. Cal. 2016. Low-order representations of the canonical wind turbine array boundary layer via double proper orthogonal decomposition. *Physics of Fluids* 28: 025103.
38. Siegel, S., J. Seidel, C. Fagley, et al. 2008. Low-dimensional modeling of a transient cylinder wake using double proper orthogonal decomposition. *Journal of Fluid Mechanics* 610: 1–42.
39. Rega, G., and H. Troger. 2005. Dimension reduction of dynamical systems: methods, models, applications. *Nonlinear Dynamics* 41: 1–15.
40. Steindl, A., and H. Troger. 2001. Methods for dimension reduction and their application in nonlinear dynamics. *International Journal of Solids and Structures* 38: 2131–2147.

41. Holmes, P.J., L. Lumley, and G. Berkooz. 2012. *Turbulence, Coherent Structures*. Dynamical Systems and Symmetry: Cambridge University Press.
42. Lu, K., H. Yu, Y.S. Chen, et al. 2015b. A modified nonlinear POD method for order reduction based on transient time series. *Nonlinear Dynamics* 79: 1195–1206.
43. Yu, H., Y.S. Chen, and Q.J. Cao. 2013. Bifurcation analysis for nonlinear multi-degree-of-freedom rotor system with liquid-film lubricated bearings. *Applied Mathematics and Mechanics* 34: 777–790.
44. Lu, K., Y.L. Jin, Y.S. Chen, et al. 2015a. Stability analysis of reduced rotor pedestal looseness fault model. *Nonlinear Dynamics* 82: 1611–1622.
45. Lu, K., Y.S. Chen, Y.L. Jin, and L. Hou. 2016. Application of the transient proper orthogonal decomposition method for order reduction of rotor systems with faults. *Nonlinear Dynamics* 86 (3): 1913–1926.
46. Adiletta, G., A.R. Guido, and C. Rossi. 1996. Chaotic motions of a rigid rotor in short jurnal bearings. *Nonlinear Dynamics* 10: 251–269.

Chapter 4
Nonlinear and Linear Phenomenon Investigation of Coupled Vibration of a Multi-disc Rotor Based on Multi-mistuned Blades Length or Multi-disordered Staggle Angle Blades

Yi-Jui Chiu, Ya-Zheng Zhao, Xiao-Yun Li, Chia-Hao Yang, Guo-Fei Yu, and Cheng-Wei Ye

4.1 Introduction

The industrial and scientific circles are concerned about the problem of rotor vibration for more than one hundred years. The high speed rotors appeared with the development of industry. The rotors often work in a flexible state, so the problem of vibration and stability is more important. Due to the error in manufacturing, the center of mass of each segment of the rotor has slight deviation from the axis of rotation. As the rotor rotates, the centrifugal force induced by the deviation will lead to the rotor transverse vibration. This vibration is unusually strong at some speeds. The mistune and disorder problems of the rotor are hot topics for many scholars.

As there are many researches in this field, this chapter only introduces the research of this century. Gennaroand Leonardo [1] used artificial neural networks of genetic algorithms and applied it to analyze structural mistuned configuration. They found that leading to the maximum amplitude of the blade vibration with the structural mistuned. Chen and Shen [2] analyzed cyclic symmetric rotors with slight mistune. They found that some mode localization phenomenons is similar as the authors' works. Li et al. [3] adopted FEM to establish the models of shaft-disk-blades, blade-disk and blade. Their results shows that some modes are excited by the strongly split and coupled vibration. The mistunes could cause and affect different coupled vibrations of blades between the disk-blades and the shaft-disk-blades model. Yasharth and Alok [4] developed a multistage rotor with geometric

Y.-J. Chiu (✉) · Y.-Z. Zhao · X.-Y. Li · G.-F. Yu · C.-W. Ye
School of Mechanical and Automotive Engineering, Xiamen University of Technology, Xiamen, P. R. China
e-mail: chiuyijui@xmut.edu.cn

C.-H. Yang
Department of Mechanical Engineering, Taipei City University of Science and Technology, Taipei, Republic of China

J. Zhang (ed.), *Dynamics and Fault Diagnosis of Nonlinear Rotors and Impellers*, Nonlinear Systems and Complexity 34, https://doi.org/10.1007/978-3-030-94301-1_4

mistuning via model. They found that two interesting results about mistuned blade. First, the blades which have geometrically mistuned could get the reduced order model. Second, natural frequencies of the two-stage rotor and the statistical distributions of peak maximum amplitudes are produced from simulations of Monte Carlo for different patterns of geometric mistuning. Paolo and Giovanni [5] applied the homogenization theory to explore a continuous model of the in-plane vibrations of a mistuned bladed rotor. They also studied localization phenomena and frequency split arising in the imperfect structure. Alok [6] took an accurate reduced order model for the development of geometrically mistuned disks-blades. The method is based on the proper orthogonal decomposition of coordinate measuring machine data on blade geometry and the vibration modes of various tuning systems. Hiroki and Akihiro [7] investigated the effects of the complex pattern on the natural frequencies and the frequency response of mistuned disk-bladed system. They obtained that the excitation with the wave number k excites only the vibration modes with the same wave number in the tuned system, and the excitation with the wave number k excites all vibration modes in the mistuned system, using sensitivity analysis which achieved the variations of the natural frequencies due to mistuning pattern. Han [8] used Floquet theory and harmonic balance method to analyze isotropic rotor, anisotropic rotor, asymmetric rotor and general rotor. They found that two results, which are the modes corresponding critical speeds are secondly dominant over the synchronous modes and the modal line due to weight crosses over the weak modal lines. Lee et al. [9] utilized 2-D unsteady vortex lattice method to predict the gust excitation to reach higher engine-order excitation. They put forward forced response characteristics of mistuned bladed disks. Raeisi and Ziaei-Rad [10] aimed to predict the worst response of mistuned bladed disk and to develop an integrated method by genetic algorithms and artificial neural networks. They investigated mistuning in modulus of elasticity and length of blades effect in the mistuned blades disk systems. Bai et al. [11] exploited hybrid interface substructure component modal synthesis method to investigate the vibration characteristics of mistuned blisk. They observed that the localization of modal shape is more risky because of geometric dimensioning mistuned. Bai et al. [12] increased the computational efficiency of mistuned disk bladed system under the condition of meeting the computational accuracy for a large amount of calculation the classical based on IHISCMS (hybrid interface substructure component modal synthesis) method. Kwon and Yoo [13, 14] explored the vibration localization phenomenon of a multi-bundle blades rotor system. They found that the mistune may cause a significant increase in the forced vibration response of the blade in the multi package blade system. Critical fatigue problems often occur in mistuned systems and the forced vibration response of mistune systemis often significantly larger than that of tuned system. Cui et al. [15, 16] studied the problems related to the rotor system. The authors also have two researches, which adopted assumed mode method to study a mono-flexible disk system which has a mistuned blades length [17] or a disordered staggle angle blade [18]. Lately, the authors [19, 20] added FEM to investigate a multi/single flexible disc rotor system with springs, and compared authors' previous discussions.

This research will investigate the coupled phenomenon among blade-bending, disk-bending and shaft-torsion. Based on the authors previous researches [15–18], this chapter focuses on coupled vibration of a multi-disc system based on mistuned blades length or disordered staggle angle. The blades are considered as Eular type with stagger angle, and the disc are supposed to be flexible in this turbine rotor system. When using the finite element method to analyze it, we employed ANSYS engineering software. Then we discussed the rotor change regulations of mode shapes (MS) and natural frequencies (NF). This research aimed at supplying a qualitative and quantitative overview for the multi flexible disc system.

4.2 Theoretical Analysis

A general rotor system with mistuned and disorder blades shown as Fig. 4.1. The rotor includes four sub-units, such as shaft, multi flexible disc, blades and the blades with length mistuned blade and disorder. The blades have staggle angle. The mistuned length blades are marked with an "*" symbol and the coordinate give as,

$$r_b^* = r_b + \Delta r = r_b\,(1 + \varepsilon_l) \tag{4.1}$$

where ε_l is the mistuned blade deviation ratio, negative or positive.In the following, the blade of disorder stagger angle is asterisked to separate from the others. It is denoted as

$$\beta^* = \beta + \Delta\beta = \beta\,(1 + \varepsilon_a) \tag{4.2}$$

where ε_a is an indication of disorder error, positive or negative.The shaft-disk subsystem torsion energies equations are the followings [18],

$$T_s = \frac{1}{2}\int_0^{Ls} I_s(\dot{\varphi} + \Omega)^2 dZ + \frac{I_d}{2}(\dot{\varphi} + \Omega)^2 \left|_{Z=Z_d}\right. \tag{4.3}$$

$$U_s = \frac{1}{2}\int_0^{Ls} G_s J_s \left(\frac{\partial \varphi}{\partial Z}\right)^2 dZ \tag{4.4}$$

where $\varphi(Z, t)$ denotes the displacement of torsional in a rotating frame; the shaft length, inertia polar moment and torsion rigidity used L_s, I_s and G_sJ_s; The polar moment of inertia of disk used I_d. The upper "dot" is the time derivative and subscripts d and s are appointed for disk and shaft.Figure 4.2 shows a disk, free outside and fixed inside. Rotating at a constant speed Ω. The outer and inner radius of disk used r_d and r_s. h_d denotes the disk's thickness.

The disk's transverse vibration energies equations are governed as following [18],

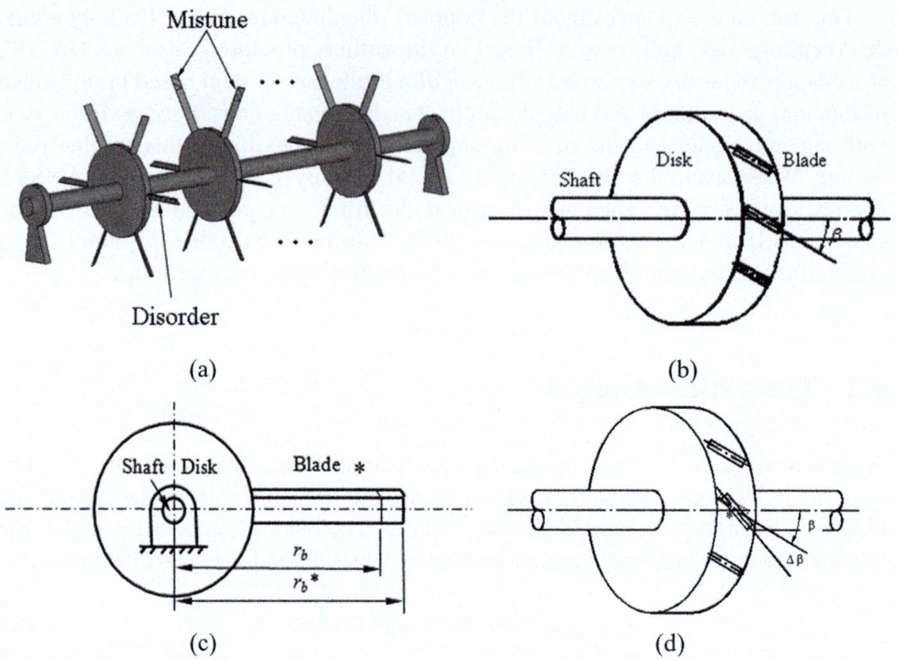

Fig. 4.1 (**a**) A gerneral multi-disc rotor with mistuned and disorder blade, (**b**) the staggle angle, (**c**) length mistuned blade (**d**) angle disorder blade

Fig. 4.2 Coordinates and geometry of the rotating disk

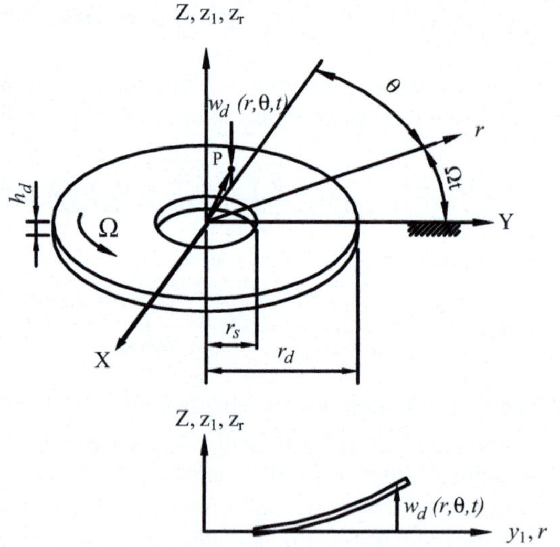

$$T_d = \frac{\rho_d h_d}{2} \int_{r_s}^{r_d} \int_0^{2\pi} \left(\dot{w}_d + \Omega \frac{\partial w_d}{\partial \theta} \right)^2 r d\theta dr \tag{4.5}$$

$$\begin{aligned} U_d = \frac{D}{2} \int_{r_s}^{r_d} \int_0^{2\pi} \Big\{ \left(\nabla^2 w_d\right)^2 - (1-\nu) \frac{\partial^2 w_d}{\partial r^2} \Big[\left(\frac{1}{r} \frac{\partial w_d}{\partial r} + \frac{1}{r^2} \frac{\partial^2 w_d}{\partial \theta} \right) \\ - \left(\frac{1}{r} \frac{\partial^2 w_d}{\partial r \partial \theta} - \frac{1}{r^2} \frac{\partial w_d}{\partial \theta} \right)^2 \Big] \Big\} r d\theta dr \\ + \frac{h_d}{2} \int_{r_s}^{r_d} \int_0^{2\pi} \left[\sigma_r \left(\frac{\partial w_d}{\partial r} \right)^2 + \sigma_\theta \left(\frac{1}{r} \frac{\partial w_d}{\partial \theta} \right)^2 \right] r d\theta dr \end{aligned} \tag{4.6}$$

Disk's transverse displacement used w_d. Bending rigidity used D. ∇^2 denotes the Laplacian; initial stresses in the radial due to Ω $_{\text{used}}$ σ_r and circumferential directions used σ_θ.

The follow is the terms,

$$\nabla^2 = \frac{\partial^2}{\partial r^2} + \frac{\partial}{r \partial r} + \frac{\partial^2}{r^2 \partial \theta^2} \tag{4.7}$$

$$D = \frac{E_d h_d^3}{12\left(1 - \nu^2\right)} \tag{4.8}$$

$$\sigma_r = \frac{3+\nu}{8} \rho_d \Omega^2 \left(r_d^2 - r^2\right) + \frac{(1-\nu)\rho_d \Omega^2 r_s^2 \left[(3+\nu) r_d^2 - (1+\nu) r_s^2\right]}{8\left[(1+\nu) r_d^2 + (1-\nu) r_s^2\right]} \left(\frac{r_d^2}{r^2} - 1\right) \tag{4.9}$$

$$\sigma_\theta = \frac{\Omega^2}{8} \left[(3+\nu) r_d^2 - (1+3\nu) r^2\right] - \frac{(1-\nu)\rho_d \Omega^2 r_s^2 \left[(3+\nu) r_d^2 - (1+\nu) r_s^2\right]}{8\left[(1+\nu) r_d^2 + (1-\nu) r_s^2\right]} \left(\frac{r_d^2}{r^2} + 1\right) \tag{4.10}$$

Figure 4.3 is shown as a rotating cantilevered blade onto a flexible disk with a stagger angle β. The (X, Y, Z) coordinate system is the inertia frame; (x_1, y_1, z_1) frame rotates at a constant speed Ω. (x_2, y_2, z_2) frame rotates a staggle angle β relative to (x_1, y_1, z_1) frame, and (x_3, y_3, z_3) frame is clamped to the root of blade.

The blade's kinetic and strain energies equations are the followings [18],

$$\begin{aligned} T_b = \frac{1}{2} \int_{r_d}^{r_b} \rho_b A_b \big\{ \dot{v}_b^2 + \dot{w}_b^2 + (v_b \cos\beta + w_b \cos\beta)^2 \Omega^2 + x^2 \Omega^2 \\ + 2x\left(\dot{v}_b \sin\beta + \dot{w}_b \sin\beta\right)\Omega \big\}\, dx + \frac{1}{2} \int_{r_d}^{r_b} I_b \left(\Omega \cos\beta + \frac{\partial \dot{v}_b}{\partial x} \right)^2 dx \end{aligned} \tag{4.11}$$

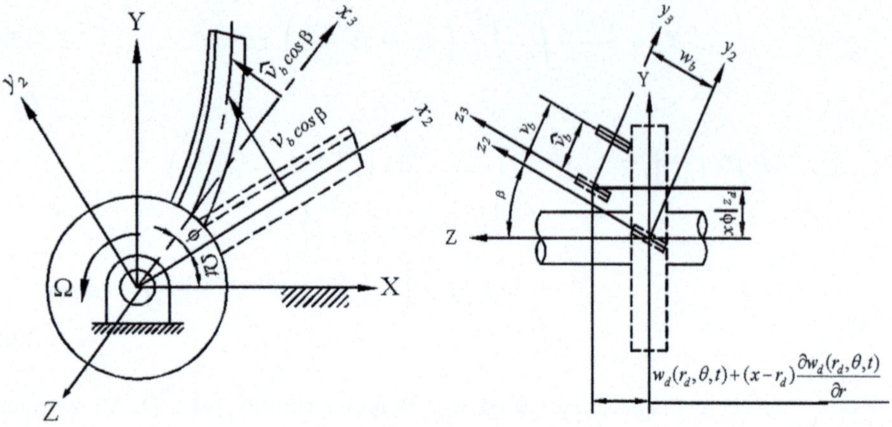

Fig. 4.3 Blade's deformation and Coordinate builds

$$U_b = \int_{r_d}^{r_b} \frac{E_b I_A}{2}\left(\frac{\partial^2 v_b}{\partial x^2}\right)^2 dx + \int_{r_d}^{r_b} \frac{1}{4}\Omega^2 \rho_b A_b \left(r_b^2 - x^2\right)\left[\left(\frac{\partial v_b}{\partial x}\right)^2 + \left(\frac{\partial w_b}{\partial x}\right)^2\right] dx \tag{4.12}$$

where, the y_2 and z_2 directions of transverse displacements used v_b and w_b. I_A denotes the area moment of inertia about the z_3 axis, and I_b denotes the polar moment of inertia.The total blade's displacements $v_b(x, t)$ and $w_b(x, t)$, are the shaft's torsion displacement $\varphi(Z_d, t)$; the transverse displacement of disk used w_d and the bending blade displacement $\hat{v}_b\,(x, t)$. The kinematic relations between these displacements are

$$v_b\,(x,t) = \hat{v}_b + x\varphi\,\big|_{Z_d} \cos\beta - \left(w_d\,\big|_{r_d} + x w_d'\,\big|_{r_d}\right)\sin\beta \tag{4.13}$$

$$w_b\,(x,t) = x\varphi\,\big|_{Z_d} \sin\beta + \left(w_d\,\big|_{r_d} + x w_d'\,\big|_{r_d}\right)\cos\beta \tag{4.14}$$

The AMM is used to discretize the continuous system, as following,

$$\varphi\,(Z,t) = \sum_{i=1}^{n_s} \Phi_i(Z)\eta_i(t) = \Phi(Z)\eta(t) \tag{4.15}$$

$$w_d\,(r,\theta,t) = \sum_{i=1}^{n_d}\left[W_i^c\,(r,\theta)\,\zeta_i^c(t) + W_i^s\,(r,\theta)\,\zeta_i^s(t)\right] = W\,(r,\theta)\,\zeta\,(t) \tag{4.16}$$

$$\hat{v}_{b_k}\,(x,t) = \sum_{i=1}^{n_b} V_i(x)\xi_{ik}(t) = V(x)\zeta_k(t) \tag{4.17}$$

The mode shapes of bending blade, transverse disk and torsion shaft is used $V_i, \{W_i^c, W_i^s\}, \Phi_i$.

$\eta_i, \zeta_i^c, \zeta_i^s, \xi_{ik}$ are the participation factors. n_s, n_d, n_b are subscripts of the corresponding sub-units, are the modes number seemed suit for permissible accuracy. The terms $(n_b, n_d, n_s) = (11, 10, 8)$ are good for yield accuracy up to 10^{-5} Hz in this chapter. These modes are

$$\Phi_i(Z) = \sin\left[\frac{(2i-1)\pi Z}{2L_s}\right] \tag{4.18}$$

$$W_i^c(r,\theta) = R_i(r)\cos[(i-1)\theta] \tag{4.19}$$

$$W_i^s(r,\theta) = R_i(r)\sin[(i-1)\theta] \tag{4.20}$$

$R_i(r)$ denotes the radial function of disk and chose the beam function.

$$V_i(x) = (\sin\tau_i x - \sinh\tau_i x) + \alpha_i(\cos\tau_i x - \cosh\tau_i x) \tag{4.21}$$

is a beam function for blade with

$$[\cos\tau_i(r_b - r_d)][\cosh\tau_i(r_b - r_d)] + 1 = 0 \tag{4.22}$$

$$\alpha_i = \frac{-\sin\tau_i(r_b - r_d) - \sinh\tau_i(r_b - r_d)}{\cos\tau_i(r_b - r_d) + \cosh\tau_i(r_b - r_d)} \tag{4.23}$$

The above equations altered the Lagrange equations yields employment and the energy expression. The following discretizated equations of motion in matrix notation as,

$$[M]\ddot{q} - \Omega[P]\dot{q} + \left([K^e] + \left[K^i\right] - \Omega^2\left[K^\Omega\right]\right)q = 0 \tag{4.24}$$

$-\Omega[P]$ induces natural frequencies bifurcation, obtained from the Coriolis effect. Note, if the disk is flexible and $[P]$ is not zero. So the disk has flexible and the frequency could be bifurcated. $[K^e]$ is observed from the elastic deflection at low rotational speed. $[K^i]$ is the stiffness form rotation with initial stress. $-\Omega^2[K^\Omega]$, observed from the rotor rotation, softening it becomes very obvious at high speed. That is very significant to the rotor stability effect. The matrices $[M]$, $[P]$, $[K^e]$, $[K^i]$, $[K^\Omega]$ are listed in Appendix.

$(n_s + 2N_d \times n_d + N_d \times N_b \times n_b) \times (n_s + 2N_d \times n_d + N_d \times N_b \times n_b)$ are the above matrices dimensions, N_b and N_d is the blade's and disk's numbers, respectively. The items of the above matrices are shown in the Appendix. q is a generalized vector as following,

$$q = \left\{ \eta^T \backslash \zeta^{cT} \zeta^{sT} \backslash \xi_1^T \xi_2^T \cdots \xi_N^T \right\}^T \tag{4.25}$$

Free vibration analysis in an ordinary way, that is a hypothetical solution which is of the form $q = \{c\}e^{\lambda t}$ with $\{c\}$, the un-known coefficient vector and λ is the eigenvalue.

4.3 Finite Element Methods

In References [19, 20], the authors have ever used three kinds of Finite element method (FEM) software to simulate the rotor system. Based on these results, the authors found ANSYS is the best one and the errors using ANSYS are less than 1%. As has been mentioned above, only the finite element method adopted ANSYS software to calculate the shaft-disk-blade system in this chapter. Table 4.1 gives the properties of geometric and material, whose parameters are utilized in the AMM and FEM. Figure 4.4 is a general FE mesh. The element types of subsystem in the rotor system were simulated for many times, which engendered the results of different frequencies, and the outstanding ones are chosen as follows. The perfect element types are that the blades, disks and shaft chose 3D hexahedral solid elements to approach. After being tested for many times, the elements and nodes number are given. Single disk rotor system utilized 70,000 nodes and 60,000 elements and the two disks rotor system utilized 110,000 nodes and 100,000 elements, respectively. The shaft torsion vibration boundary condition is clamped-free.

Table 4.1 The properties of geometric and material

Shaft	Density: ρ_s	7850 kg/m^3
	Shear modulus: G_s	75 Gpa
	Shaft length: L_s	0.6 m
	Radius: r_s	0.04 m
Disk	Density: ρ_d	7850 kg/m^3
	Young's modulus: E_d	200 Gpa
	Location: z_d	0.3 m
	Outer radius: r_d	0.2 m
	Thickness: h_d	0.03 m
	Poisson's ratio: ν	0.3
Blade	Density: ρ_b	7850 kg/m^3
	Young's modulus: E_b	200 Gpa
	Blade outer end: r_b	0.4 m
	Cross-section: A_b	1.2×10^{-4} m^2
	Area moment of inertia: I_A	1.92×10^{-9} m^2
	Staggle angle: β	30°

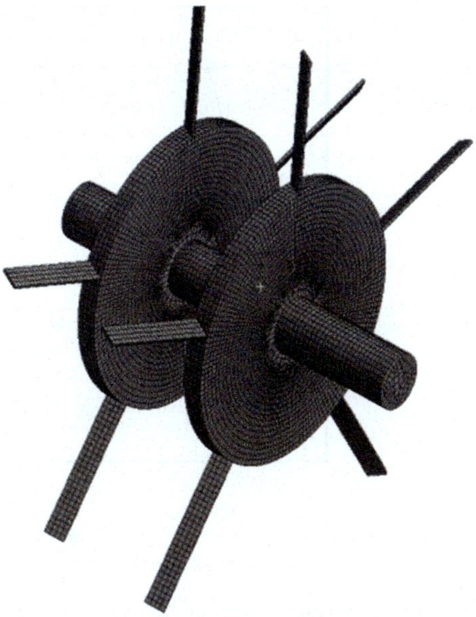

Fig. 4.4 A gerneral FE mesh of a multi-disc rotor

Table 4.2 NF (Hz) of subunits

Component's N.F.	ω_1	ω_2	ω_3
Shaft-disk(w/o blades)	207.418	2645.690	5267.204
Disk(shaft rigid)	921.227	974.922	1205.039
Clamped blade(shaft-disk rigid)	81.538	510.99	1430.788

4.4 Numerical Results

Table 4.2 lists natural frequencies (NF) of subunits on coupled vibration. These frequencies are used as the numerical results for interpretation and verification.

Its range is from root of blade (0) to tip of blade (1). Figures 4.5, 4.6, 4.7, 4.8, 4.9, 4.10, 4.11, 4.12, 4.13, 4.14, 4.15, and 4.16 and Tables 4.2, 4.3, 4.4, 4.5, and 4.6 bewrote at no rotation ($\Omega = 0$). The reasons are that the shaft torsion vibration boundary condition is clamped-free. If boundary is different, the selected modes could be transformed well. The modes of a rotating rotor system are called the traveling modes of its non-rotating modes.

This chapter is based on the research [17, 18]. Acrroding to these two paper results, this work expands the research more complex phenomenon in the multi disk system. The results that have been shown in the literature are not repeated and displayed in this chapter. And in order to make the phenomenon obviously appear, larger error is adopted in this chapter. For example, mistuned blades length adopted $\pm 10\%$ and disordered staggle angle adopted $+30^{\circ}$. Table 4.3 and 4.4 are the six/seven modes of natural frequencies of a five-/six-blades and single flexible

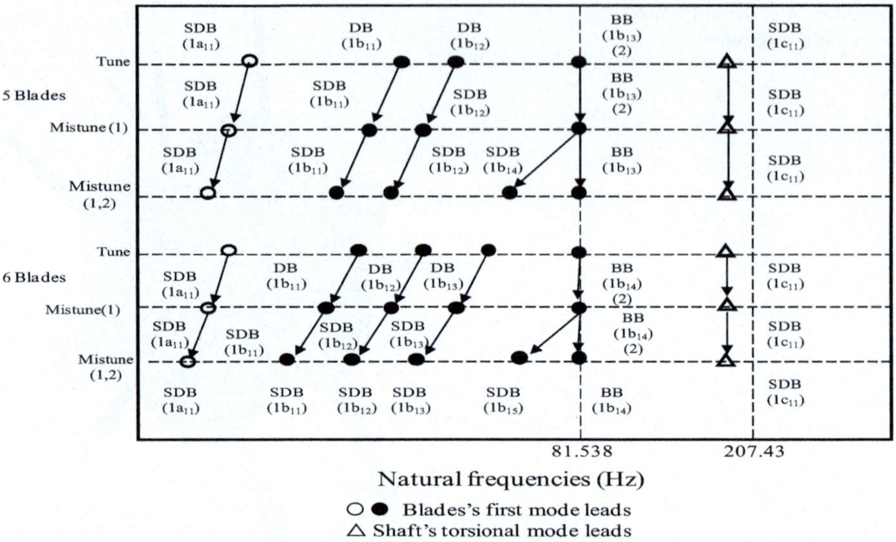

Fig. 4.5 Frequency variation based on asymmetric mistuned blade in a six-/five-blades and a single disk system

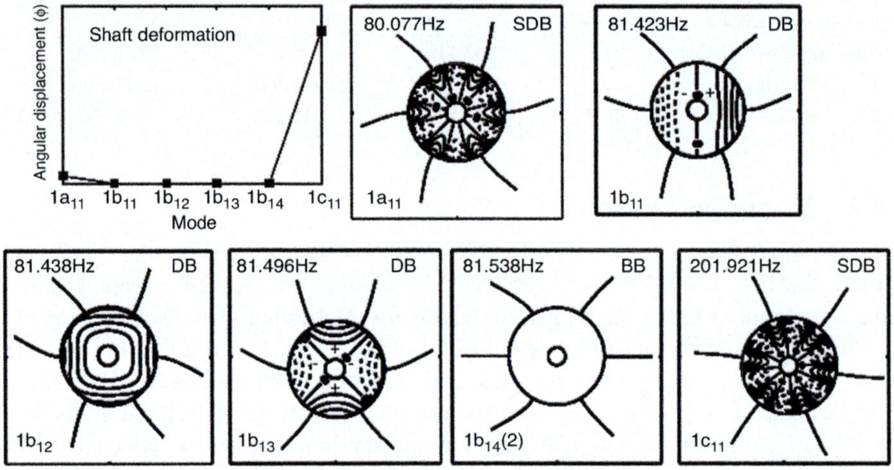

Fig. 4.6 Seven modes of a six-blades and single disk rotor using AMM

disk with mistuned blades by two methods. Figure 4.5 shows the frequency varies five- or six-blade with mistuned blades in the rotor.

For a five-blade and single flexible disk system [18], the coupling modes could be displayed in three groups, namely, SDB, DB and BB. BB repeated modes were of $N_b/2$ and $(N_b - 1)/2$ for even and odd number blades. $1a_{11}$ to $1b_{13}$ modes belong to one group which led by the blade's first mode. $1c_{11}$ mode is led by the shaft's first

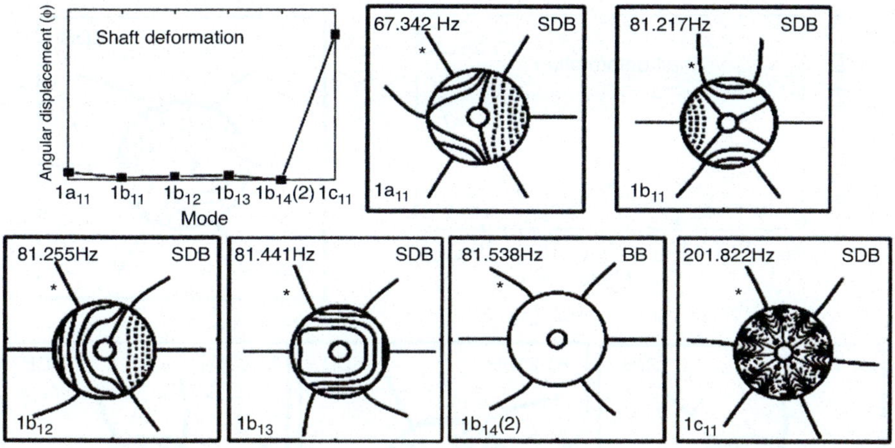

Fig. 4.7 Seven modes of a six-blades and single disk rotor with one blade mistune +10% using AMM

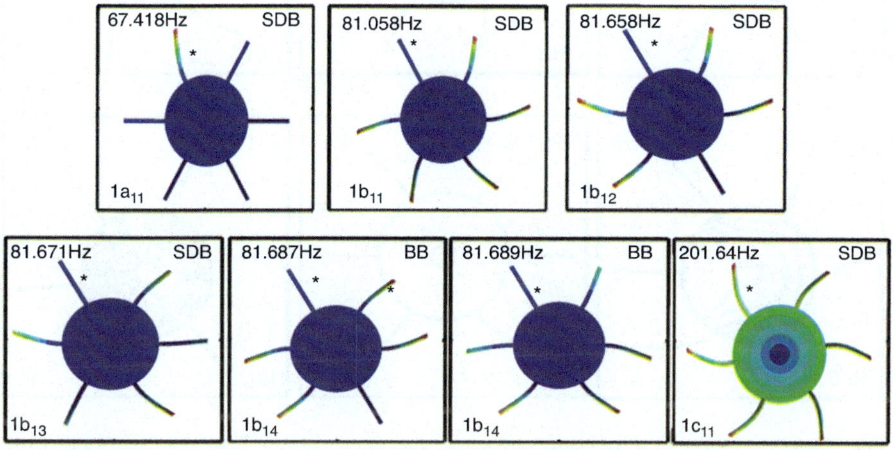

Fig. 4.8 Seven modes of a six-blades and single disk rotor with one blade mistune +10% using FEM

mode. These mistune blades locate in asymmetric place of the system. 1a–1b modes are groups where the blade's first mode predominates in a five-blade system. Note the modes change phenomenon shown in this figure is not drawn in a linear scale. And it has two reference marks, namely, ω= 81.538 and 207.43, which indicate the system frequencies predominates by cantilevered first bending of blade and first torsion of shaft (Table 4.2), respectively.

Three fascinating phenomenon were found by authors in the Fig. 4.5 and Tables 4.3 and 4.4. First, as a mistune appears in a blade, the shaft-disk-blade system only possesses two types of coupled modes, which are SDB and BB modes. Disk-blade (DB) modes vanish because the mistune destroyed the system symmetry and the

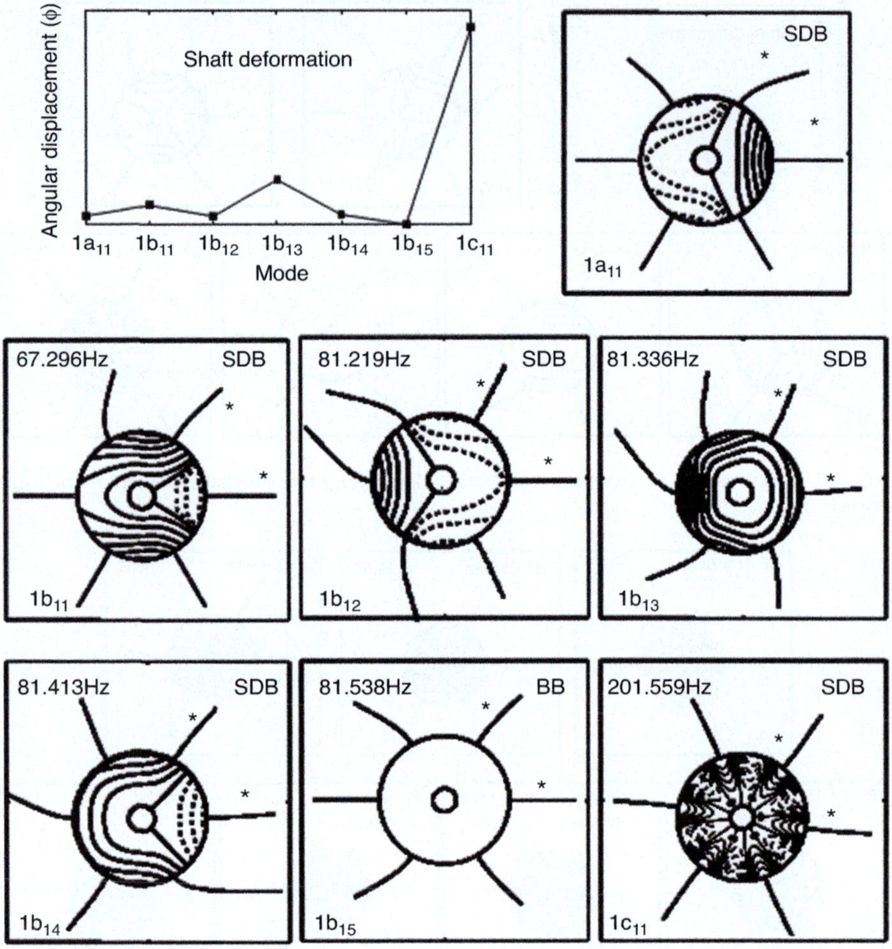

Fig. 4.9 Seven modes of a six-blades and single disk rotor with two blades (1,2) mistune +10% using AMM

balance between disks and blades no longer exists. The second is that two mistune blades appear in two asymmetric blades, the frequencies of $1a_{11}$, $1b_{11}$ and $1b_{12}$ modes drop again. The repeat frequencies of BB modes are bifurcated as one SDB mode and one BB mode. Third, if the numbers of mistune blades is more than the numbers of BB modes, the BB modes would not exist further. For example, a single disk and six-blades rotor system owns three asymmetric mistune blades, all modes are SDB modes (see Table 4.3). Fourth, different tune system, the error have negative deviation in the mistune blade rotor by using Ansys. At last, it has to be noticed that $1c_{11}$ (SDB) modes of the shaft predominates is not affected by mistune blades.

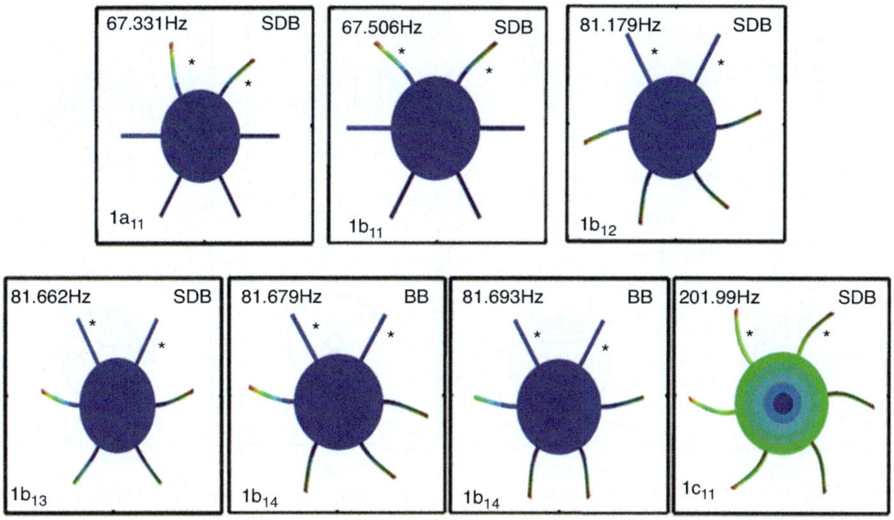

Fig. 4.10 Seven modes of a six-blades and single disk rotor with two blades (1,2) mistune +10% using FEM

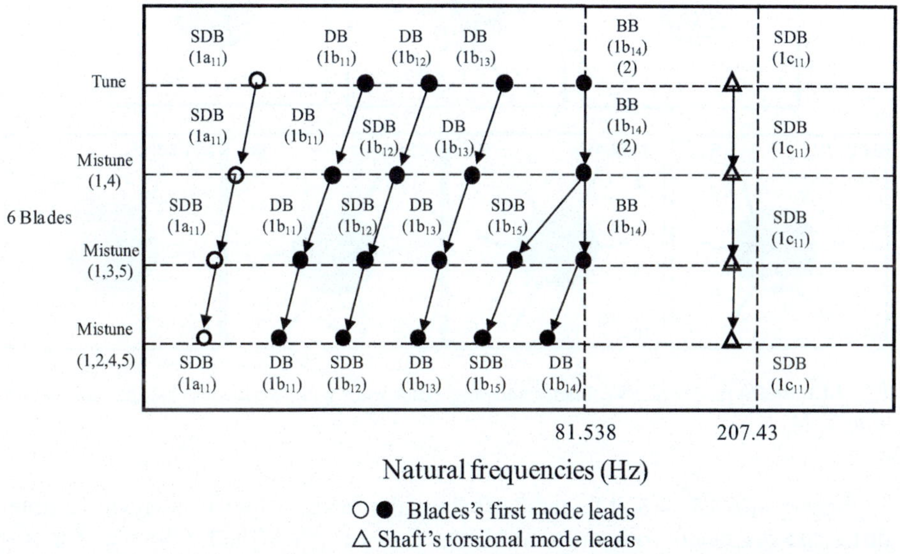

Fig. 4.11 Frequency variation based on symmetric mistuned blades in a six-blades and single disk system

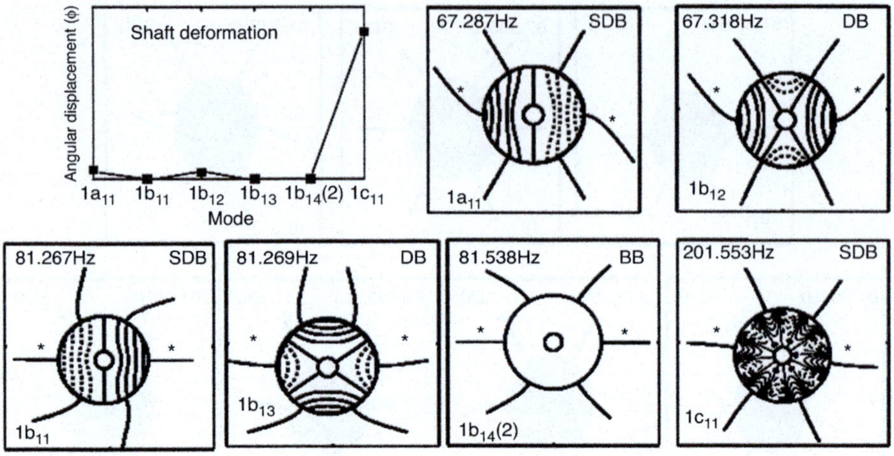

Fig. 4.12 Seven modes of a six-blades and single disk rotor with two blades (1,4) mistune +10% using AMM

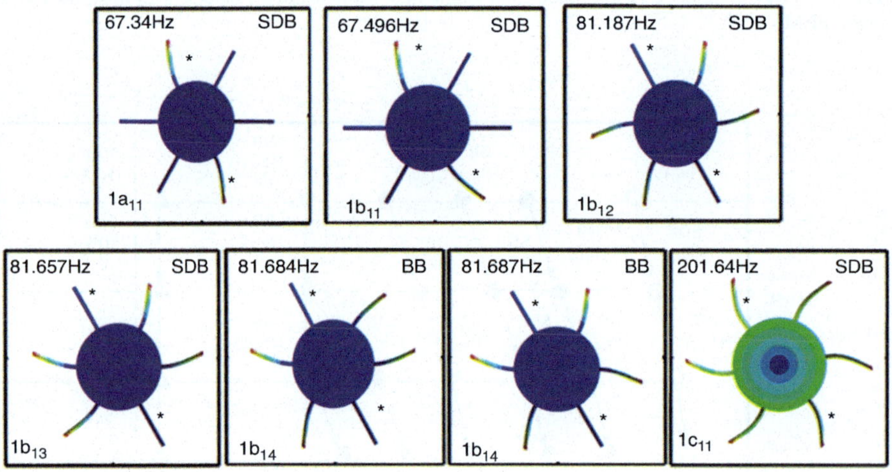

Fig. 4.13 Seven modes of a six-blades and single disk rotor with two blades (1,4) mistune +10% using FEM

Figures 4.6, 4.7, 4.8, 4.9, and 4.10 show the mode shapes of a single flexible-disks and six-blade with mistune blades case by AMM and ANSYS. The x-y figure shows the shaft torsion displacement. The deflections of blade and disk are illustrated in the diagrams. The mode type and the NF are indicated on each plot's upper right and left. Figures 4.8 and 4.10 are the modes by ANSYS, the others is the modes by AMM. After comparing results by using these figures and Tables 4.3 and 4.4, the authors obtained interesting phenomenon. The most momentous point is that all outcomes could match both results as using AMM and ANSYS.

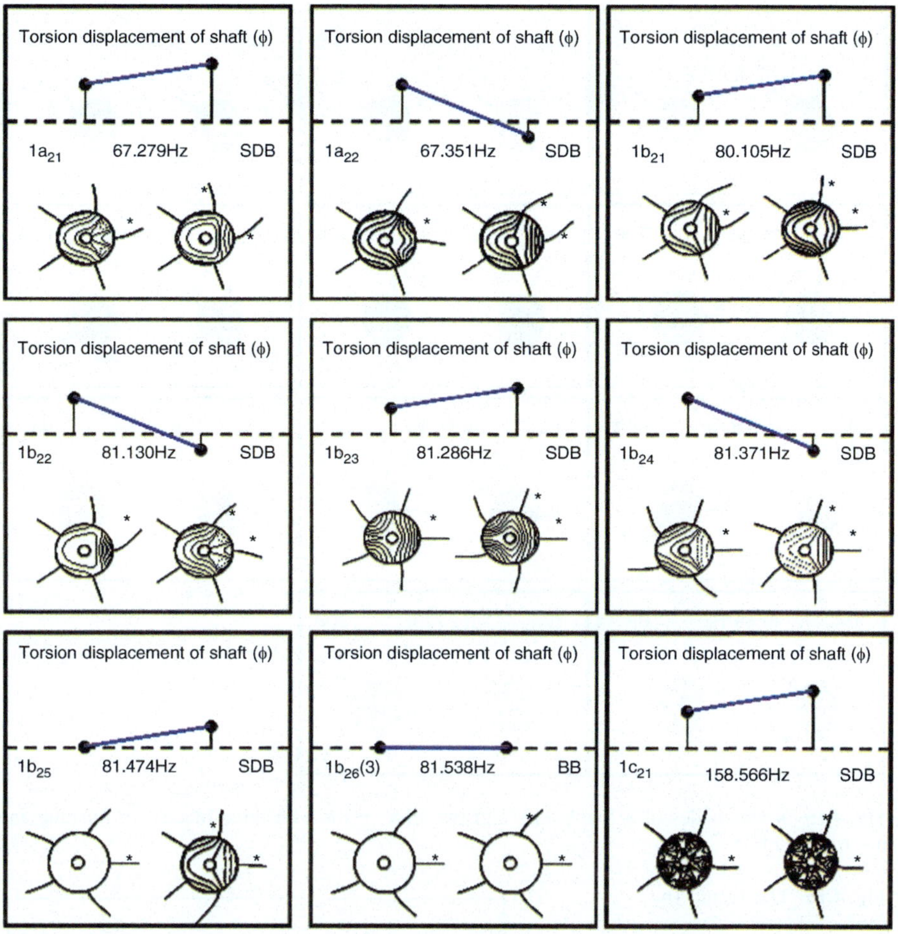

Fig. 4.14 [!t]Night modes of a five-blades and two disks rotor with three blades (*) mistune and disorder using AMM

According to references [19, 20], we understand both methods why are a bit different. The authors gave the following explanations. At first, we knew the numbers of frequencies that the blade's first mode leads is Nb and the shaft's first mode predominates $1c_{11}$ are 1 (one). Secondly, as using assumed mode method, the authors considered that the rotor is torsional shaft of fix-free, transverse disk of clamp-free and bending blade of fix-free. From the numerical results, we found that the disk deformation is much smaller than that of the blade and shaft, that are $w_d/v_b \approx 10^{-3}$~10^{-4} and $w_d/\varphi \approx 10^{-3}$~$10^{-4}$. Because of the NF changes, the disk mode can not be ignored. While the finite element method is used, the deformation is less than the error of the set and it can be ignored. So the modes are right in the usage of assumed mode method. Thirdly, we observed $1a_{11}$ and

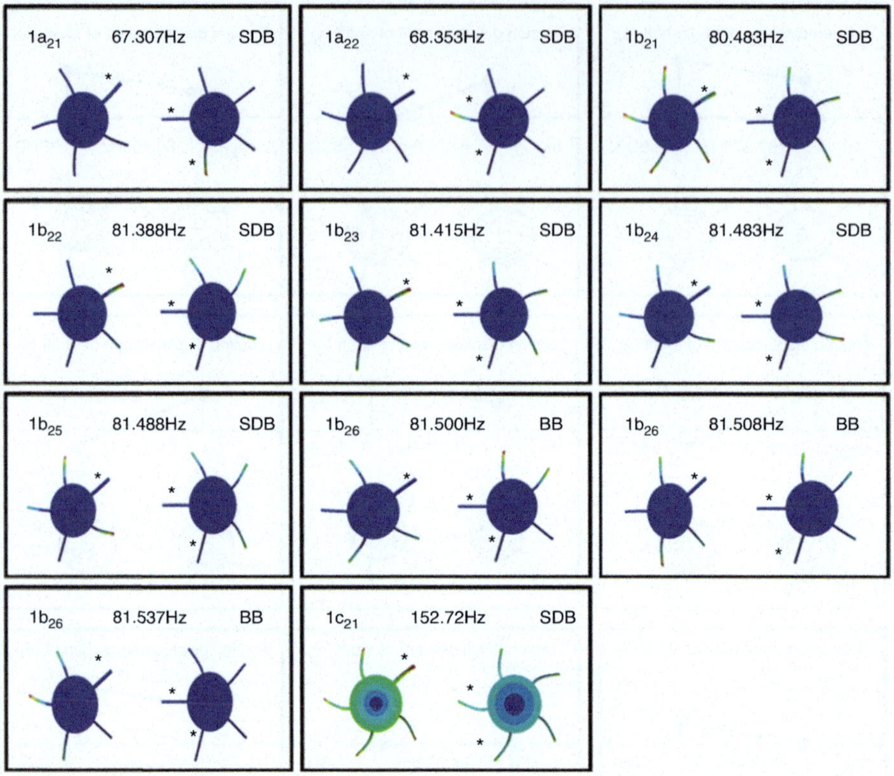

Fig. 4.15 Night modes of a five-blades and two disks rotor with three blades (*) mistune and disorder by FEM

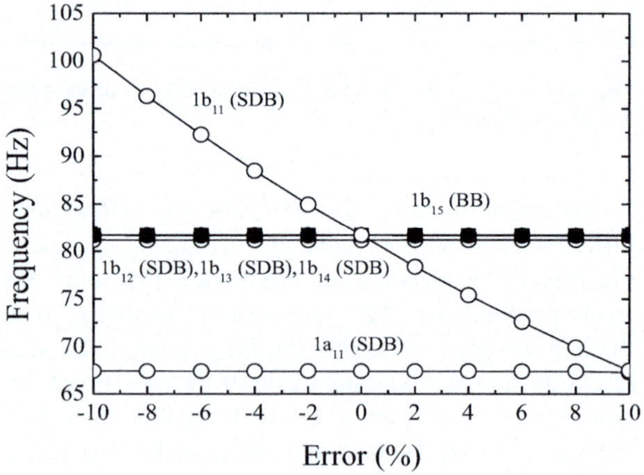

Fig. 4.16 The 1a and 1b modes frequency changes with length error of blade for a six-blade rotor

$1c_{11}$ modes, from blade deformation situation; it is convinced that there are SDB modes, although 1a11 mode has no disk deformation in the Figs. 4.8 and 4.10. Lastly, $1b_{14}$ modes are repeated modes in these figures. The NF of two modes are close and the blade deformation is so similar, we could say there are repeated BB modes in Table 4.3. After comparison, it is clear that the results are right. The other amusing phenomenon is that the numbers of mistune blades increased and the disk's deformation become more complex. $1c_{11}$ (SDB) modes of the shaft predominates, the blade joined coupled slightly and the disk keeps invariably symmetric mode. Lastly, let us see Figs. 4.8 and 4.10. We could not found different situation from the disk between tune system and mistune blade system. In other words, we could say that the finite element method is not suitable for analyzing these complicated change phenomenon of coupling vibration in a multi-disc rotor with multi mistune blades. The finite element method could capture the mistune blade deformation, but the AMM is not.

Table 4.5 is the six modes of NFs of a five-blades and single flexible disk with disorder angle blades (+30°) using two methods. The phenomenon of disorder angle blades system have same situation. But the frequencies changed slightly than the system with mistuned blades.

The next case study discussed frequency variation due to symmetric mistune blade in a six-blades and a single disk rotor is displayed in Table 4.6 and Fig. 4.11. When the mistune blades appears in the symmetric blade, the phenomenon of coupled vibration is rosy interesting and different. Firstly, the modes of predominates by cantilevered first bending of blade display order are; SDB ($1a_{11}$), DB ($1b_{11}$), SDB ($1b_{12}$), DB ($1b_{13}$) and repeat BB ($1b_{14}$) modes in mistune blades locate in number 1 and 4 blade of the rotor system. If the mistune blades locate in number 1, 3 and 5, the modes arise order are SDB, DB, SDB, DB and repeat BB (1b14) modes bifurcate SDB and BB modes. If the mistune blades locate in number 1, 2, 4 and 5, the mode sequence are SDB, DB, SDB, DB, SDB and DB and the BB mode disappears. At last, the $1c_{11}$ (SDB) modes of the shaft predominate, and are not affected by the mistune blades.

Figures 4.12 and 4.13 show the mode shapes of a six-blades and single flexible-disk with symmetric mistune blades case by AMM. Although the disk's balance is destroyed because of mistune damage, we also find new equilibrium from the symmetry of mistune, such as $1b_{11}$ and $1b_{13}$ modes. These modes always are DB modes. $1c_{11}$ (SDB) modes of the shaft predominates, the blade joined coupled slightly and the disk keeps invariably symmetric mode. Lastly we compared Fig. 4.8 with Fig. 4.10, we could find that we are hardly to distinguish mode shapes or frequencies different from mistune or tune system using ANSYS software. So FEM is not good method to complete these studies.

Figures 4.14 and 4.15 show the mode shapes of a five-blade and two flexible-disc with three mistune and disorder blades case by AMM and ANSYS. The first disk has with disorder angle blades (+30°) locate in the mark (*) blade, and the second disk has two blades (1,2) mistune +10% locate in the mark (*) blades. The upper x-y figure shows the shaft torsion displacement. The deflections of blade and disk are illustrated in the diagrams. The modes type and NF are indicated on each plot's

Table 4.3 Six modes of NF (Hz) of a five-blades and single flexible disk with mistuned blades (±10%) using two methods

Mode	AMM (Tune)	FEM (Tune)	Error (%)	AMM (+10%) (mistune,1)	FEM (+10%) (mistune,1)	Error (%)
$1a_{11}$	80.891(SDB)	81.020(SDB)	0.159	67.254(SDB)	67.411(SDB)	0.233
$1b_{11}$	81.422(DB)	81.615(DB)	0.237	81.285(SDB)	81.199(SDB)	−0.106
$1b_{12}$	81.492(DB)	81.631(DB)	0.171	81.47(SDB)	81.666(SDB)	0.241
$1b_{13}$	81.538(BB)	81.670(BB)	0.161	81.538(BB)	81.702(BB)	0.201
$1b_{13}$	81.538(BB)	81.671(BB)	0.163	81.538(BB)	81.706(BB)	0.206
$1c_{11}$	202.777(SDB)	203.250(SDB)	0.237	202.777(SDB)	202.95(SDB)	0.085
Mode	AMM(−10%)(mistune,1)	FEM(−10%)(mistune,1)	Error(%)	AMM (+10%)(mistune,1,2)	FEM(+10%)(mistune,1,2)	Error(%)
$1a_{11}$	100.35(SDB)	100.33(SDB)	−0.020	67.254(SDB)	67.327(SDB)	0.109
$1b_{11}$	81.284(SDB)	81.034(SDB)	−0.308	67.254(SDB)	67.495(SDB)	0.358
$1b_{12}$	81.47(SDB)	81.478(SDB)	0.010	81.285(SDB)	81.327(SDB)	0.052
$1b_{13}$	81.538(BB)	81.544(SDB)	0.007	81.47(SDB)	81.7(SDB)	0.282
$1b_{13}$	81.538(BB)	81.618(SDB)	0.098	81.538(BB)	81.701(BB)	0.200
$1c_{11}$	202.777(SDB)	203.68(SDB)	0.445	202.777(SDB)	202.58(SDB)	−0.097

Table 4.4 Seven modes of NF (Hz) of a six-blades and single flexible disk with mistuned blades (±10%) using two methods

Mode	AMM (Tune)	FEM (Tune)	Error (%)	AMM (+10%) (mistune,1)	FEM (+10%) (mistune,1)	Error (%)
$1a_{11}$	80.770(SDB)	81.380(SDB)	0.755	67.342(SDB)	67.418(SDB)	0.113
$1b_{11}$	81.423(DB)	81.572(DB)	0.183	81.217(SDB)	81.058(SDB)	−0.196
$1b_{12}$	81.438(DB)	81.575(DB)	0.168	81.255(SDB)	81.658(SDB)	0.496
$1b_{13}$	81.496(DB)	81.634(DB)	0.169	81.441(SDB)	81.671(SDB)	0.391
$1b_{14}$	81.538(BB)	81.668(BB)	0.159	81.538(BB)	81.687(BB)	0.183
$1b_{14}$	81.538(BB)	81.669(BB)	0.161	81.538(BB)	81.689(BB)	0.185
$1c_{11}$	201.921(SDB)	203.540(SDB)	0.802	201.822(SDB)	201.64(SDB)	−0.09
Mode	AMM(−10%)(mistune,1)	FEM(−10%)(mistune,1)	Error(%)	AMM(+10%)(mistune,1,2)	FEM(+10%)(mistune,1,2)	Error(%)
$1a_{11}$	99.97(SDB)	100.36(SDB)	0.39	67.296(SDB)	67.331(SDB)	0.052
$1b_{11}$	81.208(SDB)	81.171(SDB)	−0.046	67.349(SDB)	67.506(SDB)	0.233
$1b_{12}$	81.267(SDB)	81.664(SDB)	0.489	81.219(SDB)	81.179(SDB)	−0.049
$1b_{13}$	81.357(SDB)	81.671(SDB)	0.386	81.413(SDB)	81.662(SDB)	0.306
$1b_{14}$	81.538(BB)	81.681(BB)	0.175	81.538(BB)	81.679(BB)	0.173
$1b_{14}$	81.538(BB)	81.694(BB)	0.191	81.538(BB)	81.693(BB)	0.190
$1c_{11}$	202.505(SDB)	202.24(SDB)	−0.131	201.559(SDB)	201.99(SDB)	0.214

Table 4.5 Six modes of NF (Hz) of a five-blades and single flexible disk with disorder angle blades (+30°) using two methods

Mode	AMM (disorder,1)	FEM (disorder,1)	Error (%)	AMM (disorder,1,2)	FEM (disorder,1,2)	Error (%)
$1a_{11}$	81.265(SDB)	80.996(SDB)	−0.331	81.265(SDB)	81.085(SDB)	−0.221
$1b_{11}$	81.449(SDB)	81.403(SDB)	−0.056	81.265(SDB)	81.42(SDB)	0.191
$1b_{12}$	81.512(SDB)	81.496(SDB)	−0.020	81.449(SDB)	81.516(SDB)	0.082
$1b_{13}$	81.538(BB)	81.544(BB)	0.007	81.512(SDB)	81.567(SDB)	0.067
$1b_{13}$	81.538(BB)	81.621(BB)	0.102	81.538(BB)	81.618(BB)	0.098
$1c_{11}$	202.777(SDB)	201.93(SDB)	−0.418	202.777(SDB)	200.57(SDB)	−1.088

Table 4.6 Seven modes of NF (Hz) of a six-blades and single flexible disk with mistuned (+10%) blades in the symmetric system using assumed mode method

Mode	AMM (mistune,14)	FEM (mistune,14)	Error (%)	AMM (mistune,135)	FEM (mistune,135)	Error (%)
$1a_{11}$	67.287(SDB)	67.34(SDB)	0.079	67.281(SDB)	67.262(SDB)	−0.028
$1b_{11}$	67.318(DB)	67.496(DB)	0.264	67.313(DB)	67.496(DB)	0.272
$1b_{12}$	81.267(SDB)	81.187(SDB)	−0.098	67.394(DB)	67.496(SDB)	0.151
$1b_{13}$	81.269(DB)	81.657(DB)	0.477	81.238(SDB)	81.319(DB)	0.100
$1b_{14}$	81.538(BB)	81.684(BB)	0.179	81.327(DB)	81.671(SDB)	0.423
$1b_{14}$	81.538(BB)	81.687(BB)	0.183	81.538(BB)	81.672(DB)	0.164
$1c_{11}$	201.553(SDB)	201.64(SDB)	0.043	201.289(SDB)	201.29(SDB)	0.0005

middle line. The two cases compared to Tables 4.3 and 4.5, it is seen that the system will retain mode shapes and frequencies in a single disk system, and combine each other. In other words, they are linear relations. At the same time, we also hardly distinguished mode shapes or frequencies which are different from mistune (disk 2) and disorder (disk 1) system using ANSYS software in this case study.

The authors explored how the frequencies change with length error of blade for a six-blade system in the Fig. 4.16. We set blade (1) has 10% mistune and blade (2) has mistune from -10% to $+10\%$. From the Fig. 4.16, we found two results. First, the frequencies of $1a_{11}$ and $1b_{12}$ to $1b_{15}$ modes keep unchange in the length error of blade variation. Second, it is seemed that the frequencies of $1b_{11}$ (SDB) modes have decrease from 100.63 Hz to 67.50 Hz with blade (2) length error increase. In other words, the mistuned and disorder affect frequency and mode in the multi mistuned blades rotor system, respectively.

In last case, the authors describe the rotor's natural frequencies varying with rotation. The numerical results of this case are normalized with respect to the first order natural frequency (ω_{b1}) of the cantilever blade, i.e., $\omega^* = \omega/\omega_{b1}$, $\Omega^* = \Omega/\omega_{b1}$. Figure 4.17 shows the results of the two disks and five-blade rotor system. In Fig. 4.17, the system exist frequency bifurcating at modes (see Figs. 4.14 and 4.15). For the forward and backward frequencies close to each other, when the rotation increases, two frequencies merge and become one ultimately. At this merge point, the system undergoes a possible instability. Figure 4.17(a) is a tune system, and however Fig. 4.17(b) is a two mistuned and a disorder blades system. Compared these two figures, we could find that variation of eigenvalues with rotation speed and mistuned effect would become complex and unstable.

4.5 Conclusions

Based on the previous researches, we completely consummated the blade-bending, shaft-torsion and disk-transverse coupled vibrations phenomenon of the multi-flexible-disc rotor system. The blades subsystem has mistuned blades length or disordered staggle angle blades. We mainly used the AMM, and FEM is the complementary one to analyze the system. And we compared their different results. The researches start with the modes change resulting from a five-/six-blades and flexible disk rotor. The coupling modes could be classified into three group's types, the SDB, DB and BB. Two of the most interesting things are that symmetry of the blades which have mistuned blades or disordered staggle angle will lead to balance, that get misjudgment of frequency and model. And as the numbers of mistuned/disordered blades increases, the disk's deformation become more complex.

After comparing the results of AMM with the ones of FEM (using ANSYS), it can be concluded that the author's previous researches are right. But the results that we get by using ANSYS had some shortcomings. For example, we could not see the disk's deformation firstly. Secondly, we could not observe inter-blade (BB) modes

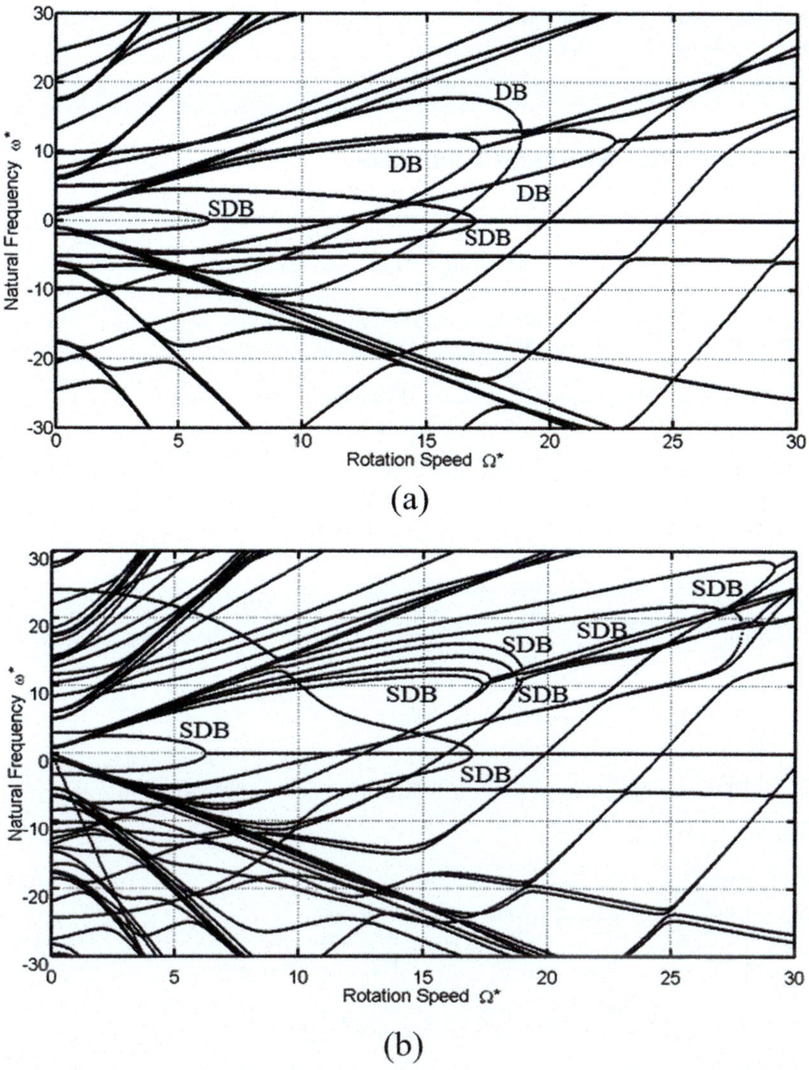

Fig. 4.17 Variation of eigenvalues with rotation speed for a five blades in the two disks rotor. (**a**) tune system (**b**) two mistuned and a disorder blades system

are repeated modes. So, we hardly distinguish mode shapes or frequencies which are different from mistune or disorder by using disk mode shapes. For these reasons, we could say that the finite element method is only complementary one as we analyze these complicated change phenomenon of coupled vibration in a multi-disc rotor with multi mistune or disorder blades. Lastly, the rotation effect is also explored in this chapter and the authors found that mistuned effect would become complex and precarious.

Acknowledgments This work was supported by the Fujian Nature foundation, No. 2016J01039; Xiamen City Project No. 3502Z20173037.

A.1 Appendix: Matrices Elements

$$[\mathrm{M}] = \begin{bmatrix} [\mathrm{M}_s]_{(n_s\times n_s)1} & [0]_{(n_s\times n_d)N_d} & [0]_{(n_s\times n_d)N_d} & [\mathrm{M}_{sb}]_{(n_s\times n_b)N_d} & [\mathrm{M}_{sb}]_{(n_s\times n_b)N_d} & \cdots & [\mathrm{M}_{sb}]_{(n_s\times n_b)N_d} \\ [0]^T_{(n_d\times n_s)N_d} & \left[\mathrm{M}^s_d\right]_{(n_d\times n_d)N_d} & [0]_{(n_d\times n_d)N_d} & \left[\mathrm{M}^s_{db}\right]_{(n_d\times n_b)N_d} & \left[\mathrm{M}^s_{db}\right]_{(n_d\times n_b)N_d} & \cdots & \left[\mathrm{M}^s_{db}\right]_{(n_d\times n_b)N_d} \\ [0]^T_{(n_d\times n_s)N_d} & [0]_{(n_d\times n_d)N_d} & \left[\mathrm{M}^c_d\right]_{(n_d\times n_d)N_d} & \left[\mathrm{M}^{*c}_{db}\right]_{(n_d\times n_b)N_d} & \left[\mathrm{M}^c_{db}\right]_{(n_d\times n_b)N_d} & \cdots & \left[\mathrm{M}^c_{db}\right]_{(n_d\times n_b)N_d} \\ \left[\mathrm{M}^*_{sb}\right]^T_{(n_b\times n_s)N_d} & \left[\mathrm{M}^{*s}_{db}\right]^T_{(n_b\times n_d)N_d} & \left[\mathrm{M}^{*c}_{db}\right]^T_{(n_b\times n_d)N_d} & \left[\mathrm{M}^*_b\right]_{(n_b\times n_b)N_d} & [0]_{(n_b\times n_b)N_d} & \cdots & [0]_{(n_b\times n_b)N_d} \\ [\mathrm{M}_{sb}]^T_{(n_b\times n_s)N_d} & \left[\mathrm{M}^s_{db}\right]^T_{(n_b\times n_d)N_d} & \left[\mathrm{M}^c_{db}\right]^T_{(n_b\times n_d)N_d} & [0]^T_{(n_b\times n_b)N_d} & [\mathrm{M}_b]_{(n_b\times n_b)N_d} & \ddots & \vdots \\ \vdots & \vdots & \vdots & \vdots & \ddots & \ddots & [0]_{(n_b\times n_b)N_d} \\ [\mathrm{M}_{sb}]^T_{(n_b\times n_s)N_d} & \left[\mathrm{M}^s_{db}\right]^T_{(n_b\times n_d)N_d} & \left[\mathrm{M}^c_{db}\right]^T_{(n_b\times n_d)N_d} & [0]^T_{(n_b\times n_b)N_d} & \cdots & [0]^T_{(n_b\times n_b)N_d} & [\mathrm{M}_b]_{(n_b\times n_b)N_d} \end{bmatrix} \tag{4.26}$$

$$[\mathrm{P}] = \begin{bmatrix} [0]_{(n_s\times n_s)1} & [0]_{(n_s\times n_d)N_d} & [0]_{(n_s\times n_d)Nd} & [0]_{(n_s\times n_b)N_d} & \cdots & \cdots & [0]_{(n_s\times n_b)N_d} \\ [0]^T_{(n_d\times n_s)N_d} & \left[\mathrm{P}^s_d\right]_{(n_d\times n_d)N_d} & [0]_{(n_d\times n_d)N_d} & [0]_{(n_d\times n_b)N_d} & \cdots & \cdots & [0]_{(n_d\times n_b)N_d} \\ [0]^T_{(n_d\times n_s)N_d} & [0]_{(n_d\times n_d)N_d} & \left[\mathrm{P}^c_d\right]_{(n_d\times n_d)N_d} & [0]_{(n_d\times n_b)N_d} & \cdots & \cdots & [0]_{(n_d\times n_b)N_d} \\ [0]^T_{(n_b\times n_s)N_d} & [0]^T_{(n_b\times n_d)N_d} & [0]^T_{(n_b\times n_d)N_d} & [0]_{(n_b\times n_b)N_d} & [0]_{(n_b\times n_b)N_d} & \cdots & [0]_{(n_b\times n_b)N_d} \\ [0]^T_{(n_b\times n_s)N_d} & [0]^T_{(n_b\times n_d)N_d} & [0]^T_{(n_b\times n_d)N_d} & [0]^T_{(n_b\times n_b)N_d} & \ddots & \ddots & \vdots \\ \vdots & \vdots & \vdots & \vdots & \ddots & \ddots & [0]_{(n_b\times n_b)N_d} \\ [0]^T_{(n_b\times n_s)N_d} & [0]^T_{(n_b\times n_d)N_d} & [0]^T_{(n_b\times n_d)N_d} & [0]^T_{(n_b\times n_b)N_d} & \cdots & [0]^T_{(n_b\times n_b)N_d} & [0]_{(n_b\times n_b)N_d} \end{bmatrix} \tag{4.27}$$

$$\left[\mathrm{K}^e\right] = \begin{bmatrix} \left[\mathrm{K}^e_s\right]_{(n_s\times n_s)1} & [0]_{(n_s\times n_d)N_d} & [0]_{(n_s\times n_d)N_d} & [0]_{(n_s\times n_b)N_d} & \cdots & \cdots & [0]_{(n_s\times n_b)N_d} \\ [0]^T_{(n_d\times n_s)N_d} & \left[\mathrm{K}^{es}_d\right]_{(n_d\times n_d)N_d} & [0]_{(n_d\times n_d)N_d} & [0]_{(n_d\times n_b)N_d} & \cdots & \cdots & [0]_{(n_d\times n_b)N_d} \\ [0]^T_{(n_d\times n_s)N_d} & [0]_{(n_d\times n_d)N_d} & \left[\mathrm{K}^{ec}_d\right]_{(n_d\times n_d)N_d} & [0]_{(n_d\times n_b)N_d} & \cdots & \cdots & [0]_{(n_d\times n_b)N_d} \\ [0]^T_{(n_b\times n_s)N_d} & [0]^T_{(n_b\times n_d)N_d} & [0]^T_{(n_b\times n_d)N_d} & \left[\mathrm{K}^e_b\right]_{(n_b\times n_b)N_d} & [0]_{(n_b\times n_b)N_d} & \cdots & [0]_{(n_b\times n_b)N_d} \\ [0]^T_{(n_b\times n_s)N_d} & [0]^T_{(n_b\times n_d)N_d} & [0]^T_{(n_b\times n_d)N_d} & [0]^T_{(n_b\times n_b)N_d} & \ddots & \ddots & \vdots \\ \vdots & \vdots & \vdots & \vdots & \ddots & \ddots & [0]_{(n_b\times n_b)N_d} \\ [0]^T_{(n_b\times n_s)N_d} & [0]^T_{(n_b\times n_d)N_d} & [0]^T_{(n_b\times n_d)N_d} & [0]^T_{(n_b\times n_b)N_d} & \cdots & [0]^T_{(n_b\times n_b)N_d} & \left[\mathrm{K}^e_b\right]_{(n_b\times n_b)N_d} \end{bmatrix} \tag{4.28}$$

$$\left[\mathrm{K}^\Omega\right] = \begin{bmatrix} \left[\mathrm{K}^\Omega_s\right]_{(n_s\times n_s)1} & [0]_{(n_s\times n_d)N_d} & [0]_{(n_s\times n_d)N_d} & \left[\mathrm{K}^\Omega_{sb}\right]_{(n_s\times n_b)N_d} & \cdots & \cdots & \left[\mathrm{K}^\Omega_{sb}\right]_{(n_s\times n_b)N_d} \\ [0]^T_{(n_d\times n_s)N_d} & \left[\mathrm{K}^{\Omega s}_d\right]_{(n_d\times n_d)N_d} & [0]_{(n_d\times n_d)N_d} & [0]_{(n_d\times n_b)N_d} & \cdots & \cdots & [0]_{(n_d\times n_b)N_d} \\ [0]^T_{(n_d\times n_s)N_d} & [0]_{(n_d\times n_d)N_d} & \left[\mathrm{K}^{\Omega c}_d\right]_{(n_d\times n_d)N_d} & [0]_{(n_d\times n_b)N_d} & \cdots & \cdots & [0]_{(n_d\times n_b)N_d} \\ \left[\mathrm{K}^\Omega_{sb}\right]^T_{(n_b\times n_s)N_d} & [0]^T_{(n_b\times n_d)N_d} & [0]^T_{(n_b\times n_d)N_d} & \left[\mathrm{K}^\Omega_b\right]_{(n_b\times n_b)N_d} & [0]_{(n_b\times n_b)N_d} & \cdots & [0]_{(n_b\times n_b)N_d} \\ \left[\mathrm{K}^\Omega_{sb}\right]^T_{(n_b\times n_s)N_d} & [0]^T_{(n_b\times n_d)N_d} & [0]^T_{(n_b\times n_d)N_d} & [0]^T_{(n_b\times n_b)N_d} & \ddots & \ddots & \vdots \\ \vdots & \vdots & \vdots & \vdots & \ddots & \ddots & [0]_{(n_b\times n_b)N_d} \\ \left[\mathrm{K}^\Omega_{sb}\right]^T_{(n_b\times n_s)N_d} & [0]^T_{(n_b\times n_d)N_d} & [0]^T_{(n_b\times n_d)N_d} & [0]^T_{(n_b\times n_b)N_d} & \cdots & [0]^T_{(n_b\times n_b)N_d} & \left[\mathrm{K}^\Omega_b\right]_{(n_b\times n_b)N_d} \end{bmatrix} \tag{4.29}$$

$$
\left[\mathrm{K}^{i}\right]=\begin{bmatrix}
[0]_{(n_s\times n_s)1} & [0]_{(n_s\times n_d)N_d} & [0]_{(n_s\times n_d)N_d} & [0]_{(n_s\times n_b)N_d} & \cdots & \cdots & [0]_{(n_s\times n_b)N_d}\\
[0]^{T}_{(n_d\times n_s)N_d} & \left[\mathrm{K}^{is}_{d}\right]_{(n_d\times n_d)N_d} & [0]_{(n_d\times n_d)N_d} & [0]_{(n_d\times n_b)N_d} & \cdots & \cdots & [0]_{(n_d\times n_b)N_d}\\
[0]^{T}_{(n_d\times n_s)N_d} & [0]_{(n_d\times n_d)N_d} & \left[\mathrm{K}^{ic}_{d}\right]_{(n_d\times n_d)N_d} & [0]_{(n_d\times n_b)N_d} & \cdots & \cdots & [0]_{(n_d\times n_b)N_d}\\
[0]^{T}_{(n_b\times n_s)N_d} & [0]^{T}_{(n_b\times n_d)N_d} & [0]^{T}_{(n_b\times n_d)N_d} & [0]_{(n_b\times n_b)N_d} & [0]_{(n_b\times n_b)N_d} & \cdots & [0]_{(n_b\times n_b)N_d}\\
[0]^{T}_{(n_b\times n_s)N_d} & [0]^{T}_{(n_b\times n_d)N_d} & [0]^{T}_{(n_b\times n_d)N_d} & [0]^{T}_{(n_b\times n_b)N_d} & \ddots & \ddots & \vdots\\
\vdots & \vdots & \vdots & \vdots & \ddots & \ddots & [0]_{(n_b\times n_b)N_d}\\
[0]^{T}_{(n_b\times n_s)N_d} & [0]^{T}_{(n_b\times n_d)N_d} & [0]^{T}_{(n_b\times n_d)N_d} & [0]^{T}_{(n_b\times n_b)N_d} & \cdots & [0]^{T}_{(n_b\times n_b)N_d} & [0]^{T}_{(n_b\times n_b)N_d}
\end{bmatrix} \tag{4.30}
$$

$$
[\mathrm{M}_s]_{ij}=\int_{0}^{Ls} I_s\Phi_i\Phi_j dZ+I_d\left[\Phi_i\Phi_j\right]_{Z=Z_d}+\rho_b A_b\sum_{1}^{N_b}\int_{r_d}^{r_b} x^2\left[\Phi_i\Phi_j\right]_{Z=Z_d}dx \tag{4.31}
$$

$$
[\mathrm{M}_{sb}]_{ij}=\rho_b A_b\int_{r_d}^{r_b} x\Phi_i\,\big|_{Z=Z_d}\; V_j\cos\beta_k dx \tag{4.32}
$$

$$
[\mathrm{M}_b]_{ij}=\rho_b A_b\int_{r_d}^{r_b} V_i V_j dx \tag{4.33}
$$

$$
\begin{aligned}
[\mathrm{M}_d]_{ij}={}&\rho_d h_d\left\{\int_{r_s}^{r_d}\int_0^{2\pi} W_i^s W_j^s r d\theta dr+\int_{r_s}^{r_d}\int_0^{2\pi} W_i^c W_j^c r d\theta dr\right\}\\
&+\rho_b A_b\sum_{1}^{N_b}\int_{r_d}^{r_b}\left\{\left[W_i^s W_j^s\right]_{\substack{r=r_d\\ \theta=\theta}}+x_k\left[W_i^s\left(\frac{\partial W_j^s}{\partial r}\right)+\left(\frac{\partial W_i^s}{\partial r}\right)W_j^s\right]_{\substack{r=r_d\\ \theta=\theta}}\right.\\
&\left.+x^2\left[\left(\frac{\partial W_i^s}{\partial r}\right)\left(\frac{\partial W_j^s}{\partial r}\right)\right]_{\substack{r=r_d\\ \theta=\theta}}\right\}dx\\
&+\rho_b A_b\sum_{1}^{N_b}\int_{r_d}^{r_b}\left\{\left[W_i^c W_j^c\right]_{\substack{r=r_d\\ \theta=\theta}}+x_k\left[W_i^c\left(\frac{\partial W_j^c}{\partial r}\right)+\left(\frac{\partial W_i^c}{\partial r}\right)W_j^c\right]_{\substack{r=r_d\\ \theta=\theta}}\right.\\
&\left.+x^2\left[\left(\frac{\partial W_i^c}{\partial r}\right)\left(\frac{\partial W_j^c}{\partial r}\right)\right]_{\substack{r=r_d\\ \theta=\theta}}\right\}dx
\end{aligned} \tag{4.34}
$$

$$[\mathrm{M}_{db}]_{ij} = -\rho_b A_b \left\{ \int_{r_d}^{r_b} \left[W_i^s + x \left(\frac{\partial W_i^s}{\partial r} \right) \right]_{\substack{r = r_d \\ \theta = \theta}} V_j \sin \beta_k dx + \int_{r_d}^{r_b} \left[W_i^c + x \left(\frac{\partial W_i^c}{\partial r} \right) \right]_{\substack{r = r_d \\ \theta = \theta}} V_j \sin \beta_k dx \right\} \tag{4.35}$$

$$\left[\mathrm{P}_d^s\right]_{ij} = \rho_d h_d \left\{ \int_{r_s}^{r_d} \int_0^{2\pi} \left[W_i^s \left(\frac{\partial W_j^s}{\partial \theta} \right) + \left(\frac{\partial W_i^s}{\partial \theta} \right) W_j^s \right] r d\theta dr + \int_{r_s}^{r_d} \int_0^{2\pi} \left[W_i^c \left(\frac{\partial W_j^c}{\partial \theta} \right) + \left(\frac{\partial W_i^c}{\partial \theta} \right) W_j^c \right] r d\theta dr \right\} \tag{4.36}$$

$$[\mathrm{K}_s^e]_{ij} = \int_0^{Ls} G_s J_s \Phi_i' \Phi_j' dZ \tag{4.37}$$

$$\begin{aligned} \left[\mathrm{K}_d^e\right]_{ij} &= D \left\{ \int_{r_s}^{r_d} \int_0^{2\pi} r \left(\nabla^2 W_i^s \right) \left(\nabla^2 W_j^s \right) dr d\theta + \int_{r_s}^{r_d} \int_0^{2\pi} r \left(\nabla^2 W_i^c \right) \left(\nabla^2 W_j^c \right) dr d\theta \right\} \\ &+ 2(1-\nu) D \left\{ \int_{r_s}^{r_d} \int_0^{2\pi} \frac{1}{r} \left[\left(\frac{\partial^2 W_i^s}{\partial r \partial \theta} - \frac{1}{r} \frac{\partial W_i^s}{\partial \theta} \right) \left(\frac{\partial^2 W_j^s}{\partial r \partial \theta} - \frac{1}{r} \frac{\partial W_j^s}{\partial \theta} \right) \right] d\theta dr \right. \\ &\left. + \int_{r_s}^{r_d} \int_0^{2\pi} \frac{1}{r} \left[\left(\frac{\partial^2 W_i^c}{\partial r \partial \theta} - \frac{1}{r} \frac{\partial W_i^c}{\partial \theta} \right) \left(\frac{\partial^2 W_j^c}{\partial r \partial \theta} - \frac{1}{r} \frac{\partial W_j^c}{\partial \theta} \right) \right] d\theta dr \right\} \\ &- (1-\nu) D \left\{ \int_{r_s}^{r_d} \int_0^{2\pi} \left[\left(\frac{\partial^2 W_i^s}{\partial r^2} \right) \left(\frac{\partial W_j^s}{\partial r} + \frac{1}{r} \frac{\partial^2 W_j^s}{\partial \theta^2} \right) + \left(\frac{\partial W_i^s}{\partial r} + \frac{1}{r} \frac{\partial^2 W_i^s}{\partial \theta^2} \right) \left(\frac{\partial^2 W_j^s}{\partial r^2} \right) \right] d\theta dr \right. \\ &\left. + \int_{r_s}^{r_d} \int_0^{2\pi} \left[\left(\frac{\partial^2 W_i^c}{\partial r^2} \right) \left(\frac{\partial W_j^c}{\partial r} + \frac{1}{r} \frac{\partial^2 W_j^c}{\partial \theta^2} \right) + \left(\frac{\partial W_i^c}{\partial r} + \frac{1}{r} \frac{\partial^2 W_i^c}{\partial \theta^2} \right) \left(\frac{\partial^2 W_j^c}{\partial r^2} \right) \right] d\theta dr \right\} \end{aligned} \tag{4.38}$$

$$\left[\mathrm{K}_b^e\right]_{ij} = E_b I_A \int_{r_d}^{r_b} V_i'' \, V_j'' dx \tag{4.39}$$

$$\left[\mathrm{K}_d^i\right]_{ij} = h_d \left\{ \int_{r_s}^{r_d} \int_0^{2\pi} \left[\sigma_r \left(\frac{\partial W_i^s}{\partial r} \right) \left(\frac{\partial W_j^s}{\partial r} \right) + \frac{\sigma_\theta}{r^2} \left(\frac{\partial W_i^s}{\partial \theta} \right) \left(\frac{\partial W_j^s}{\partial \theta} \right) \right] d\theta dr + \int_{r_s}^{r_d} \int_0^{2\pi} \left[\sigma_r \left(\frac{\partial W_i^c}{\partial r} \right) \left(\frac{\partial W_j^c}{\partial r} \right) + \frac{\sigma_\theta}{r^2} \left(\frac{\partial W_i^c}{\partial \theta} \right) \left(\frac{\partial W_j^c}{\partial \theta} \right) \right] d\theta dr \right\} \tag{4.40}$$

$$\left[\mathrm{K}_s^\Omega\right]_{ij} = \frac{1}{2} \rho_b A_b \sum_1^{N_b} \int_{r_d}^{r_b} \left(r_b^2 - 3x^2 \right) \left[\Phi_i \Phi_j \right]_{Z = Z_d} dx \tag{4.41}$$

$$\left[\mathrm{K}_{sb}^{\Omega}\right]_{ij} = \rho_b A_b \int_{r_d}^{r_b} x\Phi_i \left|_{Z=Z_d} V_j \cos\beta_k dx - \frac{1}{2}\rho_b A_b \int_{r_d}^{r_b} \left(r_b^2 - x^2\right) \Phi_i \right|_{Z=Z_d} V_j' \cos\beta_k dx \tag{4.42}$$

$$\left[\mathrm{K}_{d}^{\Omega}\right]_{ij} = \rho_d h_d \left\{ \int_{r_s}^{r_d} \int_0^{2\pi} \left(\frac{\partial W_i^s}{\partial \theta}\right) \left(\frac{\partial W_j^s}{\partial \theta}\right) r d\theta dr + \int_{r_s}^{r_d} \int_0^{2\pi} \left(\frac{\partial W_i^c}{\partial \theta}\right) \left(\frac{\partial W_j^c}{\partial \theta}\right) r d\theta dr \right\} \tag{4.43}$$

$$\left[\mathrm{K}_{b}^{\Omega}\right]_{ij} = \rho_b A_b \int_{r_d}^{r_b} V_i V_j \cos^2\beta_k dx - \frac{1}{2}\rho_b A_b \int_{r_d}^{r_b} \left(r_b^2 - x^2\right) V_i' \, V_j' dx \tag{4.44}$$

References

1. Gennaro, S., and L. Leonardo. 2004. Non deterministic approaches for the evaluation of the mistune effects on the rotor dynamics. *Collection of Technical Papers - AIAA/ASME/ASCE/AHS/ASC Structures, Structural Dynamics and Materials Conference* 4: 2579–2590.
2. Chen, Y.F., and I.Y. Shen. 2015. Mathematical insights into linear mode localization in nearly cyclic symmetric rotors with mistune. *Journal of Vibration and Acoustics, Transactions of the ASME* 137 (4).
3. Li, C.F., S.H. Zhou, S.H. Yang, X. Ren, and B.C. Wen. 2014. Dynamic characteristics of blade-disk-rotor system with structural mistuned features. *Open Mechanical Engineering Journal* 8 (1): 138–143.
4. Yasharth, B., and S. Alok. 2012. Reduced order model of a multistage bladed rotor with geometric mistuning via modal analyses of finite element sectors. *Journal of Turbomachinery* 134 (4).
5. Paolo, B., and C. Giovanni. 2012. A continuous model for the dynamical analysis of mistuned bladed rotors. *International Journal of Rotating Machinery* 2012.
6. Alok, S. 2007. Reduced-order model of a bladed rotor with geometric mistuning. *Proceedings of the ASME Turbo Expo* 5: 307–314.
7. Hiroki, T., and M. Akihiro. 2012. Vibration characteristics of mistuned bladed disk by sensitivity analysis. *Transactions of the Japan Society of Mechanical Engineers, Part C* 78 (785): 12–26.
8. Han, D.J. 2015. Complex harmonic modal analysis of rotor systems. *Journal of Mechanical Science and Technology* 29 (7): 2735–2746.
9. Lee, I., S.J. Shin, and Y.R. Kim. 2011. Mistuned bladed disk forced vibration analysis based on standing wave formulation. *Aerospace Science and Technology* 1: 1–8.
10. Raeisi, E., and S. Ziaei-Rad. 2013. The worst response of mistuned bladed disk system using neural network and genetic algorithm. *Meccanica* 48: 367–379.
11. Bai, B., G.C. Bai, and C. Li. 2014. Application of improved hybrid interface substructural component modal synthesis method in vibration characteristics of mistuned blisk. *Chinese Journal of Mechanical Engineering* 27 (6): 1219–1231.
12. Bai, B., G.C. Bai, C. Li, H.Y. Zhao, and W. Yao. 2015. Vibratory characteristic analysis of integral mistuned bladed disk assemblies for aeroengine. *Proc IMechE Part C: J Mechanical Engineering Science* 229 (16): 2921–2938.

13. Kwon, S.M., and H.H. Yoo. 2011. Vibration localization of a mistuned rotating multi-packet blade system undergoing external cyclic harmonic force. *Journal of Mechanical Science and Technology* 25 (11): 2769–2774.
14. ———. 2015. Mode and transient response localization occurred in rotating multi-packet blade systems due to random mistuning. *International Journal of Precision Engineering and Manufacturing* 16 (10): 2063–2071.
15. Cui, L.L., J.F. Huang, and F.B. Zhang. 2017. Quantitative and localization diagnosis of a defective ball bearing based on vertical–horizontal synchronization signal analysis. *IEEE Transactions on Industrial Electronics* 64 (11): 8695–8706.
16. Cui, L.L., N. Wu, C.Q. Ma, and H.Q. Wang. 2016. Quantitative fault analysis of roller bearings based on a novel matching pursuit method with a new step-impulse dictionary. *Mechanical Systems and Signal Processing*. 68-69: 34–43.
17. Chiu, Y.J., and S.C. Huang. 2007. The influence on coupling vibration of a rotor system due to a mistuned blade length. *International Journal of Mechanical Sciences* 49 (4): 522–532.
18. ———. 2008. The influence of a mistuned blade's staggle angle on the vibration and stability of a shaft-disk-blade assembly. *Shock and Vibration* 15 (1): 3–17.
19. Zhou, S.T., Y.J. Chiu, G.F. Yu, C.H. Yang, H.W. Huang, and S.R. Jian. 2017. An assumed mode method and finite element method investigation of the coupled vibration in a flexible-disk rotor system with lacing wires. *Journal of Mechanical Science and Technology* 31 (2): 577–586.
20. Chiu, Y.J., S.R. Jian, C.H. Yang, and G.F. Yu. 2017. Based on lacing wires influence of coupling vibration of a multi flexible disks turbine rotor system by two methods. *Journal of Vibroengineering* 19 (2): 1314–1331.

Chapter 5
Lateral-Torsional-Coupled Model Based Dynamic Analyses of Spur Gears Under Time-Varying External Load Conditions with Surface Wear

Jun Zhang, Jian Wang, Xike Li, Ligang Yao, and Xianzeng Liu

5.1 Introduction

Gear systems have found their wide applications in power transmission circumstances due to the merits of compact structure, strong load capacity and high transmission efficiency. During the gearing applications, vibration and noise are considered as two crucial issues that affect the gear system's operational performances. As a parametrically excitable mechanical system, a typical gear system is often suffered from both internal excitation and external excitation. From the point of view of condition monitoring and fault diagnosis, the external excitations aroused by real operational conditions and the internal excitations induced by gear faults will alter the gear system's vibrational signals greatly. For a real-world gear system that works in complicated and uncertain environments, the external excitations are usually denoted as time-varying loads for simplicity during the dynamic modeling and analysis stage. Meanwhile, the internal excitations evoked by the gear faults (for instance, tooth profile deviation and tooth crack) are often represented as time-varying meshing stiffness and displacement excitation functions as conducting a theoretical dynamic prediction. Following this track, a plenty of investigations have been carried out to reveal the influences of time-varying loads and/or gear faults on vibrational characteristics of gear systems.

To reveal the vibrational characteristics of gear systems operating under time-varying loading conditions, numerous dynamic models have been proposed in the past decades. For example, in 1990 Kahraman and Singh [1] established a 1-DOF

J. Zhang (✉) · J. Wang · X. Li · L. Yao
School of Mechanical Engineering and Automation, Fuzhou University, Fuzhou, P. R. China
e-mail: zhang_jun@fzu.edu.cn

X. Liu
School of Electrical Engineering and Automation, Anhui University, Hefei, P. R. China

J. Zhang (ed.), *Dynamics and Fault Diagnosis of Nonlinear Rotors and Impellers*,
Nonlinear Systems and Complexity 34, https://doi.org/10.1007/978-3-030-94301-1_5

nonlinear dynamic model of a spur gear pair to study the frequency responses, including the effects of the internal excitation and the external excitation as well. The predicted dynamic characteristics matched satisfactorily with the experimental data. They further considered the effects of bearing clearance and proposed a 3-DOF dynamic model to study the dynamics of a geared rotor-bearing system [2, 3]. Taking into account the effects of the internal and external excitations, Kahraman and Blankenship also conducted many experiments to study the dynamics of a spur gear system [4]. Recently, Li, et al. [5], adopted a similar method to investigate the nonlinear dynamics of a spur gear pair. Their analyses proved that the dynamical behaviors of a spur gear pair with both external and internal excitations were quite different from those with only internal excitations. More recently, Jiang, et al. [6], considered the multi-frequency excitation forces to investigate the dynamic behaviors of cracked gear systems. Khabou developed an 8-DOF dynamic model of a single-stage spur gear reducer to investigate the dynamical behaviors during transient regime [7]. With the same dynamic model, Chaari, et al. [8], studied the effects of time-varying loads on the amplitude and frequency modulations of vibrational signals of a spur gearbox. Recently, Zhou, et al. [9], also developed a lateral-torsional model to investigate the nonlinear dynamical behaviors of a coupled gear-rotor-bearing systems subject to combined internal and external excitations.

Besides the above studies on the dynamic analysis of single-stage geared systems, some investigations focused on the effect of time-varying loads on the dynamics in multi-stage gearbox and planetary gear train can also be traced. For example, Bartelmus, et al. [10], proposed two types of dynamic models for a fixed-axis two-stage gearbox and a planetary gearbox in the time-varying nonstationary loading condition to analyze the relationships between operational conditions and diagnostic features. The analysis results proved that there exists a strong correlation between the load conditions and the diagnostic features. Chaari, et al. [11], established a dynamical model of a planetary gear train and analyzed the dynamics under time-varying load conditions. The simulation results showed that the vibration amplitude of the gear system increases with the load level. Zhu, et al. [12], established a dynamical model and estimated the dynamical features of the wind turbine drive train subjected to the external excitation, due to the measured wind load spectrum.

Different from the above studies that concentrated on the effects of time-varying external excitations on dynamic performances of gear systems, there are another group of researches that aimed to reveal the influences of the defect-aroused internal excitations (tooth crack and errors) on the vibration characteristics of geared systems. For example, in 2001 Howard, et al. [13], investigated the influences of cracked gear on the frequency spectra of a gearbox. Fakhfakh, et al. [14], studied the dynamic characteristics of a gear train with teeth defects through numerical simulations and experimental tests. A 6-DOF dynamic model with localized tooth imperfection was proposed by PAREY, et al., for the identification of nascent gear fault [15]. Wu, et al. [16], established a 6-DOF dynamical model to study the influences of cracked gear on the vibrational responses of a gearbox. Mohammed and Rantatalo [17] also proposed a similar dynamical model to study the quantitative relationships between the crack size and eigenfrequencies for detecting the tooth crack and estimating its size. Ma and Chen [18] developed a 3-DOF dynamical model of a spur gear train to study the effects of tooth crack on meshing stiffness and dynamic responses. The experimental results agreed well with their theoretical analyses. Recently, Ma, et al. [19], established a finite element model to study the influences of cracked gear on vibrational features in a profile shifted gear rotor system. Besides these studies on tooth crack, the effects of tooth deviations on gear dynamics also attracted intensive attentions from the research society. For instance, Velex and Maatar [20] proposed a comprehensive model to study the effects of deviation on the dynamics of gears. Recently, Fernández, et al. [21], proposed a nonlinear model to investigate the influences of tooth profile deviations as well as support flexibility on the spur gear transmission system dynamics. A dynamical model of a spur gear system was recently developed by Inalpolat, et al. [22], to study the influence of tooth indexing errors on the system's dynamic responses. More recently, Hu, et al. [23], proposed a finite element model to investigate the influences of the profile modifications on the dynamics of a high speed gear-rotor-bearing system. Except for the aforementioned studies on single gear pairs, the dynamic characteristics of planetary gear trains affected by gear defects have also been analyzed in the past years [24, 25].

Compared with abundant studies focused on the influences of tooth crack and deviations on gear dynamics, the investigations on the influences of gear wear on dynamic behaviors of gear systems are much less. Gear wear contributes to an imperfection from theoretical tooth profiles, leading to the variation of meshing stiffness and meshing displacement, thus results in great differences between the dynamical characteristics of gear systems with gear wear and its counterpart without wear. Choy, et al. [26], proposed a dynamic model to reveal the influences of gear wear on the vibration in a geared system. Later, Kuang and Lin [27] developed a 1-DOF dynamical model of a spur gear to investigate the influences of gear wear on the vibrational signals. Based on a similar method, several researchers also conducted some investigations on the influences of gear wear on gear dynamics [28–31]. Kahraman and Ding [32, 33] investigated the interactive effects between gear wear and dynamic characteristics of the single-stage gear train and the planetary gear train. Wu, et al. [34], introduced the accumulated tooth wear depth into a dynamical model and investigated the influences of worn-surface-generated backlash on dynamics of a compound planetary gear train. Recently, the authors proposed a wear prediction model for the spur gear trains [35]. And more recently, the proposed wear model was combined with a 3-DOF nonlinear dynamical model to study the interactive influences between gear wear and gear dynamic behaviors [36].

From the above reviews, it can be roughly classified that the previous researches either focused on the influences of time-varying external excitations on dynamics of gear systems or concentrated on the effects of defect-aroused internal excitations on gear dynamics. In other words, only a few studies considered both the effects of external and internal excitations on gear dynamic characteristics simultaneously. To be more specific, the dynamics of gear systems under time-varying loads with tooth surface wear have not been investigated yet to the authors' best knowledge. For such a gear system that subjects to both the external excitations induced by time-varying loads and the internal excitations evoked by worn surfaces, it is, however, quite difficult to deal with the condition monitoring and fault diagnosis. In order to distinguish the gear vibration signal caused by time-varying loads from those caused by surface wear, it needs to indicate the vibrational mechanism of the gear systems that subjects to time-varying external loads and worn surface aroused internal excitations.

In view of this, the dynamic behaviors of a spur gear train under time-varying external load conditions with surface wear will be studied in the present study from the perspective of condition monitoring and fault diagnosis. It is believed that this paper can contribute to identify the influences of time-varying external loads as well as surface wear on gear vibration mechanism and time domain as well as frequency domain features.

In this paper, a lateral-rotational dynamical model of a spur gear system was proposed which incorporated the effects of gear wear and time-varying external loads. The effects of worn surfaces on the gear dynamics were represented in different Fourier formulations of static transmission error (STE) function and time-varying meshing stiffness function with various wear depths. Then, an example of a spur gear transmission was used to reveal the coupling influences of time-varying external loads and gear wear on the system's dynamics.

5.2 Dynamical Model

As shown in Fig. 5.1, a lateral-torsional-coupled dynamical model of a spur gear system within the influences of surface wear and time-varying external loads is developed [36]. And all the symbols used are given in the Appendix.

Based on the 2nd Newtonian law, the dynamic equations of the spur gear system can be derived.

Fig. 5.1 Lateral-torsional-coupled model

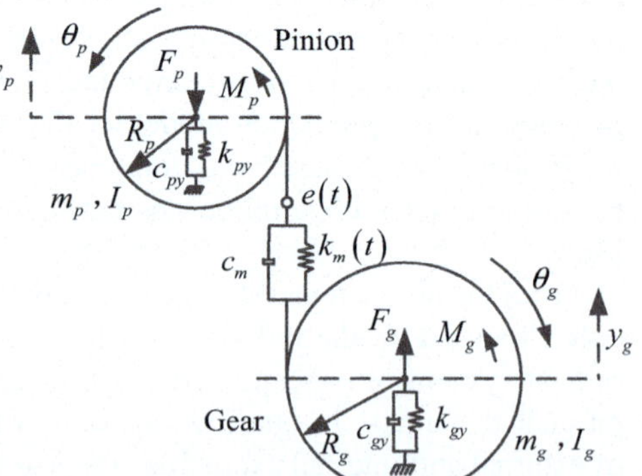

$$
\begin{aligned}
& m_p \ddot{y}_p + c_{py} \dot{y}_p + k_{py} y_p + k_m(t) f(u) + c_m \dot{u} = -F_p, \\
& m_g \ddot{y}_g + c_{gy} \dot{y}_g + k_{gy} y_g - k_m(t) f(u) - c_m \dot{u} = F_g, \\
& m_c \left(\ddot{u} - \ddot{y}_p + \ddot{y}_g\right) + k_m(t) f(u) + c_m \dot{u} = F_m + F_{aT}(t) - m_c \ddot{e}(t),
\end{aligned} \tag{5.1}
$$

$$
u = R_p \theta_p - R_g \theta_g - e + y_p - y_g, \tag{5.2}
$$

$$
m_c = I_p I_g / \left(I_g R_p^2 + I_p R_g^2\right), \tag{5.3}
$$

$$
F_m = M_{pm} / R_p = M_{gm} / R_g, \tag{5.4}
$$

$$
F_{aT}(t) = m_c R_p M_{pa}(t) / I_p + m_c R_g M_{ga}(t) / I_g, \tag{5.5}
$$

where M_{pm}/M_{gm} and $M_{pa}(t)$/$M_{ga}(t)$ are the constant and fluctuation parts of the external torques $M_p(t)$ and $M_g(t)$, respectively. M_{pa} and M_{ga} can be formulated as

$$
\begin{aligned}
& M_{pa}(t) = M_{pam} \cos\left(\omega_{pa} t + \varphi_{pa}\right), \\
& M_{ga}(t) = M_{gam} \cos\left(\omega_{ga} t + \varphi_{ga}\right),
\end{aligned} \tag{5.6}
$$

where M_{pam} and M_{gam} are the amplitudes of the external torque fluctuations; φ_{pa} and φ_{ga} are the phase angles of the external torque fluctuations; ω_{pa} and ω_{ga} are the fundamental frequencies of torque excitations.The nonlinear displacement function $f(u)$ is given by

$$
f(u) = \begin{cases} u - b, & u > b, \\ 0, & -b \leq u \leq b, \\ u + b, & u < -b. \end{cases} \tag{5.7}
$$

where $2b$ is the backlash value of the gear pair.The time-varying stiffness $k_m(t)$ is expressed in terms of the Fourier series [4]

$$
k_m(t) = k_{mm} + \sum_{r=1}^{n} \left[k_{ar} \cos\left(r \omega_m t\right) + k_{br} \sin\left(r \omega_m t\right)\right], \tag{5.8}
$$

Similarly, the STE is denoted as

$$
e(t) = e_0 + \sum_{r=1}^{n} \left[e_{ar} \cos\left(r \omega_e t\right) + e_{br} \sin\left(r \omega_e t\right)\right]. \tag{5.9}
$$

where ω_e is the fundamental frequency of STE; e_0 is the mean component of STE; e_{ar} and e_{br} are the Fourier coefficients.The above parameters of STE and meshing stiffness should be pre-obtained by loaded tooth contact analysis (LTCA) for each gear pair with or without gear wear [35, 36].

The steady-state dynamical response of a spur gear system with gear wear and time-varying external load excitation can be determined by applying the Runge-Kutta method.

5.3 Numerical Simulation and Discussion

With the lateral-torsional-coupled model developed in Sect. 5.2, a numerical example in Ref. [36] is used to demonstrate the analyses of dynamic characteristics for a spur gear train operating in time-varying external load condition with surface wear. It is expected that through the analyses, the coupling effects of surface wear and time-varying external loads on the dynamics can be revealed. For the sake of clarity, Table 5.1 lists the major design parameters of the example system. The mesh damping c_m can be set as 1000 Ns/m. The input torque M_{pm} is set as 340 N m.

5.3.1 *Effects of Wear on Meshing Stiffness and STE*

As aforementioned, surface wear not only alters the meshing stiffness but also changes STE. Therefore, the quantitative relationships between the surface wear depth and the meshing stiffness as well the STE must be prior identified to predict the dynamics of gear systems with worn surfaces accurately. For this purpose, the influences of gear wear on STE and meshing stiffness will be investigated in this subsection. The detailed computational methodology for surface wear depth of gears may be refereed to our previous investigation [36]. In the present study, the maximum wear depth is indicated as $h_{\max}$ to depict the wear degree of the tooth

Table 5.1 Design parameters

Parameters	Pinion	Gear
Module/mm	3.0	3.0
Teeth	50	50
Width/mm	20	20
Pressure angle/(°)	20	20
Density/(kg/m^3)	7830	7830
Material	42CrMo	42CrMo
Poisson's ratio	0.3	0.3
Young's modulus/GPa	206	206
Moment of inertia/(kg mm^2)	7300	7300
Mass/kg	2.8	2.8
Supporting damping/(Ns/m)	5360	5360
Supporting stiffness/(N/μm)	1150	1150
ISO precision grade	5	5

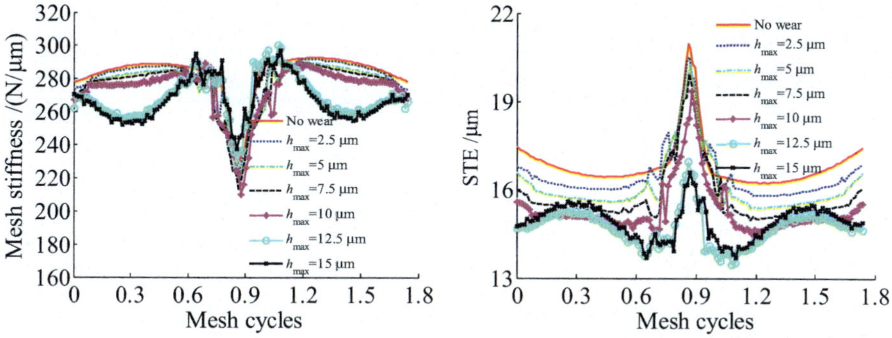

(a) Meshing stiffness curves (b) STE curves

Fig. 5.2 Variations of meshing stiffness and STE with respect to the maximum wear depth

Table 5.2 Fourier coefficients of time-varying meshing stiffness function

$k_m(t)$/(N/μm)	Without wear	With wear
k_{mm}	283.7	277.3
k_{a1}	8.589	10.74
k_{a2}	−8.448	−14.33
k_{a3}	2.707	6.6
k_{a4}	−0.02305	−5.707
k_{b1}	1.069	−0.6786
k_{b2}	−7.761	−4.263
k_{b3}	5.872	3.162
k_{b4}	−6.759	−4.355

contact surfaces. Once the wear depths are determined, an LTCA is performed to determine the meshing stiffness and STE with different wear amounts.

Figure 5.2(a) and (b) demonstrates the variations of meshing stiffness curves and those of STE curves with respect to the maximum wear depth of the mating surfaces, respectively.

As can be clearly observed from Fig. 5.2, the meshing stiffness reduces with the gear wear depth in the double contact zone while the STE reduces with the gear wear depth in the most of the meshing cycles. Nevertheless, the effects of gear wear on the meshing stiffness and STE are very intricate in the transition region from double contact zone to single contact zone.

The above LTCA obtained meshing stiffness curves and STE curves can be further expressed as time-varying meshing stiffness function and STE function in the Fourier series form defined by Eqs. (5.8) and (5.9). According to the studies in Ref. [3], the first four Fourier coefficients of STE function and meshing stiffness function are deemed to be sufficient here. For example, the corresponding Fourier coefficients of meshing stiffness and STE for the gear pair defined in Table 5.1 without surface wear and with worn surface of $h_{\max} = 5$ μm are shown in Tables 5.2 and 5.3.

Table 5.3 Fourier coefficients of STE function

$e(t)/\mu m$	Without wear	With wear
e_0	16.87	16.16
e_{a1}	−0.4554	−0.7487
e_{a2}	0.617	0.8186
e_{a3}	−0.2243	−0.384
e_{a4}	0.08364	0.2683
e_{b1}	−0.03435	−0.05636
e_{b2}	0.4778	0.3944
e_{b3}	−0.3435	−0.3072
e_{b4}	0.4668	0.3711

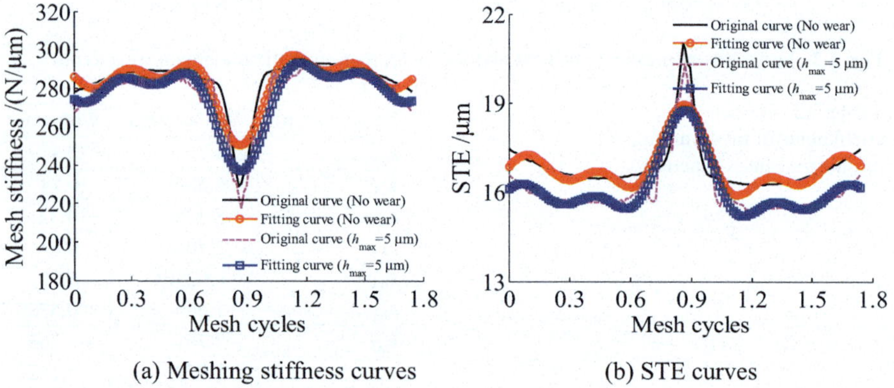

Fig. 5.3 Comparison of fitting curves and LTCA obtained curves

To be more clarity, Fig. 5.3(a) and (b) compare the fitting curves and the original curves of meshing stiffness and STE.

From Fig. 5.3, it can be observed that the fitting curves are in well agreement with the original LTCA obtained curves, manifesting the Fourier series form fitting method can describe the internal excitations arousing from worn surfaces of the gear pair with satisfactory accuracy.

With the above fitting method, the effects of worn surfaces can be introduced into the gear dynamical model through the analytical formulations of meshing stiffness function and STE function as shown in Eqs. (5.8) and (5.9).

5.3.2 *Effects of Time-Varying External Loads and Surface Wear*

In this subsection, the dynamical features of the example system including time-varying external load excitation and worn surfaces effects will be predicted by using the proposed dynamical model in Sect. 5.2.

For simplicity of the dynamics computation, only the input torque is set as time-varying in the present study though the output torque can also be set as time-varying. Specially, the fluctuant part of input torque $M_{pa}(t)$ is included and M_{pam} is assumed to be 0.5 M_{pm}, while the fluctuant part of output torque $M_{ga}(t)$ is set as 0. The external excitation is assumed to fluctuate in a cosinusoidal shape with an exciting frequency $f_{pa} = f_p = f_m/Z_p$ ($T_{pa} = T_p = Z_p T_m$). Herein, f_p is the rotational frequency of the pinion shaft and f_m is the meshing frequency; T_m and T_p are the meshing period and the pinion rotation period. Meanwhile, the tooth surface wear is accumulated with a maximum wear depth of $h_{\max} = 5$ μm.

With the above settings, the following dynamic simulations can be carried out. Here, the studies on the influences of time-varying external loads and surface wear are conducted at three speeds: (1) $n_1 = 800$ rpm (low speed condition away from any resonance), (2) $n_1 = 1860$ rpm (near the first super-harmonic resonance peak), (3) $n_1 = 3720$ rpm (near the primary resonance peak). For each operating speed, three cases are analyzed, i.e., (1) constant external loads without surface wear; (2) time-varying external loads without surface wear; (3) time-varying external loads with wear. The system responses in the three cases at each operating speed are predicted and discussed as the follows.

5.3.2.1 Dynamic Characteristics of System Operates at $n_1 = 800$ rpm

1. Case 1: Constant external load without surface wear

   When the example, gear system without surface wear, operates at $n_1 = 800$ rpm and is subjected to a constant external load of $M_p = 340$ N m, the predicted time evolutions and spectra of the dynamic meshing force (DMF) are depicted in Fig. 5.4(a) and (b), respectively.

   From Fig. 5.4(a), it can be observed that the amplitude modulation induced by the internal meshing excitation with an exciting frequency f_m (i.e., $1/T_m$) obviously occurs in the time domain response of the DMF. Correspondingly, the meshing frequency harmonics dominate the spectra in frequency domain of the DMF issued from the constant loading case, especially at $2f_m$, $3f_m$ and $5f_m$ resonances.

2. Case 2: Time-varying external loads without surface wear

   Figure 5.5 demonstrate the predicted time evolutions and spectra of the DMF when the system without surface wear operates at $n_1 = 800$ rpm and suffers from a time-varying loading condition, respectively.

   By comparison between Figs. 5.4 and 5.5, it is clear that the time-varying external load condition alters the dynamics of gears dramatically. In the time-varying loading case, the time domain signals of the DMF are modulated not only by the internal meshing excitation but also the time-varying external load excitation with an exciting frequency f_{pa} (i.e, $1/T_{pa}$). In general, the time-varying external load condition enhances the dynamical response of the gear system, yielding to a noticeable increment in the oscillation amplitudes of the DMF. To be specific, the oscillation amplitudes of the DMF are 5405.27 N and 7818.18 N

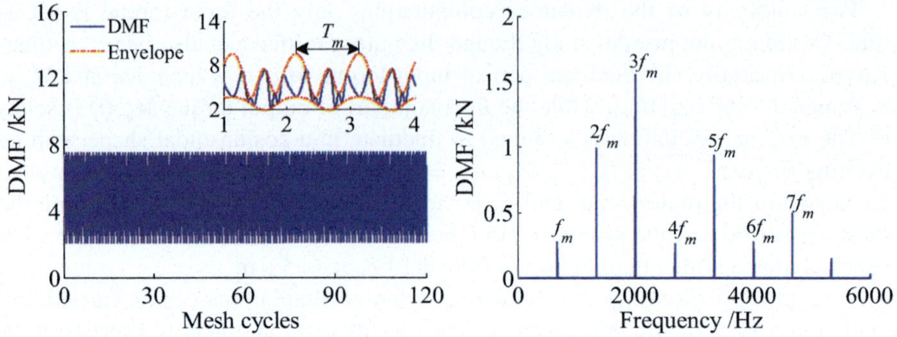

(a) Time evolutions (b) Spectra

Fig. 5.4 DMF of the system without surface wear operates at $n_1 = 800$ rpm under constant external load

in Figs. 5.4(a) and 5.5(a), respectively. In addition, from Fig. 5.5(b) it is clearly noticed that there occurs resonance modulation band with the exciting frequency f_{pa} ($f_{pa} = f_p = 13.3$ Hz) and multiplicities of sidebands (with modulation frequency f_{pa}) around the meshing frequency harmonics. These sidebands are the consequences of the external load fluctuations of the gear system. Moreover, the amplitudes of these modulation sidebands around the meshing frequency harmonics are slightly asymmetric due to the influence of system damping.

3. Case 3: Time-varying external loads with surface wear

   As the system with worn surfaces of $h_{max} = 5$ μm works at the same speed and is subjected to a time-varying external load condition, its dynamic behaviors become different from the above two situations. Figure 5.6 shows the time evolutions and spectra of DMF for this situation.

By comparison between Fig. 5.6 and 5.5, the time-domain DMF signals share the same tendency for both conditions with and without worn surfaces when the gear system operates under a time-varying load condition. Interestingly, the frequency contents of the DMF spectra are not changed by the presence of surface wear. Nevertheless, the surface wear enhances the effects of modulations induced by the internal and the external excitations and increases the oscillation amplitudes of the DMF. Further calculations show that the oscillation amplitude of the DMF increases 38.75% with the introduced surface wear.

5.3.2.2 Dynamic Characteristics of System Operates at $n_1 = 1860$ rpm

With the same procedure, the dynamic characteristics of the system operating at $n_1 = 1860$ rpm can be predicted as the follows.

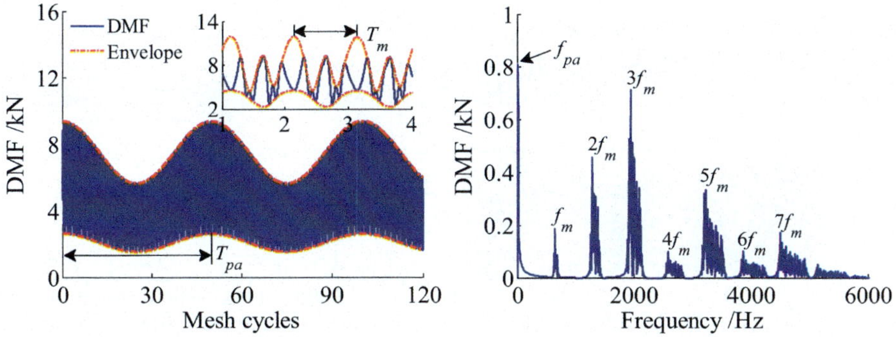

(a) Time evolutions (b) Spectra

Fig. 5.5 DMF of the system without surface wear operates at $n_1 = 800$ rpm under time-varying external loads

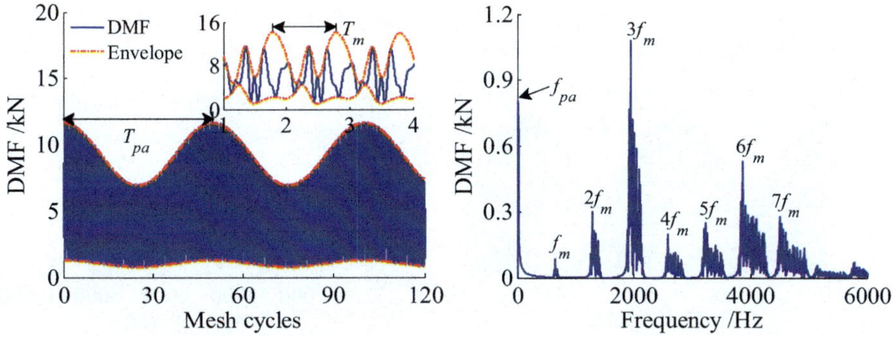

(a) Time evolutions (b) Spectra

Fig. 5.6 DMF of the system with surface wear of $h_{\max} = 5$ μm operates at $n_1 = 800$ rpm under time-varying external loads

1. Case 1: Constant external load without surface wear

   As the example gear system without surface wear is subjected to a constant external load of $M_p = 340$ N m, the predicted time evolutions and spectra of DMF are depicted in Fig. 5.7(a) and (b), respectively.

   As shown in Fig. 5.7(a), the time-domain DMF responses are modulated by the internal excitation with an exciting frequency f_m (i.e, $1/T_m$). Similarly, the meshing frequency harmonics dominate the spectra in frequency domain of the DMF issued from the constant loading case, especially at $3f_m$ resonance.
2. Case 2: Time-varying external loads without surface wear

   Figure 5.8(a) and (b) demonstrates the predicted time evolutions and spectra of the DMF when the system without surface wear operates at $n_1 = 1860$ rpm and suffers from a time-varying loading condition, respectively.

   By comparison between Figs. 5.7 and 5.8, the time-varying external load condition alters the dynamic behaviors of the gear system dramatically. In the

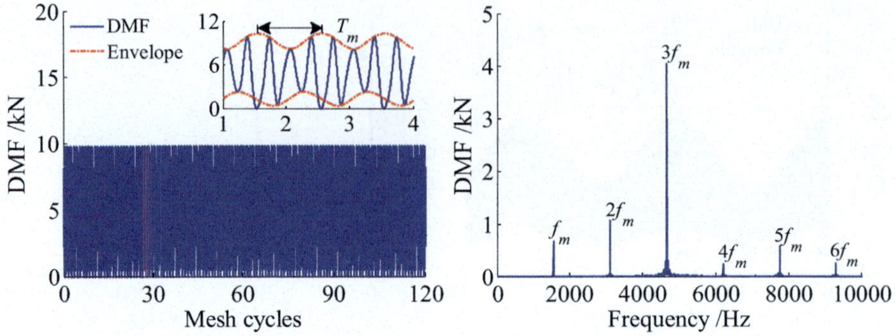

(a) Time evolutions (b) Spectra

Fig. 5.7 DMF of the example system without surface wear operates at $\eta_1 = 1860$ rpm under constant external load

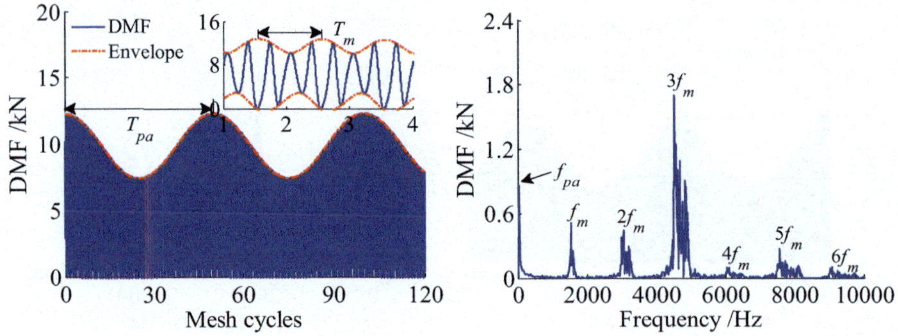

(a) Time evolutions (b) Spectra

Fig. 5.8 DMF of the system without surface wear operates at $n_1 = 1860$ rpm under time-varying external loads

time-varying loading case, the time domain signals of the DMF are modulated by both the internal meshing excitation and the time-varying external load excitation. From the comparison between Figs. 5.8(a) and 5.7(a), it can be calculated that the oscillation amplitude of DMF increases 24.62% with the introduced time-varying load. Accordingly, the resonance modulation band with the exciting frequency f_{pa} ($f_{pa} = 31$ Hz) and multiplicities of the sidebands (with modulation frequency f_{pa}) around the mesh frequency harmonics are clearly shown in the spectra of the DMF.

3. Case 3: Time-varying external loads with surface wear

   Figure 5.9(a) and (b) demonstrates the predicted time evolutions and spectra of the DMF when the system with surface wear of $h_{\max} = 5$ μm operates at the same speed and suffers from a time-varying loading condition.

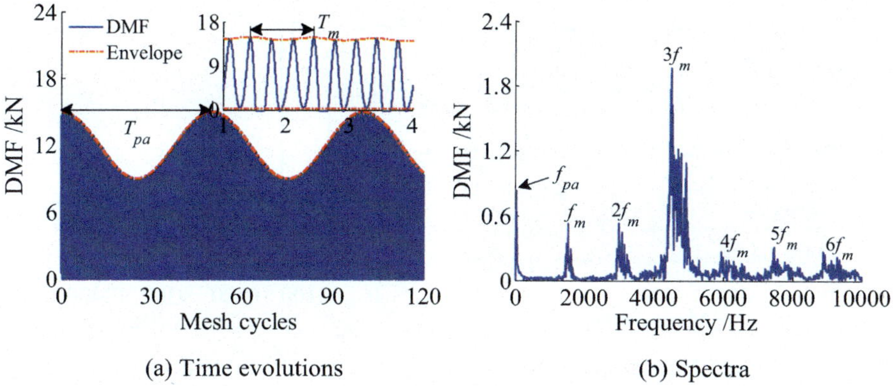

Fig. 5.9 DMF of the example system with surface wear of $h_{\max} = 5\ \mu$m operates at $n_1 = 1860$ rpm under time-varying external loads

By comparison between Figs. 5.8 and 5.9, the variation tendency of the time-domain DMF signals and the DMF spectra contents are not changed by the presence of surface wear. Nevertheless, the surface wear increases the oscillation amplitudes of the DMF dramatically. To be specific, the oscillation amplitude of the DMF increases 22.41% with the introduced surface wear.

5.3.2.3 Dynamic Characteristics of System Operates at n_1 = 3720 rpm

As the example system operates at $n_1 = 3720$ rpm, the dynamic characteristics of the system are predicted as the follows for the three cases.

1. Case 1: Constant external load without surface wear

   When the example gear system without surface wear operates at $n_1 = 3720$ rpm and is subjected to a constant external load of $M_p = 340$ N m, the predicted time evolutions and spectra of the DMF are depicted in Fig. 5.10(a) and (b), respectively.

   As can be observed from Fig. 5.10, the amplitude modulation caused by the internal meshing excitation is obvious in time domain signals of the DMF for the constant load condition without surface wear. In the corresponding frequency spectra, the fractional-octave of the meshing frequencies $mf_m/3$ ($m = 1, 2, 3, 4, 5, 6, 7, 8, 9$) are obviously shown duo to the fact that the period of system's response is $3T_m$. And compared with other resonance peaks, the amplitudes at $2f_m/3$, $4f_m/3$ and $2f_m$ are larger.
2. Case 2: Time-varying external loads without surface wear

   Figure 5.11(a) and (b) demonstrate the predicted time evolutions and spectra of the DMF when the system without surface wear operates at $n_1 = 3720$ rpm and suffers from a time-varying loading condition, respectively.

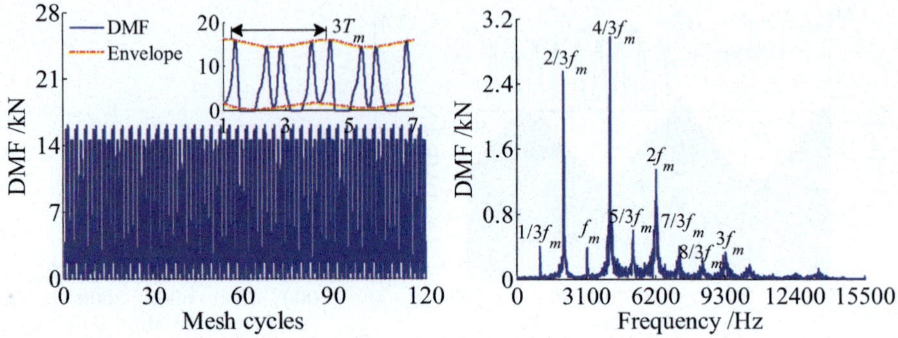

(a) Time evolutions (b) Spectra

Fig. 5.10 DMF of the system without surface wear operates at $n_1 = 3720$ rpm under constant external load

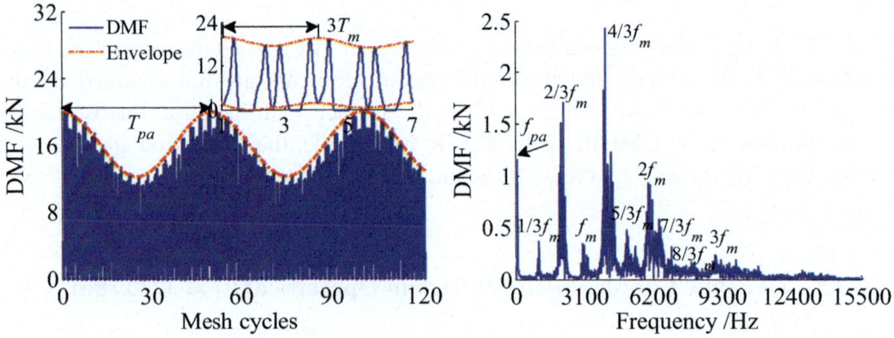

(a) Time evolutions (b) Spectra

Fig. 5.11 DMF of the system without surface wear operates at $n_1 = 3720$ rpm under time-varying external loads

Compared Fig. 5.11 with Fig. 5.10, it can be obviously observed that there appear amplitude modulations in the time domain signals of the DMF under a time-varying external load condition. Meanwhile, an exciting frequencies f_{pa} ($f_{pa} = 62$) is obviously occurred in the lower frequency band. Further comparisons indicate that the period of the DMF and the contents of the meshing frequency harmonics are not changed by the time-varying external load excitation. Nevertheless, the time-varying external excitation dramatically increases the oscillation amplitude of the DMF. To be specific, the oscillation amplitude of DMF increases 24.24% with the time-varying external load.

3. Case 3: Time-varying external loads with surface wear

When the system with worn surfaces of $h_{max} = 5$ μm works at the same speed and is subjected to a time-varying external load condition, its dynamic behaviors

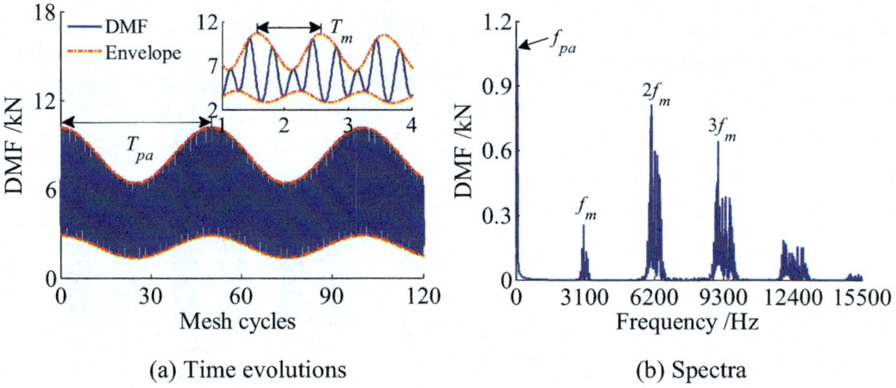

Fig. 5.12 DMF of the system with surface wear of $h_{\max} = 5\ \mu\text{m}$ operates at $n_1 = 3720$ rpm under time-varying external loads

become quite different from the above two situations. Figure 5.12 shows the time evolutions and spectra of DMF in this situation.

By comparison between Fig. 5.12 and Fig. 5.11, the DMF period turns from $3T_m$ to T_m with the introduction of surface wear. The tooth separations of the gear pair are eliminated in Fig. 5.12(a) and the oscillation amplitude of the DMF decreases 56.46% with the introduced surface wear. Meanwhile, the meshing frequency and its harmonics dominate the spectra except for the exciting frequency. What's more, the fractional-octave of the meshing frequencies $mf_m/3$ ($m = 1, 2, 4, 5, 7, 8$) are suppressed. This is very interesting that it may lead to a hypothesis that a small amount of wear depth helps to reduce the oscillation amplitude of the DMF and helps to eliminate tooth separations.

5.3.2.4 Comparisons of Dynamic Factor and Dominated Amplitudes

Based on the above analyses, it can be found that the time-varying external loads and gear wear affect the dynamical responses of the gear pair in both time and frequency domains. Therefore, the following will further compare the influences of time-varying load excitation and surface wear on the dynamics in the time domain and frequency domain.

The dynamic factor of the gear pair is defined as DF = $(\text{DMF})_{\max}/\text{SMF}$ to study the effects of time-varying external loads and gear wear more visibly, where $(\text{DMF})_{\max}$ is the maximum value of the dynamic meshing force and SMF is the static meshing force. With this definition, the DFs of the example system under different working conditions can be calculated and compared as shown in Fig. 5.13.

As can be clearly observed from Fig. 5.13 that the value of DF increases monotonously with the increment of the operating speed for the gear system without worn surfaces no matter it is subjected to a constant or a time-varying external

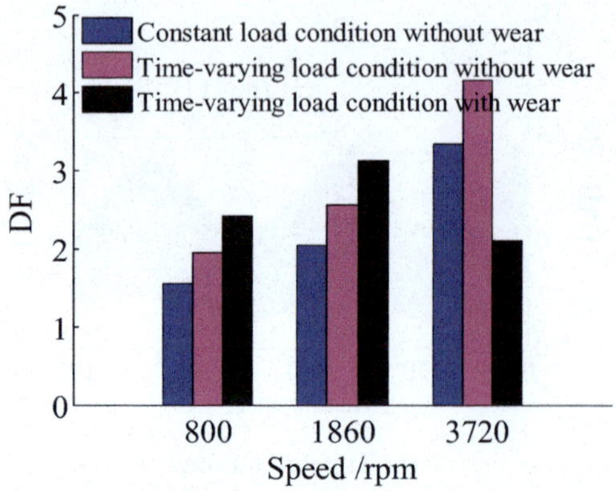

Fig. 5.13 Variations of DF with surface wear and time-varying external loads

load condition. After introducing the surface wear into the gear system, however, it becomes complicated that the DF value at $n_1 = 3720$ rpm is lower than that of the other two speeds. Further observations reveal that the time-varying external load increases the value of DF for each operating speed, indicating an external excitation may deteriorate the dynamic performance of the gear system. What's more, this effect of time-varying external loads may become more distinct as the gear system works near the primary resonance condition.

From the perspective of fault diagnosis, the frequency domain signals may be more preferred to be adopted. Hence, the dominated amplitudes of the DMF spectra are extracted and compared as the follow.

As can be observed from Fig. 5.14, for the system without gear wear, the time-varying external loading condition reduces the harmonic (and the fractional-octave) amplitudes of the DMF signals at each operating speed. However, for the system with wear, the effects of time-varying external load on the harmonic amplitudes of the DMF signals are very complicated. To be specific, Fig. 5.14(a) indicates that the first five orders of the harmonic amplitudes decrease with the time-varying load condition whereas the sixth order of the harmonic amplitudes increases with the time-varying external load when the system operates at low-speed condition away from any resonance ($n_1 = 800$ rpm). Figure 5.14(b) shows that the time-varying load condition increases the fourth order harmonic amplitude of the DMF signals while decreases the other orders of harmonic amplitudes as the gear system works near the first super-harmonic resonance peak at $n_1 = 1860$ rpm. Figure 5.14(c) indicates that at the primary resonance condition ($n_1 = 3720$ rpm), the time-varying external load condition decreases the lower order harmonic amplitudes (the first and second) while increases the higher order harmonic amplitude (the third). Once again, it can be observed that the fractional-octave of the meshing frequencies $mf_m/3$ ($m = 1, 2, 4, 5, 7, 8$) excitations are suppressed.

As the gear system works under a time-varying load condition, the influences of gear wear on the harmonic amplitudes of the DMF signals are distinct for different speeds. To be specific, Fig. 5.14(a) indicates that the first, the second and the fifth

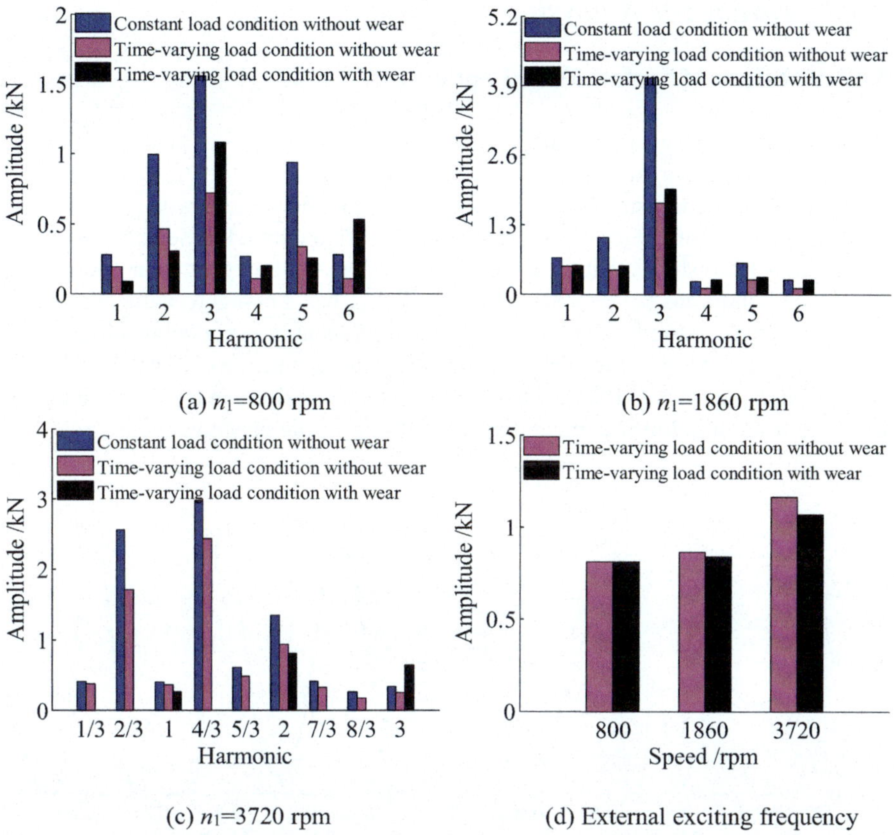

Fig. 5.14 Dominated amplitudes of the DMF spectra

orders of harmonic amplitudes decrease with the introduced surface wear, whereas the third, the fourth and the sixth orders of harmonic amplitudes increase with wear when the system operates at $n_1 = 800$ rpm. Figure 5.14(b) shows that the surface wear increases the first six orders of harmonic amplitudes of the DMF signals when the system works at $n_1 = 1860$ rpm. Figure 5.14(c) indicates that at the primary resonance condition ($n_1 = 3720$ rpm), the surface wear decreases the lower order harmonic amplitudes (the first and second) while increases the higher order harmonic amplitude (the third). It is worthy pointing out that the time-varying external load also evokes a dominant amplitude around the exciting frequency. From Fig. 5.14(d), it can be found that the introducing tooth surface wear can decrease the exciting amplitude aroused by the time-varying external load, which is more distinct when the gear system operates near the primary resonance conditions.

5.3.3 Parametric Analysis

5.3.3.1 Effects of Surface Wear Depth

As aforementioned, gear wear affects the dynamical behaviors of gear systems. The following will investigate its effects on the dynamical characteristics.

Figure 5.15 demonstrates the DMF of the example system with various wear depths under the time-varying external load condition defined in Sect. 5.3.2.

Figure 5.15(a) shows that for the example system without gear wear operating at low speed conditions ($n_1 = 800$), its DMF always has a non-zero value, indicating that there exists no tooth separation. With the increment of introducing surface wear depth, the fluctuation of the DMF is gradually getting 'intensive'. As the maximum wear depth $h_{\max}$ reaches 10 μm, an obvious tooth separation can be observed though the separation duration is comparatively short in a whole meshing cycle.

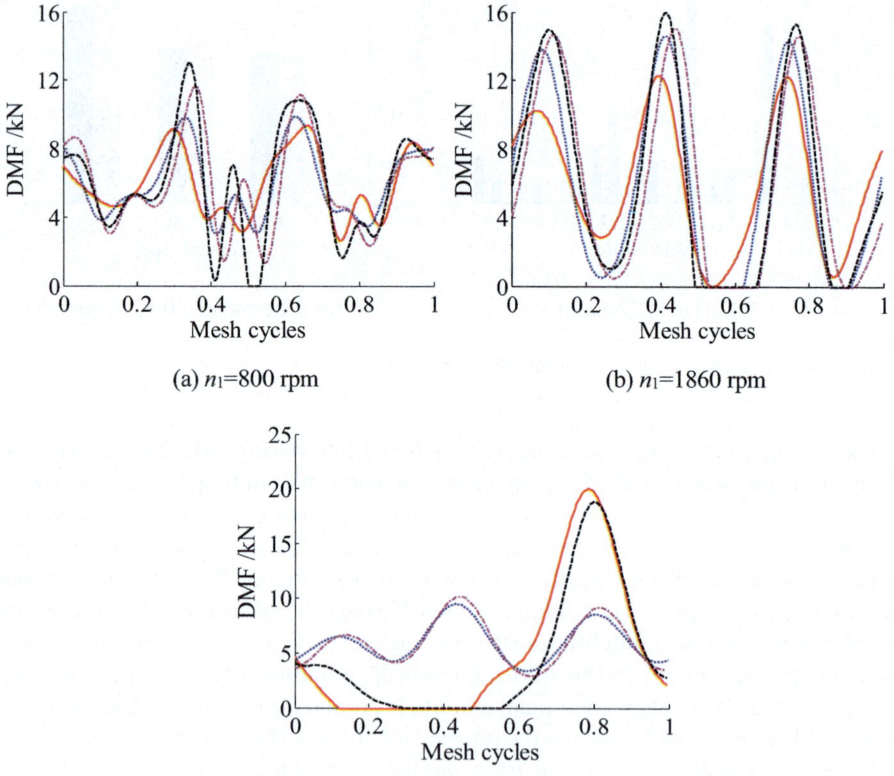

Fig. 5.15 Time domain DMF signals with various surface wear depths. (—) no wear, (···) $h_{\max} = 2.5$ μm, (-·-) $h_{\max} = 5$ μm, (– –) $h_{\max} = 10$ μm

Figure 5.15(b) shows that there is only one location in the entire meshing period where the gear pair reaches the 'threshold' of tooth separation for the condition without wear, as the gear system works at $n_1 = 1860$ rpm. With the increase of introducing wear depth, the tooth separations take place at two positions during a whole mesh period. What's more, the separation durations in each mesh cycle monotonously increase with the surface wear depth. Similarly, the fluctuation amplitudes of the DMF increase monotonously with the surface wear depth though the incremental tendency declines marginally when the wear depth getting greater.

Figure 5.15(c) shows that the DMF value becomes zero for about 35.3% of the meshing cycle caused by tooth separations, as the example system without worn surfaces operates at a high speed of $n_1 = 3720$ rpm,. It is very interesting that the tooth separations appear in the unworn gear pair are eliminated after introducing the surface wear of $h_{\max} = 2.5$ μm and $h_{\max} = 5$ μm and reappear again when the maximum wear depth $h_{\max}$ reaches 10 μm. It can be roughly summarized that the early gear wear may help to reduce the DMF fluctuation and eliminate tooth separation when the gear train runs near a resonance peek. With more running times, the incremental wear depth will enhance the DMF fluctuation and evoke tooth separation again.

Besides the above observations, the amplitude modulations and phase modulations are also clearly demonstrated by comparing the DMF signals in Fig. 5.15. In general, the effects of modulation are more distinct with a larger wear depth for different operational speeds. For example, significant phase differences of the DMF signals in Fig. 5.15(a) can be observed for the gear system with various wear depths.

Figure 5.16 further demonstrates the influence of gear wear on the dynamical responses in frequency domain.

As can be seen from Fig. 5.16, the peaks related to the exciting frequency f_{pa} are obviously shown in the DMF spectra for the three speeds with different wear depths, which agrees well with the analyses in Sect. 5.3.2. For the system works at operational speeds of $n_1 = 800$ rpm and $n_1 = 1860$ rpm, the frequency contents of the DMF spectra are not changed by the increase of surface wear depth, while the dominant amplitudes are affected significantly by the wear depth. However, the influence of accumulated gear wear on the spectra of the DMF is quite complicated as the example system works at $n_1 = 3720$ rpm. To be specific, the fractional-octave of meshing frequencies $mf_m/3$ ($m = 1, 2, 3, 4, 5, 6, 7, 8, 9$) are dominant in the frequency domain for the condition without wear due to the fact that the period of the system's response is $3T_m$. However, the dominant frequencies turn to be the meshing frequency harmonics with a small surface wear amounts (i.e., $h_{\max} = 2.5$ μm and $h_{\max} = 5$ μm). With the accumulated surface wear reaches 10 μm, the fractional-octave of meshing frequencies $mf_m/3$ reappear as dominant frequencies in the DMF spectra.

The effect of gear wear on the dominant amplitudes in the spectra of DMF is further studied and depicted in Fig. 5.17.

From Fig. 5.17(a), it can be observed that the surface wear significantly affects the harmonic amplitudes of the DMF signals when $n_1 = 800$ rpm. The first, the second and the fifth orders of the harmonic amplitudes firstly decrease and then

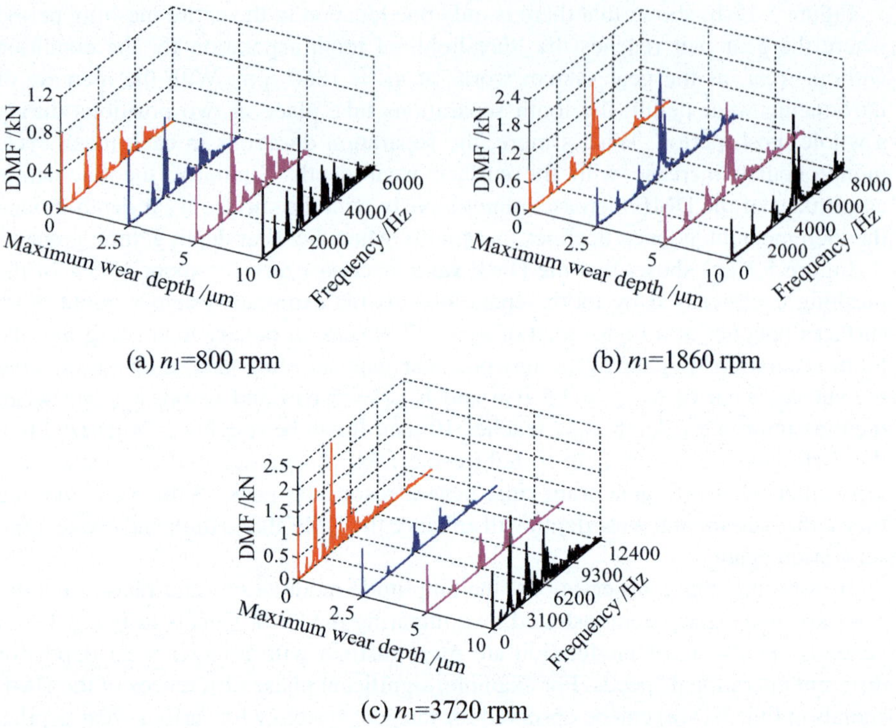

Fig. 5.16 Frequency domain DMF signals with various surface wear depths

increase with surface wear depth, whereas the third and the sixth orders of the harmonic amplitudes firstly increase with wear then slightly decrease when $h_{\max}$ reaches 10 μm. The fourth order of the harmonic amplitude increases monotonously with wear depth.

For the system working at $n_1 = 1860$ rpm, the first order harmonic amplitude firstly reduces and then rises with surface wear, while the fourth, the fifth and the sixth orders of the harmonic amplitudes increase with wear depth. The variation tendencies of the second and the third orders of the harmonic amplitudes with respect to wear depth are comparatively complex in that it comes through an increase, a decrease and an increase sequentially.

For the gear system works at primary resonance condition ($n_1 = 3720$ rpm), the accumulated surface wear significantly affects the frequency contents and corresponding amplitudes of the DMF spectra. The fractional-octave of meshing frequencies $mf_m/3$ ($m = 1, 2, 3, 4, 5, 6, 7, 8, 9$) are dominant for the conditions without wear and with a wear depth of $h_{\max} = 10$ μm. Nevertheless, the dominant frequencies turn to be the meshing frequency harmonics with wear depths of $h_{\max} = 2.5$ μm and $h_{\max} = 5$ μm. Compared to the condition without wear, the amplitudes at $f_m/3$ and $4f_m/3$ are smaller for the condition with $h_{\max} = 10$ μm, while

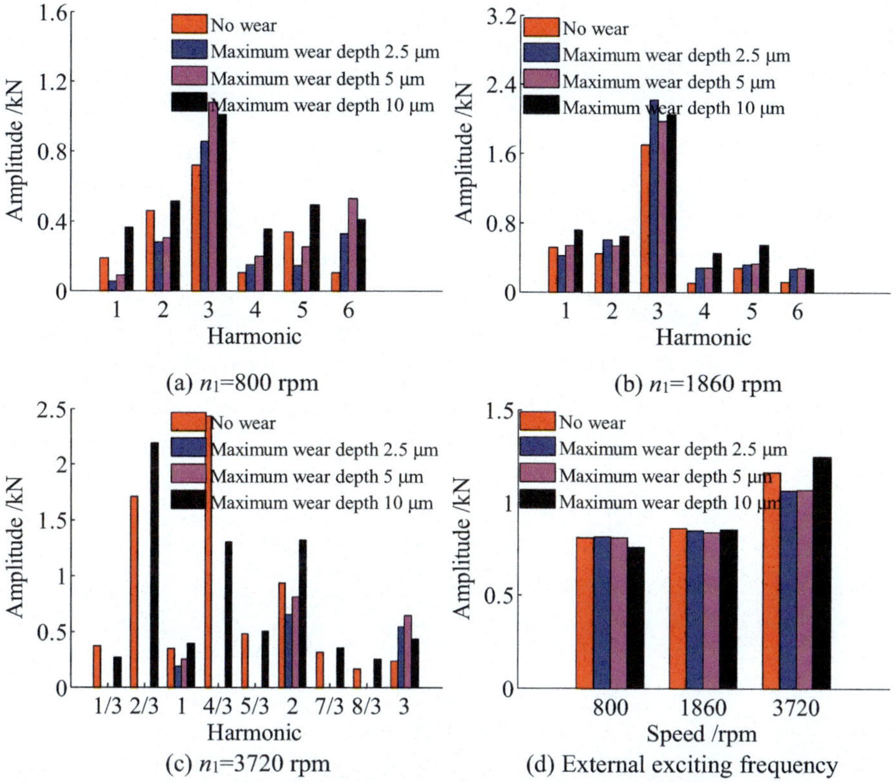

Fig. 5.17 Dominant harmonic amplitudes in the DMF spectra with various wear depths

the amplitudes at $2f_m/3$, $5f_m/3$, $7f_m/3$ and $8f_m/3$ are larger. The first and the second orders of the harmonic amplitudes firstly decrease and then increase with surface wear, while the third order harmonic amplitude firstly increases and then decreases with surface wear.

As to the time-varying external load excited amplitudes shown in Fig. 5.17(d), the influences of gear wear on the amplitudes at exciting frequencies are very trivial. To be specific, the amplitudes of exciting frequencies slightly decrease with wear depth when the operating speed is $n_1 = 800$ rpm and $n_1 = 1860$ rpm; the amplitude of exciting frequencies drops after an initial wear and then increases with the wear accumulations at an operating speed of $n_1 = 3720$ rpm.

5.3.3.2 Coupling Effects of Time-Varying Load and Gear Wear

To illustrate the coupling effect of gear wear and time-varying load on dynamics of the example system, the following investigations can be carried out.

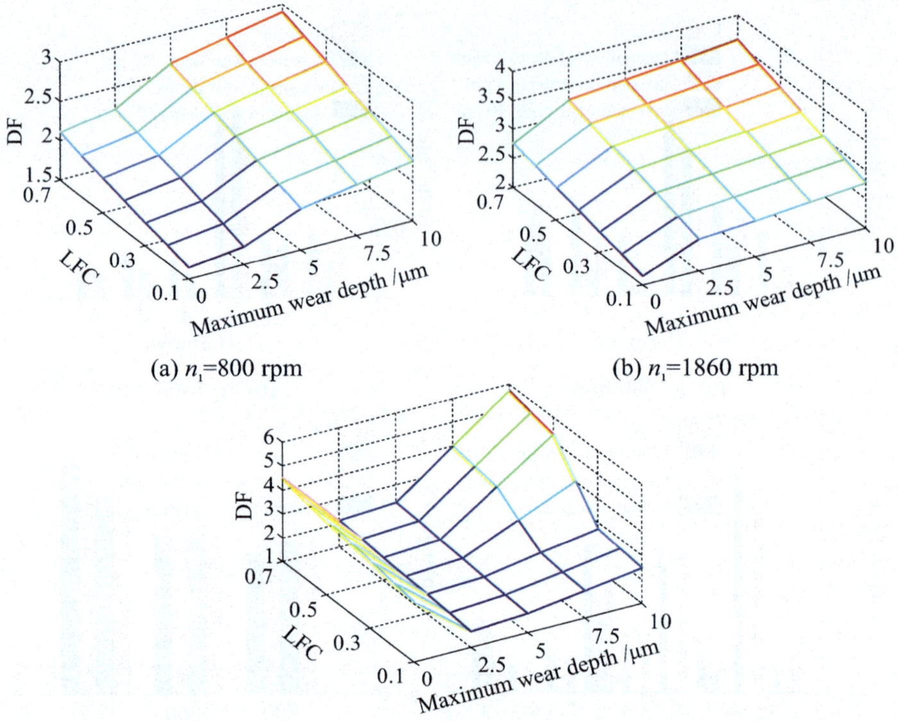

(a) n_1=800 rpm (b) n_1=1860 rpm

(c) n_1=3720 rpm

Fig. 5.18 DF variations with respect to LFC and $h_{\max}$

To quantify the influences of time-varying external load excitation, we define two variables named as the external load fluctuation coefficient (LFC) and the exciting frequency coefficient (EFC). And we have LFC $= M_{pam}/M_{pm}$ and EFC $= f_{pa}/f_p$, where M_{pm} is kept to be constant values listed in Table 5.1 and $f_p = n_1/60$.

The coupling effects of LFC and maximum wear depth $h_{\max}$ on the DF value of the example system are shown in Fig. 5.18 when the exciting frequency is $f_{pa} = f_p$.

As shown in Fig. 5.18(a) and (b), the DF value monotonously increases with LFC and $h_{\max}$ when the system operates at $n_1 = 800$ rpm and $n_1 = 1860$ rpm. However, the influences of LFC and $h_{\max}$ on the DF value are obviously different as the system operates at $n_1 = 3720$ rpm. As can be seen from Fig. 5.18(c), the DF value increases monotonously with LFC while firstly decreases and then increases with $h_{\max}$. To be more specific, the DF value dramatically decreases during the interval of $h_{\max} =$ [0, 2.5] μm and then slightly increases during the interval of $h_{\max} =$ [2.5, 10] μm. Meanwhile, the DF value is very sensitive to the variations of LFC and $h_{\max}$ during the interval of LFC = [0.3, 0.7] and $h_{\max}$= [5, 10] μm.

The coupling effects of EFC and $h_{\max}$ on the DF value are shown in Fig. 5.19 with $M_{pam} = 0.5M_{pm}$.

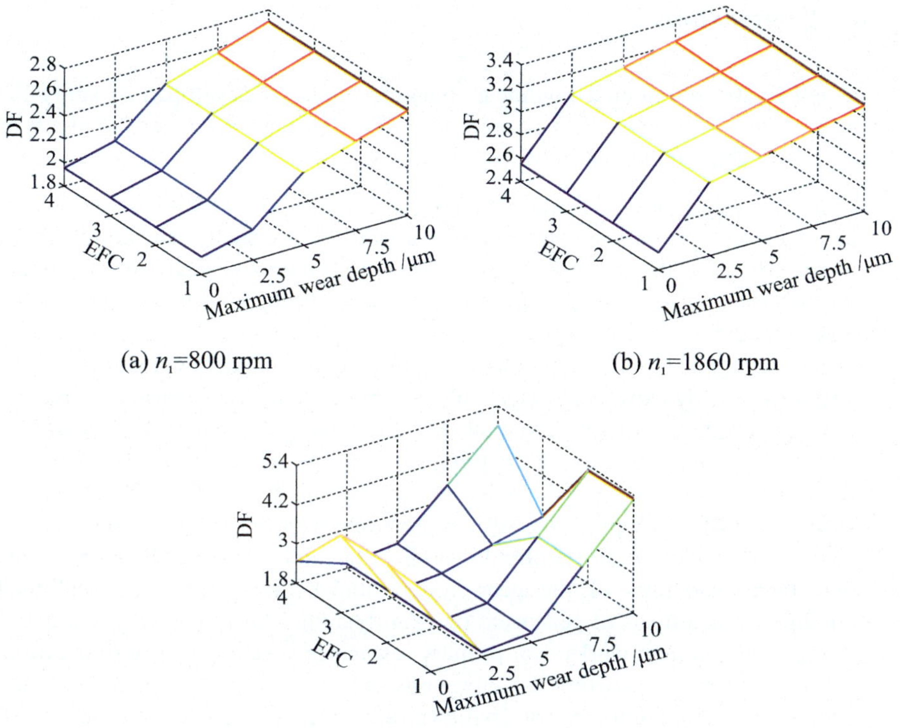

(a) n_1=800 rpm (b) n_1=1860 rpm

(c) n_1=3720 rpm

Fig. 5.19 DF variations with respect to EFC and $h_{\max}$

As shown in Fig. 5.19(a) and (b), the DF value monotonously increases with $h_{\max}$ whereas the effects of EFC are negligible as the system operates at $n_1 = 800$ rpm and $n_1 = 1860$ rpm. Nevertheless, the influences of EFC and $h_{\max}$ on the DF value are very complicated when the system operates at $n_1 = 3720$ rpm. For a given EFC value, the DF value firstly decreases and then increases with $h_{\max}$. To be more specific, the DF value dramatically decreases during the interval of $h_{\max} = [0, 2.5]$ μm, then slightly increases during the interval of $h_{\max} = [2.5, 5]$ μm, and finally increases during the interval of $h_{\max} = [5, 10]$ μm. The effect of EFC on the DF value is negligible when $h_{\max} = 5$ μm. For other circumstances, however, no general rule can be extracted. When EFC = 4 ($f_{pa} = 4f_p$) with $h_{\max} = 0$ μm and EFC = 3 ($f_{pa} = 3f_p$) with $h_{\max} = 10$ μm, the DF value is abruptly changed. Our further investigation reveals that this abrupt reduction in DF value is induced by the system nonlinearity that the response period turns from $3T_m$ to be T_m.

5.4 Conclusions

A lateral-rotational-coupled nonlinear dynamic model is developed to study the coupling effect of time-varying external load and gear wear on dynamics in a spur gear system. The conclusion is drawn as follows.

1. Gear wear alters the STE and meshing stiffness of the gear pair, which may aggravate the system's dynamical characteristics dramatically. The influences of worn surfaces on the gear dynamics are incorporated into the gear dynamic model through the Fourier formulations of time-varying meshing stiffness function and STE function.
2. Time-varying external load induces additional amplitude modulation in the gear system's dynamic responses, enhancing the excitation intensity of internal meshing excitation of the gear pair. For the spur gear system, the dynamical factor (DF) value increases monotonously with the increment of external load fluctuation coefficient (LFC) while the DF value is barely affected by the variation of EFC except for the system operating near the primary resonance.
3. Gear wear evokes both amplitude modulation and phase modulation in the dynamical responses of the gear system under time-varying external load condition. The effects of gear wear on gear dynamics turn to be very distinct as the spur gear system works near primary resonance speed. Under this condition, early gear wear helps to diminish vibrations and eliminate tooth separations while excessive wear amount can worsen the vibrational characteristics and cause tooth separations again.
4. For the spur gear system, the coupling influences of time-varying external load and gear wear on the dynamical response are complicated, the quantitative relationships between the dynamic factor and operating parameters as well as surface wear depth still need to be revealed.

The present study provides an in-depth understanding on the dynamical behaviors of spur gear trains under time-varying load conditions with tooth surface wear.

Appendix

$2b$: Backlash value of gear pair
c_{py}, c_{gy}: Damping of bearings for pinion and gear
c_m: Mesh damping
e: Static transmission error
F_{aT}: Fluctuating force
F_m: Average mesh force
F_p, F_g: External radial preloads of bearing
$h_{\max}$: Maximum wear depth
I_p, I_g: Mass moments of inertia for pinion and gear

k_m: Time-varying stiffness
k_{py}, k_{gy}: Supporting stiffness of pinion and gear
m_p, m_g: Mass of pinion and gear
m_c: Equivalent mass of gear pair
M_p, M_g: Input torque and output torque
R_p, R_g: Radii of base circles for pinion and gear
t: Time
T_m: Meshing period
T_p: Rotational period of the pinion shaft
u: Relative displacement along the line of action
n_1: Speed of pinion
y_p, y_g: Transverse displacements of pinion and gear
θ_p, θ_g: Torsional displacements of pinion and gear
ω_{pa}, ω_{ga}: Fundamental frequencies of torque excitations
ω_e: Fundamental frequency of STE
ω_m: Fundamental frequency of mesh stiffness

References

1. Kahraman, A., and R. Singh. 1990. Non-linear dynamics of a spur gear pair. *Journal of Sound and Vibration* 142 (1): 49–75.
2. Kahraman, A., and R. Singh. 1991. Non-linear dynamics of a geared rotor-bearing system with multiple clearances. *Journal of Sound and Vibration* 144 (3): 469–506.
3. Kahraman, A., and R. Singh. 1991. Interactions between time-varying mesh stiffness and clearance non-linearities in a geared system. *Journal of Sound and Vibration* 146 (1): 135–156.
4. Kahraman, A., and G.W. Blankenship. 1997. Experiments on non-linear dynamic behavior of an oscillator with clearance and periodically time-varying parameters. *Journal of Applied Mechanics* 64 (1): 217–226.
5. Li, Y.G., T.N. Chen, and X.P. Wang. 2016. Non-linear dynamics of gear pair with dynamic backlash subjected to combined internal and external periodic excitations. *Journal of Vibration and Control* 22 (6): 1693–1703.
6. Jiang, Y., H. Zhu, Z. Li, et al. 2016. The nonlinear dynamics response of cracked gear system in a coal cutter taking environmental multi-frequency excitation forces into consideration. *Nonlinear Dynamics* 84 (1): 203–222.
7. Khabou, M.T., A.N. Bouchaal, F. Chaari, et al. 2011. Study of a spur gear dynamic behavior in transient regime. *Mechanical Systems and Signal Processing* 25 (8): 3089–3101.
8. Chaari, F., W. Bartelmus, R. Zimroz, et al. 2012. Gearbox vibration signal amplitude and frequency modulation. *Shock and Vibration* 19 (4): 635–652.
9. Zhou, S., G. Song, Z. Ren, et al. 2016. Nonlinear dynamic analysis of coupled gear-Rotor-bearing system with the effect of internal and external excitations. *Chinese Journal of Mechanical Engineering* 29 (2): 281–292.
10. Bartelmus, W., F. Chaari, R. Zimroz, et al. 2010. Modelling of gearbox dynamics under time-varying nonstationary load for distributed fault detection and diagnosis. *European Journal of Mechanics* 29 (4): 637–646.
11. Chaari, F., M.S. Abbes, F.V. Rueda, et al. 2013. Analysis of planetary gear transmission in non-stationary operations. *Frontiers of Mechanical Engineering* 8 (1): 88–94.

12. Zhu, C., S. Chen, H. Liu, et al. 2014. Dynamic analysis of the drive train of a wind turbine based upon the measured load spectrum. *Journal of Mechanical Science and Technology* 28 (6): 2033–2040.
13. Howard, I., S.X. Jia, and J.D. Wang. 2001. The dynamic modelling of a spur gear in mesh including friction and a crack. *Mechanical Systems and Signal Processing* 15 (5): 831–853.
14. Fakhfakh, T., F. Chaari, and M. Haddar. 2005. Numerical and experimental analysis of a gear system with teeth defects. *International Journal of Advanced Manufacturing Technology* 25 (5): 542–550.
15. Parey, A., M.E. Badaoui, F. Guillet, et al. 2006. Dynamic modelling of spur gear pair and application of empirical mode decomposition-based statistical analysis for early detection of localized tooth defect. *Journal of Sound and Vibration* 294 (3): 547–561.
16. Wu, S.Y., M.J. Zuo, and A. Parey. 2008. Simulation of spur gear dynamics and estimation of fault growth. *Journal of Sound and Vibration* 317 (3–5): 608–624.
17. Mohammed, O.D., and M. Rantatalo. 2016. Dynamic response and time-frequency analysis for gear tooth crack detection. *Mechanical Systems and Signal Processing* 66–67: 612–624.
18. Ma, R., and Y. Chen. 2012. Research on the dynamic mechanism of the gear system with local crack and spalling failure. *Engineering Failure Analysis* 26 (12): 12–20.
19. Ma, H., R. Feng, X. Pang, et al. 2015. Effects of tooth crack on vibration responses of a profile shifted gear rotor system. *Journal of Mechanical Science and Technology* 29 (10): 4093–4104.
20. Velex, P., and M. Maatar. 1996. A mathematical model for analyzing the influence of shape deviations and mounting errors on gear dynamic behaviour. *Journal of Sound and Vibration* 191 (5): 629–660.
21. Fernández, A., M. Iglesias, A. De-Juan, et al. 2014. Gear transmission dynamic: Effects of tooth profile deviations and support flexibility. *Applied Acoustics* 77 (3): 138–149.
22. Inalpolat, M., M. Handschuh, and A. Kahraman. 2015. Influence of indexing errors on dynamic response of spur gear pairs. *Mechanical Systems and Signal Processing* 60–61: 391–405.
23. Hu, Z., J. Tang, J. Zhong, et al. 2016. Effects of tooth profile modification on dynamic responses of a high speed gear-rotor-bearing system. *Mechanical Systems and Signal Processing* 76–77: 294–318.
24. Cheon, G.J., and R.G. Parker. 2004. Influence of manufacturing errors on the dynamic characteristics of planetary gear sys-tems. *Journal of Mechanical Science and Technology* 18 (4): 606–621.
25. Parra, J., and C.M. Vicuña. 2017. Two methods for modeling vibrations of planetary gearboxes including faults: Comparison and validation. *Mechanical Systems and Signal Processing* 92: 213–225.
26. Choy, F.K., V. Polyshchuk, J.J. Zakrajsek, et al. 1996. Analysis of the effects of surface pitting and wear on the vibration of a gear transmission system. *Tribology International* 29 (1): 77–83.
27. Kuang, J.H., and A.D. Lin. 2001. The effect of tooth wear on the vibration spectrum of a spur gear pair. *Journal of Vibration and Acoustics* 123 (3): 311–317.
28. Wojnarowski, J., and V. Onishchenko. 2003. Tooth wear effects on spur gear dynamics. *Mechanism and Machine Theory* 38 (2): 161–178.
29. Lin, A.D., and J.H. Kuang. 2008. Dynamic interaction between contact loads and tooth wear of engaged plastic gear pairs. *International Journal of Mechanical Sciences* 50 (2): 205–213.
30. Yuksel, C., and A. Kahraman. 2004. Dynamic tooth loads of planetary gear sets having tooth profile wear. *Mechanism and Machine Theory* 39 (7): 695–715.
31. Osman, T., and P. Velex. 2010. Static and dynamic simulations of mild abrasive wear in wide-faced solid spur and helical gears. *Mechanism and Machine Theory* 45 (6): 911–924.
32. Ding, H., and A. Kahraman. 2007. Interactions between nonlinear spur gear dynamics and surface wear. *Journal of Sound and Vibration* 307 (3): 662–679.
33. Kahraman, A., and H. Ding. 2010. A methodology to predict surface wear of planetary gears under dynamic conditions. *Mechanics Based Design of Structures and Machines* 38 (4): 493–515.

34. Wu, S.J., H.B. Zhang, X.S. Wang, et al. 2017. Influence of the backlash generated by tooth accumulated wear on dynamic behavior of compound planetary gear set. *Proceedings of the Institution of Mechanical Engineers Part C Journal of Mechanical Engineering Science* 231 (11): 2025–2041.
35. Zhang, J., and X.Z. Liu. 2015. Effects of misalignment on surface wear of spur gears. *Proceedings of the Institution of Mechanical Engineers Part J Journal of Engineering Tribology* 229 (9): 1145–1158.
36. Liu, X.Z., Y.H. Yang, and J. Zhang. 2016. Investigation on coupling effects between surface wear and dynamics in a spur gear system. *Tribology International* 101: 383–394.

Chapter 6
Experimental and Numerical Studies on Compressor Nonlinear Behaviors with Inlet Distortion and Their Interaction

Jun Li, Youtian Zhou, and Guoxing Song

6.1 Introduction

Nowadays, the modern aircrafts' flight envelope has expanded a lot, the low observability and high maneuverability of an aircraft are lay more stress on. Generally, the aircraft inlet system design concept is basically determined by two aspects, namely, the upper limits of the conditions within which the aircraft will operate and the specific roles of the aircraft. For the compressor stability is greatly reduced by the non-uniform inflow conditions. It is important for designers to evaluate the distortion influences on the compressor stall line. Therefore, a plenty of aero-engine/compressor stability assessment tests with various distortion coefficients has been established to solve this problem [1]. However, for the lack of understanding of the interaction between turbofans and inlet distortion, the earlier stability assessment works are roughly and conservative. The underlying physics of compressor performance with inlet distortion needs to be obtained from more delicate experiments or high fidelity CFD capabilities. When compressor suffers the inlet distortion, the axisymmetric characteristic no longer exists, which means the whole annulus needs be considered. This is a challenge for experimental measurement methods and computational capabilities. Furthermore, extracting useful and meaningful information from volumes of experiments and CFD data is also challenging.

J. Li (✉) · Y. Zhou · G. Song
Science and Technology on Plasma Dynamics Lab, AFEC, Xi'an, P. R. China

J. Zhang (ed.), *Dynamics and Fault Diagnosis of Nonlinear Rotors and Impellers*, Nonlinear Systems and Complexity 34, https://doi.org/10.1007/978-3-030-94301-1_6

6.1.1 Experimental Investigations

The earlier investigations of engine/inlet compatibility focused on total pressure distortion. In the 70s, industry and government agencies have carried numerous distortion experiments in the J85-GE-13 turbojet and TF30-P-3 turbofan engine [2, 3], to solve the engine/inlet compatibility problem. Aerospace Recommended Practice (ARP) 1420 [4] and Aerospace Information Report (AIR) 1419 [5] was established to assess total pressure distortion effect at the aerodynamic interface plane (AIP). At that time, a number of investigations [6–9] gave some preliminary and simplified descriptions of the compressor behavior with inlet distortion. And their analytical methods mainly based on experimental statistics and some linear assumption of distortion effect on compressor characteristic. Further studies [10] by Hynes and Greitzer offered a formulation to account for the circumferential distortion. Compared with other type of distortion, circumferential total-pressure-distortion pattern was considered as the most representative factor to engine stability. Another importance work [11, 12] in distortion showed that the compressor behaviors would have a strong impact on the upstream inlet flow distortion. Further, Greitzer [13] highlighted that there might also exist a coupling between the rotor and the downstream components as a compressor is subject to circumferential distortion.

Since 1980s, for the complexity of modern aircraft inlet systems, such as serpentine inlet and S-inlet, swirl distortion also got much more attention. Nowadays, inlet distortion is widely divided into total pressure, swirl, and temperature distortion. And total-pressure and swirl distortions are considered as the most common form. Baffle-produced inlet distortion is a complex distortion that include total pressure, swirl and axis-unsteady distortion. For the concision and practicability of a movable baffle product, it is widely used in aero-engine stability assessment [14]. Li Fanyu [15] used baffle inlet distortion to study the applicability of stall precursor-suppressed (SPS) casing treatment on an axial-flow compressor.

6.1.2 Numerical Investigations Using CFD (Computational Fluid Dynamics)

Due to the cost and the limitation of the measuring method in distortion experiment, numerical simulations have been introduced to study the inlet distortion and compressor characteristics. The early strategies on the numerical method for the compressor distortion include parallel compressor theory [8, 16], actuator disk model [9, 17] and body force model [18–20]. For the limit of computing power, these methods generally make some assumptions to simplify full annulus simulation. Such as a 3D approximate method made by Gong in1999 [18], blade rows were replaced by body force source terms in Euler equations to simplify computational mesh numbers. More recently, a number of full annulus computations [21–25] base on Navier–Stokes equations have been published. V. Jerez Fidalgo

[21] used unsteady RANS to study the NASA rotor 67 stage, operating with a total pressure distortion covering a 120° sector of the inlet flow-field. It is found that the rotor local behavior is highly dependent on the swirl and mass flow redistribution due to the incoming distortion.

Fabian Wartzek Heinz-Peter Schiffer [25] present an experimental and URANS study about the rotor distortion interaction. He found that flow separation widely exists in local blade passage before the rotor stall and an accurate compressor simulation with inlet distortion would not be realized if only a model of the flow separation is analyzed.

RANS methods use a universal model to simulate the large-scale eddies, which makes RANS methods cannot well predict unsteady separation flows, especially the propagation of stall cell in blade rows. Hence, LES is promising to overcome such disadvantage of the RANS methods. In the method, the large energy-containing scales are directly simulated, and the small-scale eddies are considered by sub-grid scale model. However, full annulus LES simulation is still unrealistic for today's computational power, as the complexity of wall boundary conditions in the compressor simulation are considered. To reduce the CPU requirement of LES, Spalart et al. [26] developed the detached-eddy simulation (DES) method, which use RANS model within the solid wall boundary layer in the LES simulation to cut down near-wall mesh number. Gecheng Zha [27, 28] have used the full-annulus DES to get the rotating stall inception of rotor 67. Therefore, it is prospected to study compressor distortion interaction by using the DES method, with the existing computer capabilities.

The purpose of this chapter is to enhance the understanding of the interaction between compressor performance and inlet distortion flow, especially the compressor nonlinear behaviors and stall process in distorted flow. The full-annulus DES method for a compressor with inlet distortion is examined in a low speed rotor. The work presented in this chapter is part of a research project, addressing future aero-engine/inlet integrated design.

6.2 Experimental and Simulation Setups

6.2.1 Experimental Setups

The test compressor in this chapter is a single stage, low-speed and axial-flow one. The geometric data of the test compressor are listed in Table 6.1. The detail geometric data can be obtained in reference [29].

Figure 6.1 briefly depicts the test compressor with the baffle distortion generator and the stable flow field measuring system. The test rig is operated in open-loop conditions, and the distortion generator is installed on the inlet section. At the outlet of the compressor, an exit throttle enables the back-pressuring of the stage. Compressor flow and pressure rise are measured with wall static pressure holes,

Table 6.1 Design parameters of the compressor

Rotational speed (rpm)	3000
Design mass flow rate (kg/s)	6.5
Design pressure ratio	1.025
Outer diameter (mm)	600
Hub-tip ratio	0.66
Tip clearance(mm)	0.42
Rotor blade number	45
Stator blade number	60
Reynolds number (based on the mid-span chord)	1.6×10^5

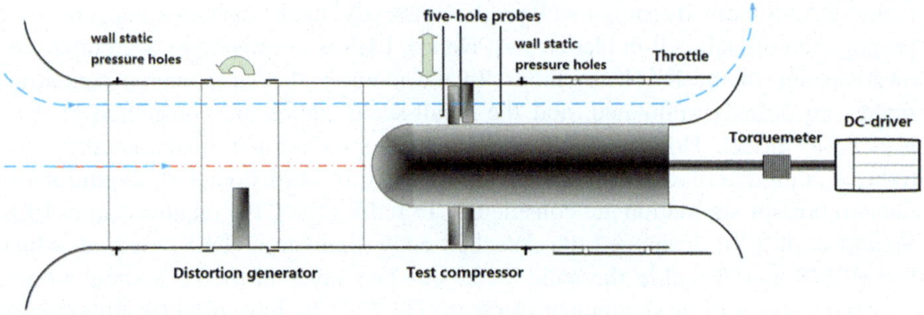

Fig. 6.1 Schematic diagram of single-stage low-speed axial compressor

and three five-hole probes are used to measure the flow field in three key planes. Compressor flow coefficient and total-to-static pressure rise coefficient are defined as follow to evaluate compressor performance,

$$\psi_{ts} = \frac{P_{static}^{out} - P_{total}^{in}}{0.5\rho U^2} \tag{6.1}$$

$$\phi = \frac{\overline{V}_x}{U} \tag{6.2}$$

where P_{static}^{out} and P_{total}^{in} respectively are the outlet static pressure and the inlet total pressure, ρ, U, $\overline{V}_x$ is density, blade speed and mess average axial velocity.The 360° high-resolution measurement is necessary for the distortion flow field. By rotating the distortion generator around an axis and the translation of the five-hole probes, the distortion flow field can be measured. Three key planes are chosen to study the interaction between the compressor and the distortion flow field, and they are the plane inlet, the plane inter-stage, and the plane outlet, respectively. There are 24×19 measure points on each plane, and measure points circumferential divided into 24 by 15° apart and 19 through blade span.

Nine high-frequency Kulite transducers are used to measure the stall pressure signals. The transducers are placed in two different type locations, respectively,

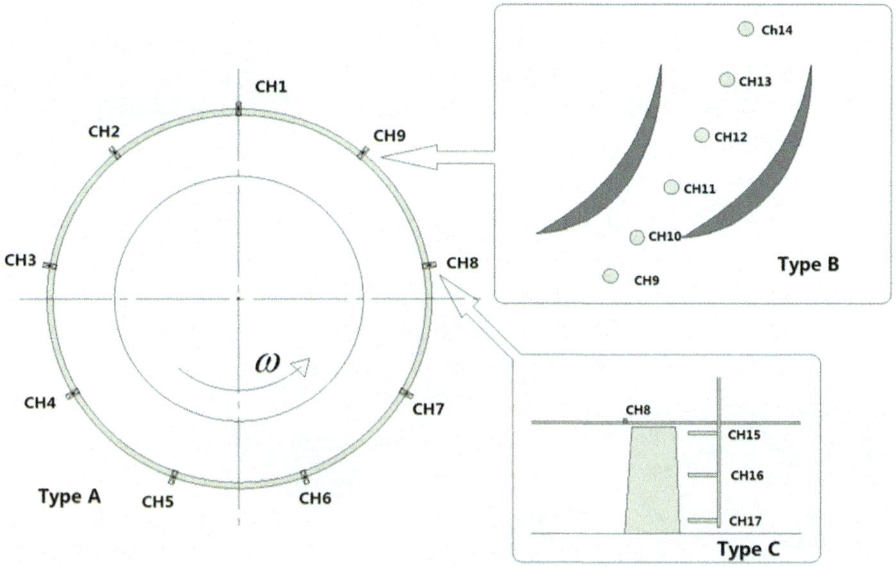

Fig. 6.2 The location of high-response transducers

Table 6.2 Computing resources for DES simulation

Number of cells	130 million
CPU	16× Intel Xeon E5-2682 V4
Partitions	256 (16 × 16)
Memory (RAM)	1 TB
Calculation period	About 370.000 CPU-hours

as shown in Fig. 6.2. Type A: nine transducers are equally spaced along the circumference by 40° apart. Type B: six transducers are spaced along the chord, from the leading edge to the trailing edge.

6.2.2 Numerical Simulation

Due to the complexity of the compressor flow and the mesh requirement of the DES model, the simulations are extremely time-consuming, compared to classical undistorted compressor simulations. Table 6.2 gives an overview of the numerical computing resources.

The simulation model contained distortion generator, inlet tunnel and full annulus of the rotor, and the stator does not contain such elements, so as to reduce computing resources. Figure 6.3 shows the simulation domain and mesh. Compared to the experiment, the simulation model is simplified, that is, the inlet dome and outlet throttle are neglected. Former studies show that the effects of inlet dome on

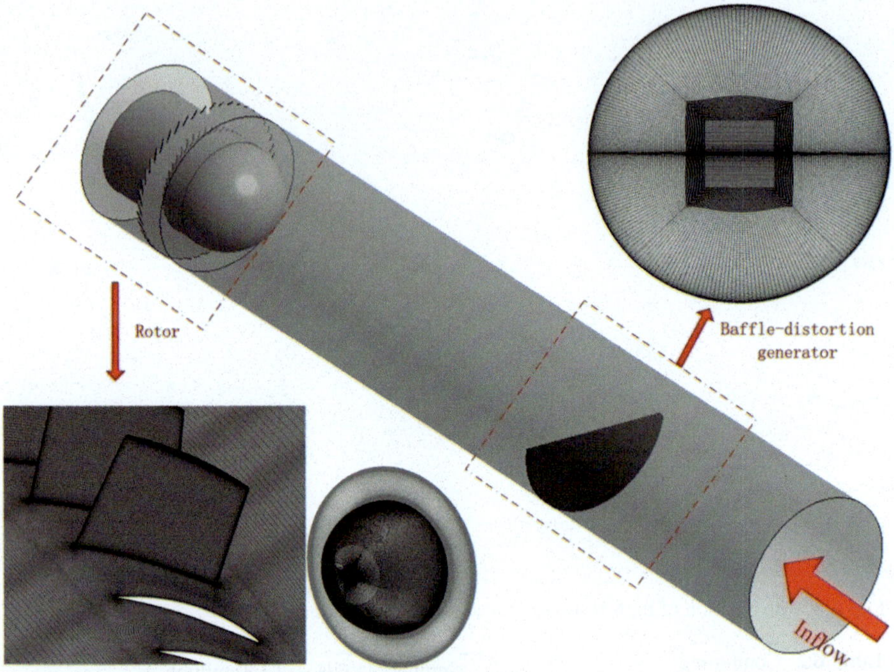

Fig. 6.3 Overview of the mesh for the rotor and the DG

the compressor is slight, compared to the distortion generator. And the effects of outlet throttle can be replaced by a throttle valve model, which is set on the outlet boundary conditions. As a result, this simplification can reduce the mesh number to save the simulation times.

The structured mesh is used to discrete the domain. The compressor meshes are generated with the NUMECA meshing tool AutoGrid5, and HOH-topology is introduced to the rotor meshes. Following the former DES mesh independency study on a compressor, the total mesh number for a single blade passage was around 2.21 million. The blade tip gap is set for 33 cells, the whole span with 145 cells. The blade passage was resolved with 77 cells in blade-to-blade and 157 in streamline direction. The y+ is set to a minimum value of 1 for the RANS-SST model.

The mesh in the inlet domain with the distortion generator is generated by ANSYS meshing tool ICEM. The grid resolution is adapted to the downstream rotor grid to minimize interfaces errors between the two domains, and wall functions were used in this domain to reduce the requirement of near-wall mesh size. The majority part of wall boundary y+ is less than 10. This enables to save a huge amount of cells, but it is known to result in some numerical errors in some near wall separation. To the baffle-produced inlet distortion simulation, the backflow behind the baffle is the dominating factor. Hence, near-wall numerical errors is acceptable to the whole calculation. The mesh number of inlet domain is about 26.1 million, and the total mesh number was $21 \times 45 + 26.1 = 125.6$ million.

The numerical simulations have been carried out using ANSYS fluid solver CFX, which is a widely recognized flow solver specialized on turbomachinery flows and internal flows. An implicit finite volume, the cell-centered approach is used to solve the Navier-Stokes equations. For time marching, the second-order dual pseudo-time-step approach is used to achieve a high convergence rate. The diffusive fluxes are discretized using a Central-Difference-Scheme. the unsteady Reynolds-averaged Navier-Stokes (URANS) method has been used to obtain the initial flow field for detached-eddy simulation (DES). For the DES in this chapter, Shear Stress Transport (SST) turbulence model is used in the RANS part and the WALE subgrid-scale model is used in the LES part.

The boundary condition at the inlet is total pressure and total temperature, and the boundary condition at the outlet is set by a throttle valve model which can offer a reasonable static pressure distribution and set the mass flow as the experimental measurements. The mixing model for the interface between rotor domain and inlet domain is the transient rotor/stator approach, and the physical time step is 1/60 blade passing time.

6.3 Results

The distortion degree of flat baffle generator mainly depends on multiple factors, the baffle height(H), the distance between baffle and compressor(S), the average inlet flow velocity ($\overline{V}_x$). One typical baffle distortion flow field is shown in Fig. 6.4. The baffle height was set as 0.5 times diameter (0.5D). And the baffle was placed at 3 times diameter away from the compressor. The distortion generator produced a backflow zone behind the baffle. To the compressor, the backflow formed a low-speed and low-pressure zone at the aerodynamic interface plane (AIP). In this section, this typical flat baffle distortion flow field will be analyzed from numerical simulations and experimental results. The content in this section includes both steady and unsteady parts.

6.3.1 Steady Part of Flat Baffle Distortion

In this steady analysis, the simulation results are taken from the time-averaged DES results, and the experimental results are taken from the five-hole probe's measurements. Due to the low-frequency response characteristic of the pneumatic probe's measurements, the experiment data is unable to catch the high-frequency signal of the flow field, such as the blade pass and transmission of stall cell. Therefore, the time-averaged and circumferential smoothing was used for the

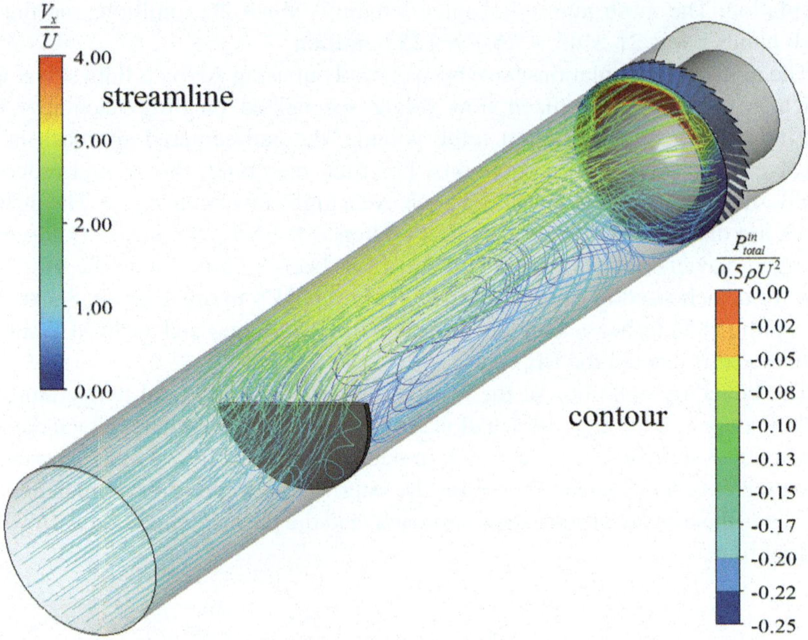

Fig. 6.4 The streamline of distortion flow field and total pressure distribution in AIP

simulation data, in order to make a better comparison with experimental data. For this purpose, the circumferential smoothing range was set as five blade area.

Figure 6.5 shows the total pressure and axial velocity distribution in the aerodynamic interface plane (AIP). (Fig. 6.5a depicts the distortion generator (DG) location and rotation direction of the rotor, and the viewing direction is from the inlet to the outlet. The viewing direction and the DG location of the other figures are the same.) The numerical results show that the distribution can be further qualified especially in the near-wall area. Simulation and experiment results show that the total pressure distribution is almost bilateral symmetrical about the baffle axis. Nevertheless, the axial velocity distribution does not have this asymmetrical features. It is because that the total pressure distribution is depended on the upstream total pressure loss, which is mainly induced by the distortion generator. However, the velocity distribution not only depends on the upstream flow but also strongly affected by downstream rotor operating condition. Therefore, the axial symmetry of the distortion flow field would be broken in front of the rotor.

Figure 6.6 shows the velocity quiver distribution in AIP. The non-uniform pressure distribution leads to the flow transverse movement in AIP, and there is a transverse flow from the high-pressure area to the low-pressure area, and this kind flow determines the rotor's attack angle. Half of the blades suffer high attack angle and half of the blades are on low attack angle. In the high attack angle zone, the

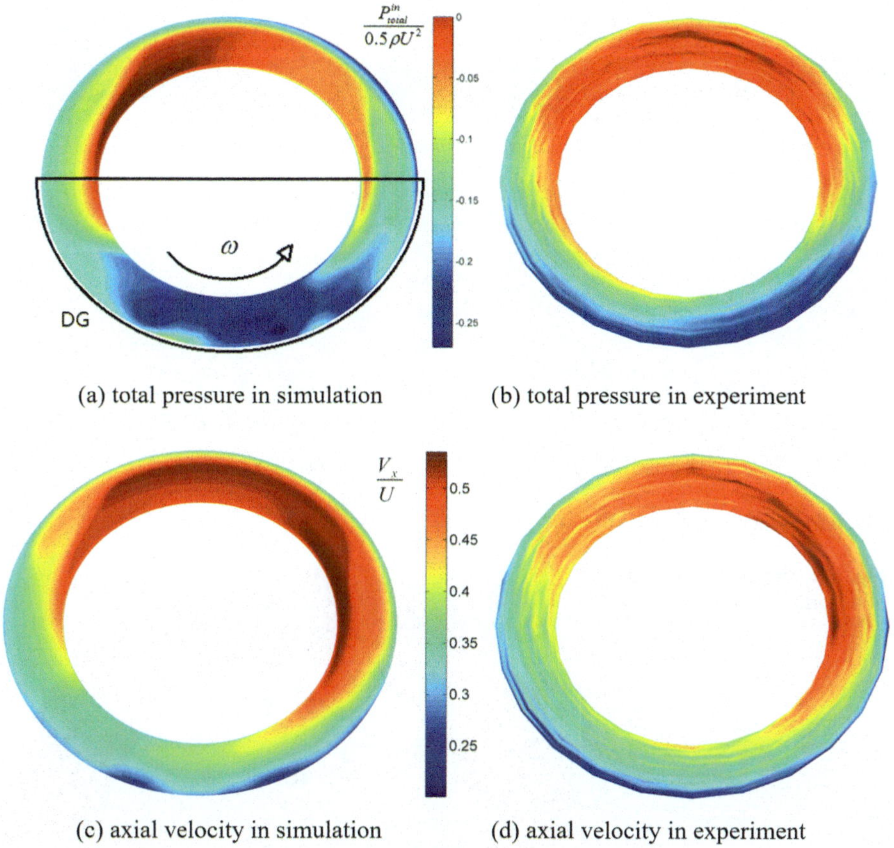

Fig. 6.5 The total pressure and axial velocity distribution in AIP

blade's load and pumping action are stronger, making the upstream flow moving faster in high attack angle zone. The symmetrical features of flow in AIP is broken by this effect.

6.3.2 Rotor Steady-State Performance with Distortion

As discussed above, this distortion flow field can be divided into 4 parts on the bases of the measuring result, namely, a high-speed and high-pressure zone with high attack angle, a high-speed and high-pressure zone with low attack angle, a low-speed and low-pressure zone with high attack angle and a low-speed and low-pressure zone with low attack angle, shown in Fig. 6.7. For the rotor, a low-speed zone means a lower flow coefficient and higher pressure rise. Figure 6.8 shows the experimental results of the total pressure and axial velocity distribution in the inter-

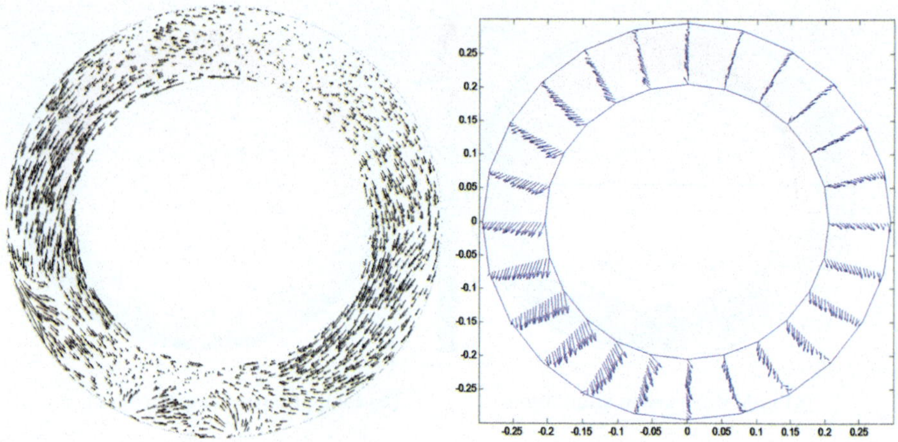

(a) Simulation (b) Experiment

Fig. 6.6 Velocity vector in AIP

Fig. 6.7 The divisional chart of distortion flow field

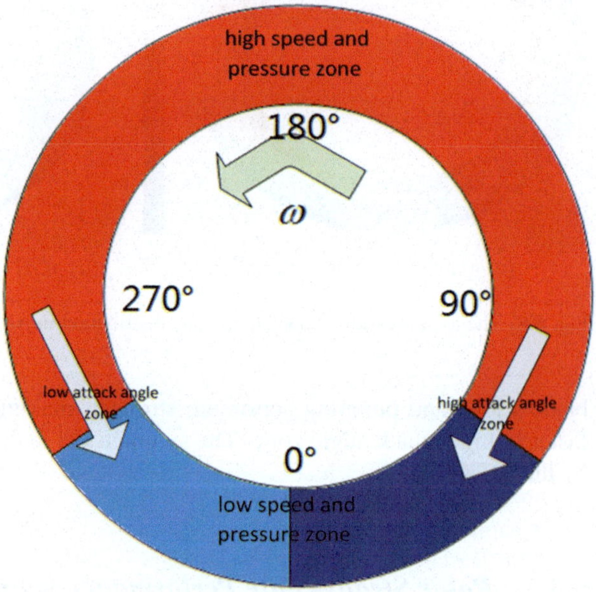

stage plane. Through the rotor's work, it can be found that the flow in the low-speed and low-pressure zone will get high total pressure. In particular, half of the low-speed zone (circumference location at 0° ~ 90°) has a higher attack angle. As rotor blades enter this region, the blade load would be maximum. Through Fig. 6.8, it can be found that there is a low axial velocity and total pressure zone in those blades' tip, indicating that the flow separation occurs in those rotor blades'. However, this

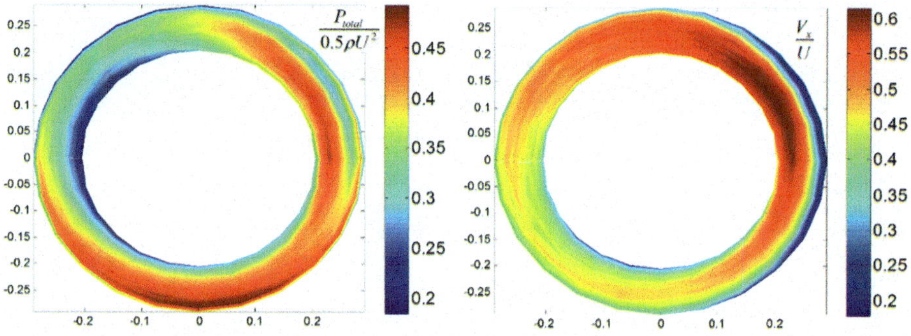

Fig. 6.8 The experiment total pressure and axial velocity distribution in inter stage plane

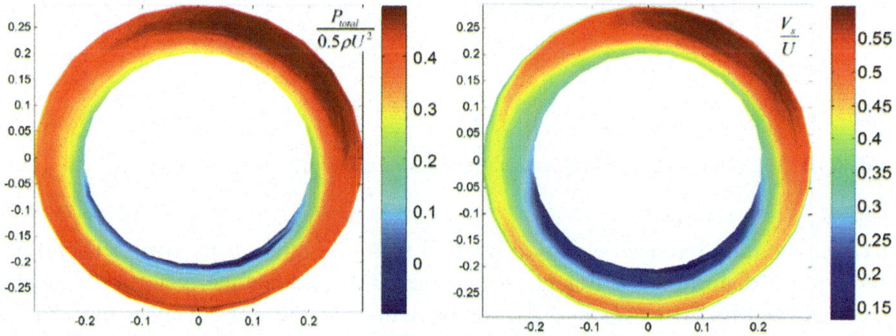

Fig. 6.9 The experiment total pressure and axial velocity distribution in outlet plane

flow separation could not be found in the stator blades behand those rotor blades through the axial velocity distribution in the outlet plane as Fig. 6.9 shows, meaning that the typical flow separation just exist in the rotor. To the stator passage, there is large flow separation around the blades in the low-speed and low-pressure zone near the hub.

For further analysis, the compressor was divided into 24 sub-compressors circumferentially to study the compressor performance and the interaction between the compressor and the distortion inflow, with the post-processing methods [21]. Figure 6.10 shows sub-compressors' local rotor operating points in the typical baffle distortion flow field. 24 local rotor operating points form an operating loop. It is quite interesting to notice that most of the local rotor operating points surround the clean operating line, despite the mean operating point falling on the clean operating line. As in the previous analysis, the transverse secondary flow created by the pressure gradient on the inlet plane is the main reason, especially the circumference flow leading to the variation of sub-compressors' attack angle. The half of local rotors (circumferential position 0° ~ 180°) with a high attack angle have higher operating points than the clean line. The distribution of flow coefficient and attack angle are the main influence factors for the compressor, and they are induced by the

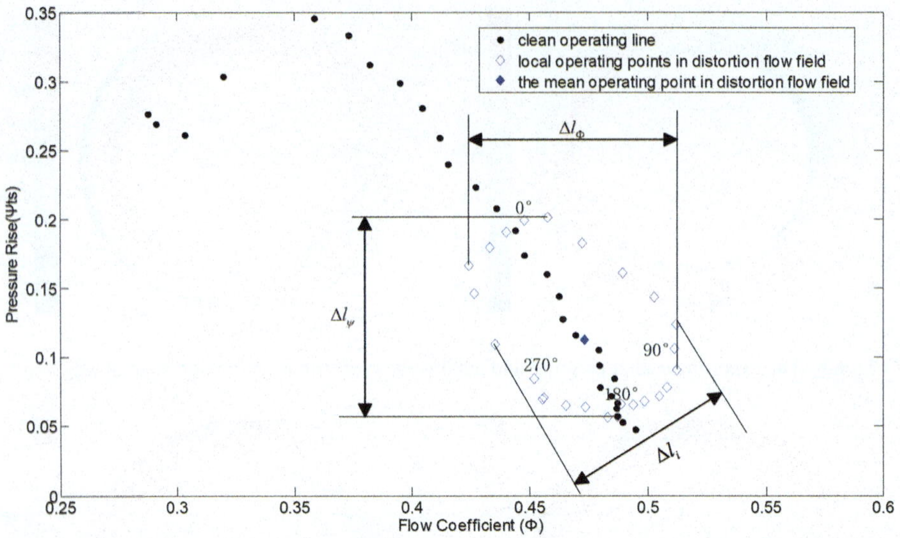

Fig. 6.10 The variation in local rotor operating conditions as operating in distortion

axial and circumference velocity distortion from the inlet, respectively. Therefore, three novel distortion coefficients are presented to evaluate the distortion features, as shown in Fig. 6.10. Δl_{Φ} and Δl_i can be defined as the consequences of the axial and circumference flow distortion, and the area of the operating loop and Δl_{ψ} can represent the comprehensive influences.

In this part, the effects of the baffle's height and distance will be considered. Other 6 experiments will be conducted to analyze the influence of various type of parameters on the distortion flow field. The three coefficients, namely, Δl_{Φ}, Δl_i, Δl_{ψ} mentioned above, is used to evaluate the distortion influences on the compressor. To determine the inlet distortion level, the distortion coefficient DC60 is widely used, which is defined as,

$$DC60 = \left(P_{0,60} - P_0\right) / \left(P_0 - P\right) \tag{6.3}$$

where P_0 and P are the average total and static pressures at the compressor inlet plane, respectively. $P_{0,\,60}$ is the area-averaged total pressure for the 60-deg segment with the lowest mean total pressure.DC60 can be considered as a coefficient to measure the inlet distortion degree and Δl_{Φ}, Δl_i, Δl_{ψ} are used to evaluate the compressor reactivity under the distortion flow flied. Table 6.3 shows the major parameters and distortion coefficients in 7 Cases, and Fig. 6.11 shows the local rotor operating points of experiment, that is, No. 1# to 5#.

It intuitively detects that the increase of the baffle height and the decrease of the baffle distance will strengthen the DC60 and Δl_{Φ}, Δl_i, Δl_{ψ}. For 2#'s baffle distance

Table 6.3 Major parameters and distortion coefficients of distortion experiments

Experiment no.	baffle height (H)	baffle distance (S)	velocity (V_x)	DC60	Δl_{Φ}	Δl_{ψ}	Δl_i
1#	30%D	2D	0.4747 U	−0.1799	0.0876	0.1448	0.0741
2#	30%D	3D	0.4742 U	−0.1207	0.0440	0.0903	0.0240
3#	30%D	2D	0.4209 U	−0.1712	0.0732	0.1042	0.0588
4#	50%D	2D	0.4597 U	−0.5097	0.1613	0.3828	0.1195
5#	50%D	2D	0.3770 U	−0.7929	0.1217	0.3777	0.1137
6#	50%D	3D	0.4478 U	−0.3828	0.1235	0.2834	0.0794
7#	50%D	3D	0.4077 U	−0.4440	0.1080	0.2409	0.0675

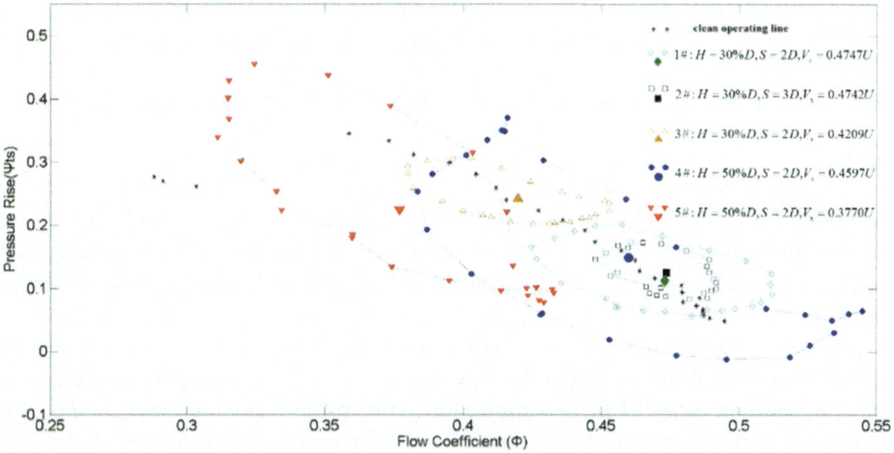

Fig. 6.11 The variation in local rotor operating conditions of different experiments

(S) is farther than 1#, and the 2# operating loop is less than 1#. By comparing experiment 1# and 4#, we can also find the influences from baffle height.

However, DC60 shows that the inlet distortion level mostly is enhanced, as the flow velocity decreases. While the Δl_{Φ}, Δl_i, Δl_{ψ} indicate the distortion influences on the compressor weaken. It should be noted that the total pressure distortion is always accompanied by some strength velocity distortion. And the velocity distortion, especially the axial and circumference flow distortion, is the direct factor to compressor performance and stability. Therefore, the distortion coefficient defined by total pressure distortion is unable to determine the inlet distortion level in some situations, and the velocity distortion plays a major role. So some swirl distortion coefficient is taken into account in the velocity distortion.

The 5# operating loop shows the compressor stall point with the baffle distortion. The 5#'s mean operating point is under the clean operating line, for most of the local rotor operating points are below the clean operating line. However, a portion of local rotors operating points are posted stall regime and a part of their pressure

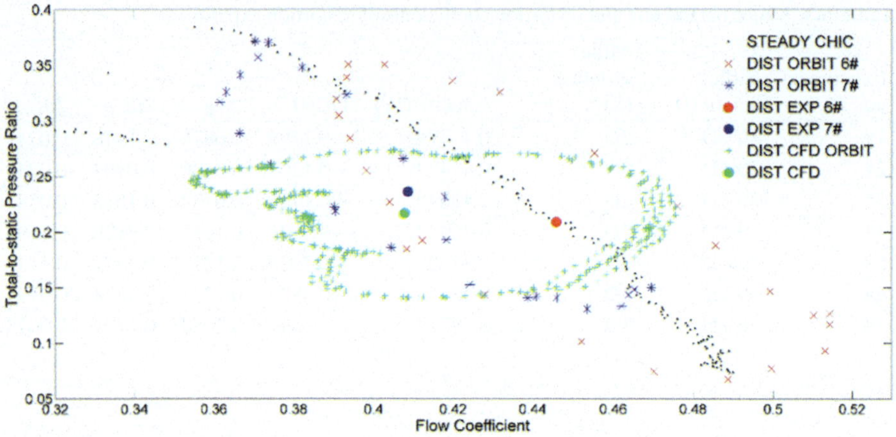

Fig. 6.12 The variation in local rotor operating conditions of experiments and simulations

rise is even above the clean operating line. On other hands, the operating points of this local rotors can't appear in uniform inlet condition.

For further analysis of the rotor performance with the distortion, several typical rotor operating points with distortion need to be simulated. Due to the time needed to simulate one single operating point, this was not feasible here. The No.7# experiment was chosen as a simulation case. Because in this case, the rotor's mean operating point is under the clean operating line, implying a dangerous phenomenon in a multistage compressor. This rotor behaviors will change the matching of stages. Figure 6.12 shows the simulation results of the rotor behaviors under the flat baffle distortion. Similar to the experimental analysis method, the rotor behavior in simulation results is described as orbit. It can be find that the axial and circumference distortion coefficient Δl_{Φ} and Δl_i in CFD are basically the same as the experiment. However, CFD overestimates the rotor pressure ratio in the high-speed zone and underestimate the rotor pressure ratio in the low-speed zone. The rotor mean operating point is also below the experiment result. Overall, the DES method can give an accurate estimate of the flat baffle distortion flow field. But to the rotor simulation with inlet distortion, the results from the DES method are still not accurate enough.

6.3.3 *Unsteady Part of Flat Baffle Distortion*

In order to complete the description of distortion flow, the unsteady analysis is carried out through high-response transducers measurement and DES result. Figure 6.13 shows the vortex structure of the baffle distortion flow field from RANS steady result and DES result. The snapshot of Q criterion iso-surface, colored by the

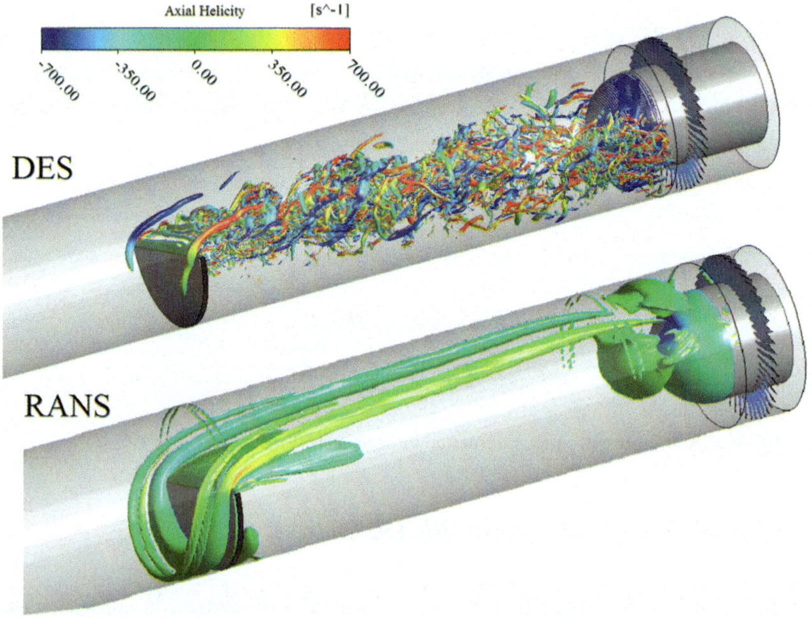

Fig. 6.13 Vortex structure of baffle distortion flow field from RANS steady result and DES result

normalized helicity, was used to describe the vortex. The Q criterion, defined as the equation below, represents the balance shear strain rate S and vorticity magnitude Ω. As it is a positive scalar, it represents locations where the rotation dominates the strain. Therefore, the Q criterion is widely used to describe the vortex structure, as following,

$$Q = \frac{1}{2}\left(\|\Omega\|^2 - \|S\|^2\right) \tag{6.4}$$

$$\Omega = \frac{1}{2}\left(\mathrm{grad}\,(\mathrm{V}) - (\mathrm{grad}\,(\mathrm{V}))^T\right) \tag{6.5}$$

$$S = \frac{1}{2}\left(\mathrm{grad}\,(\mathrm{V}) + (\mathrm{grad}\,(\mathrm{V}))^T\right) \tag{6.6}$$

For both RANS and DES results, DG could induce stream-wise vortex pair and a span-wise vortex. In RANS result, the span-wise vortex is soon dissipated and the stream-wise vortex pair could develop to influence the rotor. In the DES result, this three vortex soon breaks, and entangle with each other, developing into the low-speed and low-pressure distortion flow field to affect the rotor behavior.

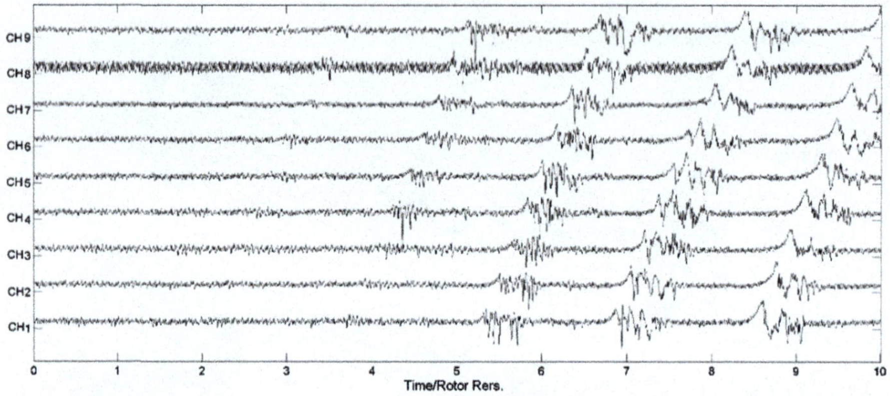

Fig. 6.14 Wall static pressure measurements without distortion

6.3.4 Rotor Performances with Distortion

In this subsection, the rotor stall process with inlet distortion is studied based on experiment and DES results. Rotating stall is usually initiated around the rotor tip, and hence the monitor located at the rotor tip, as sketched in Fig. 6.2, can catch the information on the stall inception and development. Figure 6.14 shows the experimental casing pressure traces at 9 circumferential even spacing locations. Without distortion, the rotor stall inception is a typical 'spike', and the initial rotating speed of the stall cell is 61.3% of rotor speed.

Figure 6.15 presents the distortion result. According to the stall cell development, the rotor stall process with inlet distortion can generally be divided into three stages. From Fig. 6.15a, the stall inception is observed in CH4 ~ CH9 transducers, meaning flow separation occurs in those blade tips. It is in agreement well with the steady analysis. For those blades is with high attack angle and blade loading. Suffered the dynamic part of distortion, blades in this zone are more likely to generate flow separation. Moreover, this kind of local flow separation can circumferentially propagate a few blades and disappear in low blade loading zone (CH1 ~ CH3). In this condition, the flow coefficient is 0.467. According to No 6# experiment (flow coefficient is 0.4478) and the Fig. 6.11, we can know that the rotor's mean operating point is still on the clear operating line, meaning that this local flow separation has not yet influenced the whole rotor pressure rise capability.

As the flow coefficient drops to 0.432, the local flow separation, which only exists in part area, developed into the stall cell. The stall cell can circumferentially propagate to all the passage of rotor blades. This stall cell is unsteady, deduces and disappears after several cycles, just as Fig. 6.15b shows. By comparison with Fig. 6.15c, it can be seen that the stall cell became stable as the flow coefficient drops to 0.421. It can be found that the rotor can be in a critical unstable state as suffering from flat baffle distortion. Furthermore, the stall cell rotating speed is 68.7% in Fig.

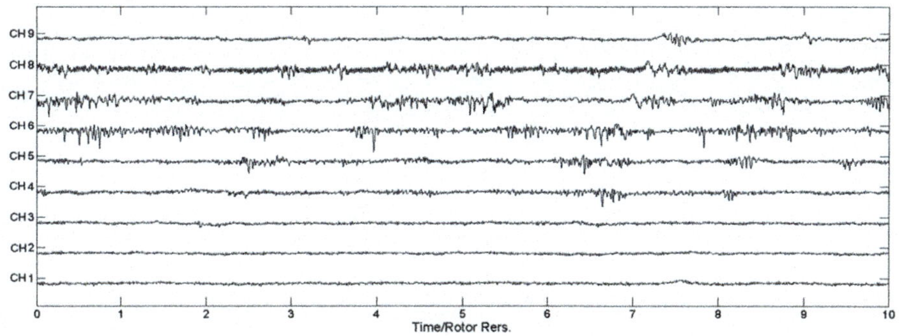

a) Flow coefficient 0.467

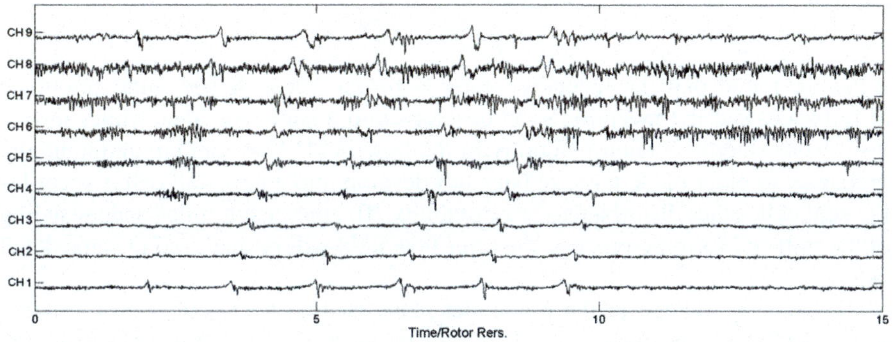

b) Flow coefficient 0.432

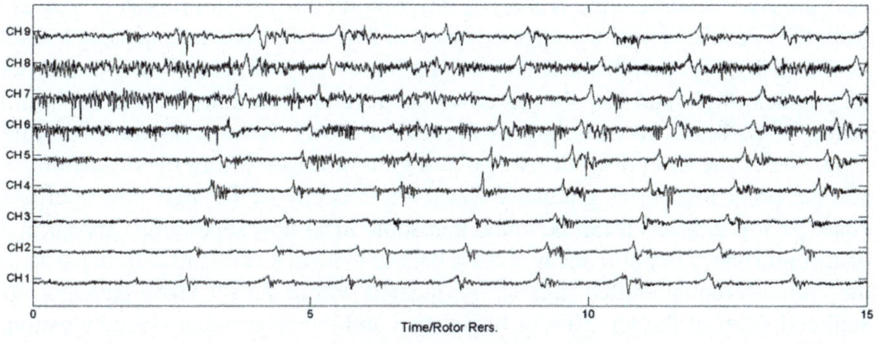

c) Flow coefficient 0.421

Fig. 6.15 Wall static pressure measurements with circumferential inlet distortion

6.15b and 66.2% in Fig. 6.15c. When the rotor suffers from distortion, the stall cell rotating speed is a slightly faster than usual.

In DES simulation, the speed of the stall cell was 66.7% as Fig. 6.16 showed, and this parameter is almost the same as the experiments. From the simulation result,

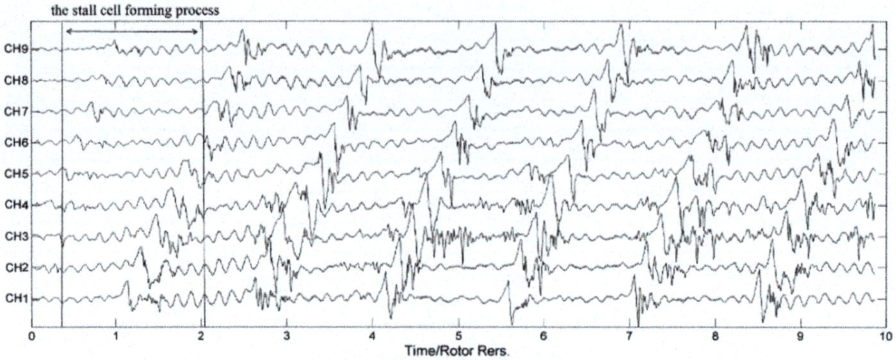

Fig. 6.16 Wall static pressure of DES result

it can be found that the inception starts from one spike-shaped finite-amplitude disturbance, and the inception soon develops into a stall cell. Meanwhile, several disturbances are generated randomly in the flow field. They can circumferentially propagate to other blade passage but can't develop into a new stall cell. Except for the stall cell, other disturbances are instability. To approach a fuller understanding of the stall cell forming process, the rotor flow field will be analyzed in some detail below.

The entropy change in the rotor inlet plane is used to describe the formation of the stall cell since entropy can stand for the degree of energy loss; hence, high entropy could reflect flow separation or stall cell zone. In Fig. 6.17, entropy change near the rotor leading edge is used to describe the formation process of stall cells from DES results. The time range is shown in Fig. 6.16. Like the experiment results as shown by Fig. 6.15b, local flow separation is easy to occur in the blades with high attack angle and low flow coefficient. Several inceptions of stall cell can be found in the blade tip zone, just as Fig. 6.15a shows. After a 2/9 rotor revolution time, one of the inceptions developed and grew stronger within the distortion zone and the other inceptions gradually weaken and decay in the high flow coefficient zone. At 1 + 2/9 revs it can be found that some local flow separations appear in the distortion zone. They still decay and are unable to form a new stall cell. In particular, the circumferential stalled regions are located within 7 to 9 blade passages. The stall cell starts at the top 50% of blade span and the span radial stalled location is changing with the stall cell circumferential location. For instance, the stall cell can impact the whole blade span after one cell revolution.

In Fig. 6.18, positive Q iso-surfaces are used to detect the vortex structure of stall cell. As Q criterion is 0.005 s^{-2}, stall cell and some local flow separation can be clearly demonstrated, meanwhile only tip leakage flow vortex can be seen in the regular blade passage. Just as Fig. 6.18 showing, there are two primary flow separation zones. One of them locates at those blades which start to enter the distortion zone. This flow separation is the inception of the stall cell, as Fig. 6.17 indicated. Through the streamline in those blade tips, reverse flows can be seen in

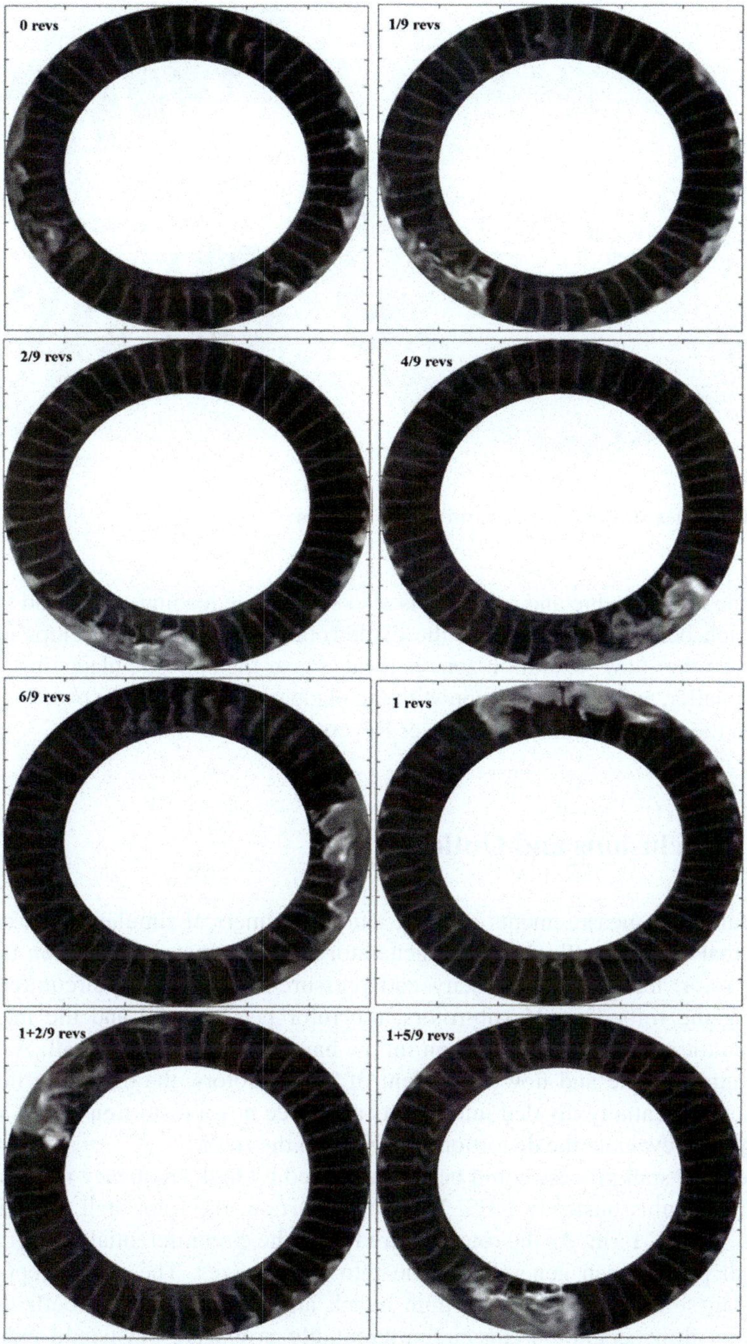

Fig. 6.17 Entropy change near rotor leading edge

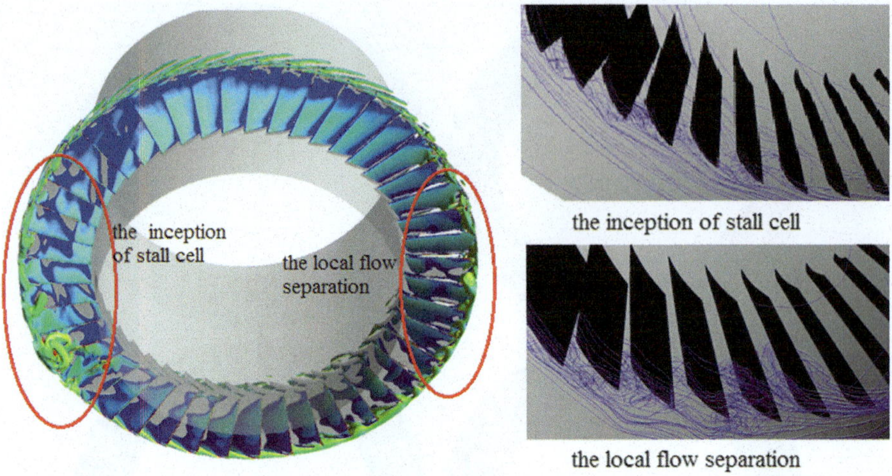

Fig. 6.18 Vortex structure and local streamline at 0 revs

the blade trailing edge and overflow is also seen in the leading edge. And this flow separation became stronger with the blade rotating and finally develops into stall cell. The other flow separation zone is located near the opposite blade tips, and it is much smaller, and dissipates soon with the blade rotating. Only the overflow can be seen on the leading edge, and however Reverse flows are not formed.

6.4 Conclusions and Outlook

With full 360° measurements and full annulus numerical simulations, it could be found that the flat baffle distortion generator could produce a separation upstream of the rotor, leading to a velocity and pressure distortion. By circumferentially dividing the rotor into 24 sub-rotors, the rotor performance and the interaction between the rotor and the distortion inflow have been studied in detail. According to the attack angle and flow coefficient of 24 sub-rotors, the distortion flow field are circumferentially divided into 4 parts and three novel distortion coefficients are presented to evaluate the distortion influence on the rotor.

The rotor stall process could be well captured by high-frequency measurements and full annulus unsteady DES. Without distortion, the rotor stall inception is a typical 'spikes' form. As the rotor suffered from the circumferential inlet distortion, the stall process can generally divide into three stages. The stall inception first occurs in some blade tips with high attack angle and low flow coefficient, and this local flow separation can circumferentially propagate but would reduce and disappear quickly as it leaves those zones. With the reduction of mass flow, the local flow separation develops into the unsteady stall cell. It circumferentially propagates to all the passage of rotor blades and disappears after several cycles. Finally, the stall cell become stable and the rotor runs into rotating stall. Indeed, this nonlinear phenomenon is the compressor complex feature.

The experiment data or results in this chapter can be used to examine the simulation models. For larger flow separation existed widely in this case, it is a challenge for CFD codes to capture the rotor characteristic and stall process with such strong inlet distortion. In this chapter, the full-annulus DES could get the rotating stall inception of rotor but could not determine precisely the rotor characteristic well. In particular, the interaction between the rotor and the distortion must be considered, as simulating real aircraft inlet with an engine.

References

1. Cousins, W.T. 2004. History, philosophy, physics, and future directions of aircraft propulsion system/inlet integration. In ASME Turbo Expo 2004: power for land, sea and air, Vienna, Austria, 14-17 June 2004.
2. Bobbala, G.A. 1979. Effect of steady-state pressu\re distortion on stall margin of a J85-21 turbine engine. NASA-TM-79123 E-9958 NASA Lewis research center.
3. Calogeras, J.E., and C.M. Mebalic. 1971. Experimental investigation of the effect of screen-induced total-pressure distortion on turbojet stall margin, NASA TMX-2239 March.
4. Gas Turbine inlet flow distortion guidelines. Aerospace Recommended practice 1420. Society of automotive engineers.
5. Inter total-pressure-distortion considerations for gas turbine engines. Aerospace information report 1419. AIR 1419.
6. Horlock, J.H., and E.M. Greitzer. 1977. The response of turbomachinery blades to low frequency inlet distortions. *Jeng Gas Turb Power* 99: 195–203.
7. Greitzer, E.M., and T. Strand. 1978. Asymmetric swirling flows in turbomachinery annuli. *ASME\Jeng Gas Turb Power* 100: 618–629.
8. Mazzawy, R.S. 1977. Multiple segment parallel compressor model for circumferential flow distortion. *ASME Jeng Gas Turb Power* 99: 288–296.
9. Hawthorne, W.R., N.A. Mitchell, and J.E. McCune. 1978. Non-axisymmetric flow through annular actuator disks: Inlet Distortion Problem. *ASME Jeng Gas Tueb Power* 100: 604–617.
10. Hynes, T.P., and E.M. Greitzer. 1987. A method for assessing effects of circumferential flow distortion on compressors stability. *ASME Journal of Turbomachinery* 109: 371–379.
11. Callahan, G.M., and A.H. Stenning. 1971. Attenuation of inlet flow distortion upstream of axial flow compressors. *Journal of Aircraft* 4: 227–233.
12. Greitzer, E.M. 1973. Upstream attenuation and quasi-steady rotor lift fluctuations in asymmetric flows in axial flow compressors. ASME Paper 73-GT-30.
13. Greitzer, E.M., R.S. Mazzawy, and D.A. Fulkerson. 1978. Flow field coupling between compression system components in asymmetric flow ASME Jeng Gas Turb. *Power* 100: 66–72.
14. Li, F.Y., J. Li, X. Dong, D.K. Sun, and X.F. Sun. 2016. Influence of SPS casing treatment on axial flow compressor subjected to radial pressure distortion. *Chinese Journal of Aeronautics* 02: 223–225.
15. Khaleghi, H., G. Doulgeris, and M. Boroomand. 2009. A method for calculating inlet distortion effects on stability of split-flow fans. *Chinese Journal of Aeronautics* 113: 591–598.
16. Horlock, J.H. 1978. *Actuator disk theory*. New York: McGraw-Hil.l.
17. Gong, Y.F. 1999. *A computational model for rotating stall and inlet distortions in multistage compressors*. Cambridge, MA: Massachusetts Institute of Technology.

18. Chima, R.V. 2006. A three-dimensional unsteady CFD model of compressor stability. ASME Turbo Expo 2006, Power for Land Sea and Air, Barcelona, Spain 8-11 May 2006.
19. Cao, T., and A.R. Smith. 2016. Fan-intake interaction under high incidence. In Proceedings of ASME Turbo Expo 2016: Turbomachinery Technical Conference and Exposition.
20. Fidalgo, V.J., C.A. Hall, and Y. Colin. 2010. A study of fan-distortion interaction within the NASA rotor 67 transonic stage, Proceedings of ASME Turbo Expo 2010: Power for Land, Sea and Air, 2010, Glasgow, Scotland.
21. Castaneda, J., A. Mehdi, and D. Cugno. 2011. A preliminary numerical CFD analysis of transonic compressor rotors when subjected to inlet swirl distortion. In Proceedings of ASME Turbo Expo 2011 June 6-10, 2011, Vancouver, British Columbia, Canada GT2011-46560.
22. Florea, R.V., D. Voytovych, and G. Tillman. 2013. Aerodynamic analysis of a boundary-layer-ingesting distortion-tolerant fan[C] Proceedings of ASME Turbo Expo 2013: Turbine Technical Conference and Exposition June 3-7, 2013, San Antonio, Texas, USA GT2013-94656.
23. Sharma, G., L. Zori, and S. Connell. 2013. Efficient computation of large pitch ratio transonic flow in a fan with inlet distortion. In Turbine Technical Conference and Exposition June 3-7, 2013, San Antonio, Texas, USA GT2013-95059.
24. Wartzek, F., H. Schiffer, J. Haug, and R. Niehuis. 2016. Investigation of engine distortion interaction. In Proceedings of ASME Turbo Expo 2016: Turbomachinery Technical Conference and Exposition June 13–17, 2016, Seoul, South Korea.
25. Spalart, P.R.., M. Strelets, and S.R. Allmaras. 1997. Comments on the feasibility of LES for wings, and on a hybrid RANS/LES approach advances in DNS/LES. In Proceedings of the First AFOSR International Conference on DNS/LES, Greyden Press, Columbus, OH.
26. Hong-Sik, I.M., X.Y. Chen, and G.C. Zha. 2010. Detached eddy simulation of unsteady stall flows of a full annulus transonic rotor. In Proceedings of ASME Turbo Expo 2010: Power for Land, Sea and Air, June 14-18, 2010, Glasgow, UK GT2010-23465.
27. Hongsik, I.M., X.Y. Chen, and G.C. Zha. 2012. Detached-eddy simulation of rotating stall inception for a full-annulus transonic rotor. *Journal of Propulsion and Power* 28 (4): 782–798.
28. Li, F.Y., J. Li, X. Dong, Y.T. Zhou, D.K. Sun, and X.F. Sun. 2016. Stall warning approach based on aeroacoustic principle. *AIAA Journal of Propulsion and Power* 32 (6): 1–12.
29. Vester, A.K., R. Örlü, and P.H. Alfredsson. 2015. POD analysis of the turbulent flow downstream a mild and sharp bend. *Experiments in Fluids* 56 (3): 57.

Chapter 7
Study on Mode Localization Induced by Material and Aerodynamic Mistunings in Impellers with Periodical and Cyclic Symmetry

Jiazhong Zhang, Yan Liu, and Pengliang Wang

7.1 Introduction

7.1.1 Background

It is shown that vibration localization plays an important role in inducing fatigue failures of impeller of centrifugal compressor with large diameter, and such kind of failure is frequently encountered in industry.

The impeller of centrifugal compressor is one typical structure, that is, it is of periodic or cyclic symmetry, and there are some distinct dynamic properties compared with normal or non-cyclic structures. Specifically, it is very sensitive to mistuning, ensuing vibration localization and spatial confinement of vibration energy. Such mistuning mentioned above is always random and includes the deviations from manufacturing tolerances, material tolerances, in-operation wear, non-uniform of flow in the passage and other causes. Following survey, it is shown that the impeller with large diameter are sensitive to the mistuning. Indeed, it is well-known that the dynamics of impeller with large diameter in centrifugal compressor behaves as that of thin-walled structure, that is, the natural frequencies become closer with each other. The mistuning could drastically destroy the symmetry pattern of impeller inducing vibration localization, though it's value is small [1,

---

J. Zhang (✉) · P. Wang
School of Energy and Power Engineering, Xi'an Jiaotong University, Xi'an, Shaanxi Province, P. R. China
e-mail: jzzhang@mail.xjtu.edu.cn

Y. Liu
School of Mechanical Engineering, Northwestern Polytechnical University, Xi'an, Shaanxi Province, P. R. China
e-mail: liuyan@nwpu.edu.cn

J. Zhang (ed.), *Dynamics and Fault Diagnosis of Nonlinear Rotors and Impellers*, Nonlinear Systems and Complexity 34, https://doi.org/10.1007/978-3-030-94301-1_7

2]. Therefore, how to control the vibration localization and fatigue failure becomes one urgent task in community of turbo-machinery.

In summary, large centrifugal compressors with high speed are used widely and play important role in industries and aerospace engineering. In particular, the dynamics and stabilities of the whole systems are the main points as the speed, pressure have increased to a higher level.

7.1.2 State of the Art

In decades, the dynamics of impeller with mistuning and its analysis have been paid more attention. In 1976, Ewin has noted the mode of the vibration of mistuning blade-disk and observed the complex mode in experiments [3], and he could not define such phenomenon as vibration localization at that time. In 1980, Fabunmi found that there exist obvious differences between mistuning and tuning impellers in modal shape of dynamics, as one experiment was carried out [4]. In fact, such mistuning modes are intrinsic modes of vibration localization. However, vibration localization is just discovered and studied initially in the field of solid-state physics, and the researchers in mechanics community could not understand such phenomenon in structures. Until 1980s, the vibration localization in the fields of structure dynamics and vibration have been studied systematically [5]. In 1982, Hodges studied the similarities between mechanics and solid-state systems, and introduced the localization founded by Anderson, and studied initially the localization in vibration of quasi-periodic structures [6, 7]. Since then, both theoretical and experimental studies have been further carried out, and the existence of vibration localization in structural dynamics and vibration has been verified. In particular, the importance of localization in dynamic stress analysis of structures has been explained further.

Since then, vibration localization induced by mistuning in engineering has attracted more attention as periodic or cyclic structures are used widely in engineering, and hence a number of works relevant to mistuned bladed disk are obtained. In 1988, Wei and Pierre applied firstly perturbation and modal analysis methods to calculating the natural frequencies, steady responses and vibration localization of mistuning impeller, and studied the influences of mistuning parameters on the vibration localization [8–10]. In 1992, Pierre and Murthy investigated mode localization considering the coupling of aeroelasticity of mistuned blade assemblies [11]. In 1995, Watson and Kamat analyzed the dynamics of mistuned cyclic systems with introducing mistune modes, and perturbation method was used to solving the response of the structure and localization factor was proposed further [12]. Happawana studied the free and forced response of mistuned linear cyclic systems based on singular perturbation method, and the vibration localization and frequency veering are captured [13]. Further, Sextro proposed blade-disk of with dry friction damping to model nonlinear mistuned systems with cyclic symmetry, and dry friction damping is used to describe the nonlinearity in the system, localization in system is investigated in detail [14]. Judge gave an experimental study to investigate the influences of stochastic mistuning from blades on mode localization and forced response for a mistuned bladed disk, and the theoretical analysis of mode localization, forced response amplitude magnification and frequency veering are verified [15, 16]. Using modal synthesis techniques, Xie studied vibration mode localization in disordered cyclic structures with single substructure mode and multiple substructure modes, and then one numerical method for localization factor was proposed [17, 18].

Since 1990s, there are a number of researchers in China whose researches focus on mode localizations and veering frequencies in cyclic structures, and some more important results are obtained. Hao Wang proposed a modal synthesis method to analyze the natural vibration of bladed disk with small mistuning, based on perturbation method [19]. Following this method, the principal modes of the tuning bladed disk are used as Ritz basis. Yunju Yan built a mathematical model with component mode synthesis method for analyzing vibration of mistuned bladed-disk, and several low order modes of each sub-structure are obtained in combination with experimental modal analysis and modifying calculation of the modes [20]. Further, the real mistuned bladed disk and several kinds of theoretical mistuned models are set up based on the small perturbation of blade modal stiffness. Jianjun Wang gave a detail review on several models of mistuned bladed disk to analyze the vibration localization, and the models include lumped parameter model, continuous parameter model, finite element model and dynamic substructure model etc. [21]. Hongjian Wang presented a method for predicting the forced response of mistuned bladed disks within entire coupling range, that is, classified the response of mistuning bladed disk as partial and entire mistuning responses, and then combination of these two models can accurately describe the vibration of mistuned bladed disks [22]. Erming Huo presented a method to study the influences of random mistuning on the forced response of nearly cyclic assemblies, with Monte Carlo simulation and N-DOF mass-spring model, and the results show that weakly coupled systems are more sensitive to mistuning than strongly coupled ones, leading to a wider range of resonant frequency and an increase in the largest amplitude, and damping ratio plays a positive role in suppressing forced response amplitudes [23]. Further, Hongjian Wang studied the influences of disorders induced by dry friction, one kind of damper with nonlinear property, on dynamics of bladed disk with Harmonic Balance Method and Fast Fourier Transform, and the results show that the disorder of dry friction can significantly increase the resonant amplitude of bladed disks, especially as system with weaker coupling and larger viscous damping ratio [24]. Jingjun Dai gave a method to study the influences of disorder of blades on vibration of bladed disk assemblies [25]. By this method, the blades are modeled as cantilever beams fixed on the disk, and the result indicates that vibration localization appears in the system due to the disorder, and the degree of blade disorder and rotating speed have important influences on the vibration mode localization.

However, all the studies stated above are relevant to the vibration localization of bladed disks, instead of the impeller of centrifugal compressors, which include shroud, blades and disk. Jiazhong Zhang and his colleagues gave explanations for the nature of the fatigue failure of impeller in centrifugal compressor from viewpoint of dynamics [26]. In his study, the dynamics of the centrifugal impeller is studied numerically using finite element method and mode synthesis analysis, and it is found that centrifugal impeller is the typical cyclic structure, and there exist many unique dynamics in comparison with non-cyclic structure, such as vibration localization, passbands and stopbands. Specifically, the amplitude and the resulting energy will be uniform in the structure as the exciting force in the range of passbands. On the other hand, the amplitude and the resulting energy will be confined to local part, and hence the amplitude and stress will be increased greatly. Under such situation, the centrifugal impeller will be destroyed as the energy is localized. Additionally, the influences of mistuning from the manufacture process, material tolerance and wear on the dynamics of the system are significant. As a conclusion, it can be drawn that such mode localization is the key to the fatigue failure of the centrifugal impeller and such fatigue failure is the result from resonance.

Recent years, the studies in our group indicate that the non-uniforms of mechanics properties of material, such as Young's modulus, Possion ratio, density, and aerodynamics in passages have significant influences on the dynamics of the impellers of centrifugal compressors, except the mistunings from manufacturing tolerances, wear and others. This chapter will give a detail introduction to the influences of mistunings induced by mechanics properties of material and aerodynamics in passages of centrifugal impellers on the dynamics of whole impeller. The Chapter is scheduled as following: In Sect. 7.2, the fundamental theories relevant to vibration localization of periodic structures are introduced briefly, as well as the analysis method and models; In Sect. 7.3, a numerical method based on finite element method is developed to consider the mistunings of materials with stochastic description, and one method to study the influences of non-uniform of aerodynamics in passages is given; In Sect. 7.4, the methods presented in Sect. 7.3 are further developed in combination with ANSYS, and some examples are given to verify the methods with stochastic description of the mistunings of materials and analyze the vibration localization of mistuned centrifugal impellers; In Sect. 7.5, the influences of non-uniform of aerodynamics in passages on vibration localization are investigated in detail; In Sect. 7.6, some conclusions are drawn and further works are suggested.

7.2 Fundamental Theories of Vibration Localizations of Periodic Structures

7.2.1 Concept of Vibration Localization and Its Developments

In 1958, Anderson maybe the first one to present the concept of localization, as he studied the influences of disorders on the electric conductivity of solid-state physics [27]. In experiment, he found that there exists a unique conductivity property in the

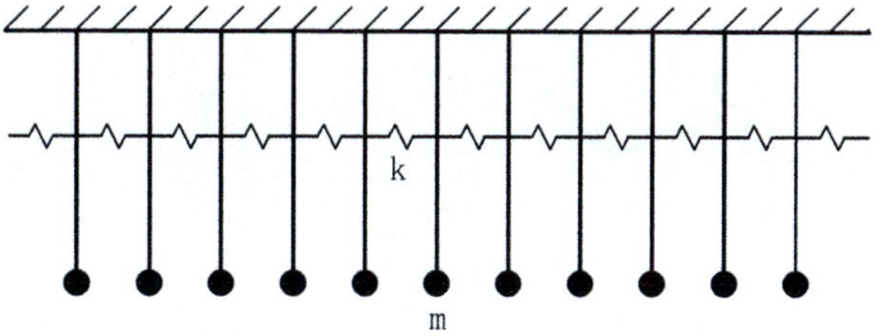

Fig. 7.1 Coupled pendulum system [27]

mistuned solids, such finding has attracted more attention of researchers. In 1982, Hodges made a comparison between mechanics system and solid-state physics, and found there are some similarities between them, and then he introduced the localization to the vibration of mistuned structures.

Since then, the problem initially studied is the coupled pendulum system, as shown in Fig. 7.1. By this model, the vibration localization and its intrinsic physical properties are obtained theoretically and experimentally [28].

In structure dynamics, there are two main points in vibration localization, namely, mode localization and vibration transferring localization. As for mode localization, there are two physical phenomena, that is, mode shape localization and frequency veering. The mean of mode shape localization is that mode shape can become confined in some partial substructures or in a small region of the structure, instead of global structure. As the results, most of the sub-structures vibrate with small amplitudes, and however few of sub-structures vibrate with relative larger magnification amplitudes since they are localized [29]. As for the vibration transferring localization, it means that the excitations have just influences on some partial region of the structure, and the energy in vibration is confined to few sub-structures of the system, resulting in oscillation of local regions which are normally closed to excitation resources. In this Chapter, only the mode localizations of centrifugal impeller are studied, with mistunings from material and aerodynamics in passages.

7.2.2 Nature of Vibration Localization and Its Effects

Vibration localization could be explained theoretically by wave propagation as following: The forced responses of specific blade can be considered as the superimposition of waves propagated circumferentially in bladed disk with different velocities, and only single wave propagation could be used to explained the nature of vibration localization if the system is linear one [30]. As depicted in Fig. 7.2, it is clear that the force transmitted from one layer to another one, as it transmitted through "multi-layers structures", can be used to simulate the wave propagated from one blade to another one, and the function of velocity c is the same as $H(\omega)$, the transmission function defined in theory of vibration.

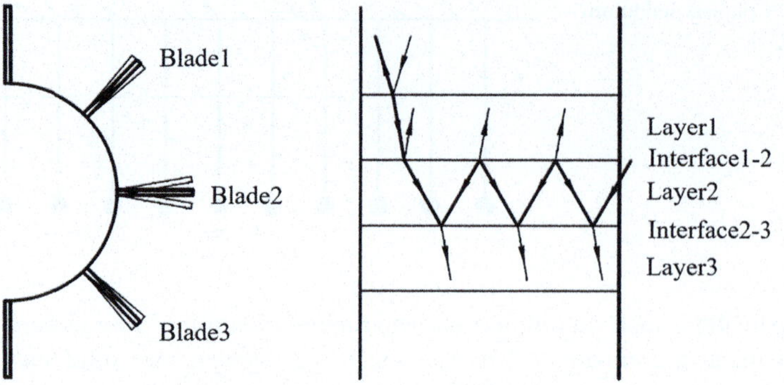

Fig. 7.2 Transmission of vibration

For tuned structures, all the blades (or layers) are the same exactly with each other, the interfaces between layers have no influences on the propagation of wave induced by vibration; As for mistuned structures, the blades are different from each other, energy relevant to vibration can be transferred and reflected partially, and the energy reflected could lead to large magnification amplitudes as the dissimilarity of frequency response functions of two layers around interface becomes higher. The wave induced by vibration will be confined to several blades or layers, as the wave propagated to an interface with high reflection, that is, there exists a notable difference between the frequency response functions of two adjoint blades [31]. As a conclusion, it can be drawn that the amplitudes of several blades become magnified as the wave induced by vibration is confined or the energy is confined to these blades, that is the physical meaning of vibration localization.

In particular, the results show that vibration localization could have positive and negative influences on the dynamics of the systems. For positive influence, a certain mistuning of the structure could confine the range which aeroelasticity will become unstable within, so that vibration localization can be considered as an available passive method to suppress the oscillation of system; On the other hand, vibration localization could confine the energy to the region where excitation resources are located, so that it is possible that some strategies could be introduced to reduce or dampen the amplitudes of the systems. For the negative influence, it can be listed as following: (1) The regularity of modes of the structures will be destroyed as localization occurs, and system become complicated and the respected values could not be captured by the sensors in experiments, ensuing the failure of experiments; (2) Vibration localizations have influences on the controlling system, the reason is that the controlling system is designed following the regular modes or global modes, the controlling system could not work as mode localization occurs; (3) The feasibility of optimization design, parameter identification, reliability depend strongly on the vibration localization; (4) Under excitation, the existence of vibration localization could lead to large magnification of maximum amplitude of local structures or regions of the whole structure, resulting in stress increases and high cycle fatigue.

7.2.3 Main Models of Vibration Localization

A number of models have been proposed and developed, since the vibration or mode localizations in mistuned bladed disks and other cyclic structures are studied, the main models include the following, (1) Lumped-parameter model, such as cyclic chains of spring-mass oscillators; (2) Distributed-parameter model, such as plate-beam-based method; (3) Reduced-order model, such as component-mode-based methods, system-mode-based methods [32].

In our study, finite element method is used to analyze numerically the vibration localization of mistuned centrifugal impellers with cyclic structure induced material tolerances and non-uniform of aerodynamics in passages. Indeed, there are a lot of research works in which finite element method is used to approach the real bladed disk systems, and analyze the dynamics induced by localization of the systems. Recent years, full models of the mistuned bladed disk are studied numerically rather than reduced the model, since finite element method and computational resource have been developed further. For the tuned cyclic structures, it is available to analyze single sector to understand the dynamics of the full system. For the mistuned bladed disk system, the cyclic symmetry has been destroyed completely, the full system must be used to analyze the natural frequencies and forced responses. To this end, finite element method is introduced to approach the full system.

7.2.4 Main Analysis Methods

There are a lot of researchers focus their researches on vibration localizations of periodic structures or cyclic structures, such as bladed disk, and hence some analysis methods are proposed and developed. The main analysis method can be listed as following, perturbation method, mode synthesis method, transfer matrix method, wave analysis method, and experiment.

In our work, mode synthesis method is used to study the vibration localization of mistuned centrifugal impellers. By this method, Chao Hu and Fengming Li investigated the vibration localization of mistuned bladed disk. In their model, the blade is modeled as cantilevered beam fixed at disk, and the governing equations are given by variational method, and the Eigen values and vectors are obtained further [33]. More, Kuang and Huang studied mode localization of a cracked bladed disk with mode synthesis method, modeling blade as distributed beam fixed at disk. In their works, the influences of crack and Coriolis effects on the mode localizations are investigated in detail, and the results show that the distributed model could give an accurate solution to the problem [34–36].

7.3 Finite Element Method Combined with Materials Mistuning

7.3.1 Finite Element Method to Model Material Mistuning

As mentioned above, the mistunings in bladed disk or centrifugal impellers could be induced by manufacturing tolerances, material tolerances, wear, non-uniform of aerodynamics in passages and others. In our study, the mistunings relevant to Young's modulus, Possion's ratio, density and non-uniform of aerodynamics in passages are mainly considered in the model.

As one example, how to model the mistuning of Young's modulus by element method is stated in detail. It is assumed that the nominal Young's modulus is E^0, and hence the Young's modulus of each element can be expressed as

$$E^e = E^0\left(1+\alpha^e\right) \quad \alpha^e \geq -1 \tag{7.1}$$

where α denotes the deviation of Young's modulus due to material mistuning. $\alpha = 0$ implies there is no mistuning, and $\alpha \geq -1$ since the value of Young's modulus could be negative. As elements are generated, the stochastic sequences of Young's modulus governed by probability distribution function are allocated to each element, and then the distribution of mistuning of Young's modulus can be obtained following the specified probability statistics distribution. For uniform state, the elasticity matrix can be obtained in the following form.

For stress problem,

$$\boldsymbol{D} = \frac{E}{1-\nu^2}\begin{bmatrix} 1 & \nu & 0 \\ \nu & 1 & 0 \\ 0 & 0 & \frac{1-\nu}{2} \end{bmatrix} \tag{7.2}$$

For strain problem,

$$\boldsymbol{D} = \frac{E\left(1-\nu\right)}{\left(1+\nu\right)\left(1-2\nu\right)}\begin{bmatrix} 1 & \frac{\nu}{1-\nu} & 0 \\ \frac{\nu}{1-\nu} & 1 & 0 \\ 0 & 0 & \frac{(1-2\nu)}{2(1-\nu)} \end{bmatrix} \tag{7.3}$$

Under consideration of mistuning, elasticity matrix for each element can be expressed as,

$$D^e = f\left\{E^e\right\} = f\left\{E^0\left(1+\alpha^e\right)\right\} \tag{7.4}$$

Following Taylor series, yields,

$$\boldsymbol{D}^e = \boldsymbol{D}_0 + \alpha^e \boldsymbol{D}_0 \tag{7.5}$$

For plane stress problem,

$$D_0 = \frac{E^0}{1-\nu_0^2}\begin{bmatrix} 1 & \nu_0 & 0 \\ \nu_0 & 1 & 0 \\ 0 & 0 & \frac{1-\nu_0}{2} \end{bmatrix} \tag{7.6}$$

For plane strain problem,

$$D_0 = \frac{E^0\,(1-\nu_0)}{(1+\nu_0)\,(1-2\nu_0)}\begin{bmatrix} 1 & \frac{\nu_0}{1-\nu_0} & 0 \\ \frac{\nu_0}{1-\nu_0} & 1 & 0 \\ 0 & 0 & \frac{1-2\nu_0}{2(1-\nu_0)} \end{bmatrix} \tag{7.7}$$

Further, the stiffness matrix of each element can be obtained as follow,

$$\boldsymbol{K}^e = \boldsymbol{B}^T \boldsymbol{D}^e \boldsymbol{B} t A \tag{7.8}$$

where $\boldsymbol{B}$ is the strain matrix of element, $\boldsymbol{D}^e$ the elasticity matrix of element. After assembling, the global stiffness matrix can be obtained,

$$\boldsymbol{K} = \sum_e \boldsymbol{G}^{\mathrm{T}} \boldsymbol{K}^e \boldsymbol{G} \tag{7.9}$$

Finally, the governing equation following finite element method is given as following,

$$\boldsymbol{Ka} = \boldsymbol{P} \tag{7.10}$$

where $\boldsymbol{a}$ is displacement vector, $\boldsymbol{P}$ load vector.As for 3D problems, elasticity matrix of each element can be expressed as,

$$\boldsymbol{D}^e = f\left\{E^e\right\} = f\left\{E_0\left(1+\alpha^e\right)\right\} \tag{7.11}$$

Using Taylor series, Eq. (7.11) can be rewritten as,

$$\boldsymbol{D}^e = \boldsymbol{D}_0 + \alpha^e \boldsymbol{D}_0 \tag{7.12}$$

where

$$D_0 = \frac{E^0\,(1-\nu_0)}{(1+\nu_0)\,(1-2\nu_0)} \times \begin{bmatrix} 1 & \frac{\nu_0}{1-\nu_0} & \frac{\nu_0}{1-\nu_0} & 0 & 0 & 0 \\ \frac{\nu_0}{1-\nu_0} & 1 & \frac{\nu_0}{1-\nu_0} & 0 & 0 & 0 \\ \frac{\nu_0}{1-\nu_0} & \frac{\nu_0}{1-\nu_0} & 1 & 0 & 0 & 0 \\ 0 & 0 & 0 & \frac{1-2\nu_0}{2(1-\nu_0)} & 0 & 0 \\ 0 & 0 & 0 & 0 & \frac{1-2\nu_0}{2(1-\nu_0)} & 0 \\ 0 & 0 & 0 & 0 & 0 & \frac{1-2\nu_0}{2(1-\nu_0)} \end{bmatrix} \tag{7.13}$$

The stiffness matrix of each element can be given as following,

$$\boldsymbol{K}^e = \boldsymbol{B}^T \boldsymbol{D}^e \boldsymbol{B} t A \tag{7.14}$$

where $\boldsymbol{B}$ is the strain matrix of element, $\boldsymbol{D}^e$ the elasticity matrix of element. After assembling, the global stiffness matrix can be obtained,

$$\boldsymbol{K} = \sum_e \boldsymbol{G}^{\mathrm{T}} \boldsymbol{K}^e \boldsymbol{G} \tag{7.15}$$

Finally, the governing equation following finite element method is given as following,

$$\boldsymbol{K a} = \boldsymbol{P} \tag{7.16}$$

where $\boldsymbol{a}$ is displacement vector, $\boldsymbol{P}$ load vector [37].

7.3.2 Mistuning from Aerodynamics

Previous studies on the influences of aerodynamics on dynamic performance of turbomachinery mainly focus on unsteady flows induced by the deformation of blades [38, 39]. However, the study shows that non-uniform of steady flows in each passage could lead to mistuning, ensuing vibration localization. How to model such kind of mistuning is the key point. In our work, the inflow is disturbed to model the non-uniforms in each passage, so that mode localization induced by aerodynamics could be investigated.

7.3.2.1 Stress Stiffening Effects

The nature of stress stiffening effect is that the structure behaves as stiffening increasing under steady stress, and hence such phenomena is referred to as geometric stiffening, or pre-stressing stiffening. Such kind of stiffening effect could lead to varying of natural frequencies and modal shapes, and has an influence on response of the system.

For arbitrary infinitesimal segment *ds*, the relationship between strain and displacement can be described as

$$\varepsilon_x = \frac{\partial u}{\partial x} + \frac{1}{2}\left[\left(\frac{\partial v}{\partial x}\right)^2 + \left(\frac{\partial w}{\partial x}\right)^2\right] \tag{7.17}$$

where subscript x denotes direction of infinitesimal segment.Following this geometric relationship, the prestress matrix could be transformed into a pre-stress stiffness matrix with additional stress as following,

$$\mathbf{K_{pre}} = \iint_A \mathbf{G^T}\sigma_{\mathbf{pre}}\mathbf{G}dxdyt \tag{7.18}$$

where $\boldsymbol{\sigma}_{\mathbf{pre}}$ is the prestress matrix, $\mathbf{G}$ differential matrix relevant to shape function. Therefore, the actual stiffness matrix includes two parts, original one and additional stiffness matrix, as following,

$$\mathbf{K_T} = \mathbf{K} + \mathbf{K_{pre}} \tag{7.19}$$

7.3.2.2 Aerodynamics Effects

Under aerodynamics load, the governing equation of dynamics of impeller derived in traveling wave-based coordinate system could be expressed as,

$$\left[-\omega^2\mathbf{I} + \boldsymbol{\Lambda}_{\mathbf{K}}\right]\mathbf{Y} = \mathbf{F_a} \tag{7.20}$$

where $\mathbf{F_a}$ is the aerodynamic load. The aerodynamic load on blade is unsteady due to the vibration induced by the coupling between structure and flow in passages, and hence the aerodynamic load can be described as $\varphi = \overline{\varphi} + \varphi'$, $\overline{\varphi}$ is the averaged or steady part, φ' the fluctuation part. At first, the perturbating part of aerodynamic load is not included in the following, as to study the influences of aerodynamics mistuning on modes of impeller,

$$\mathbf{F_a} = \overline{\mathbf{F_a}} \tag{7.21}$$

The stress matrix reduced by aerodynamic load $\mathbf{F_a}$ on blades can be transformed into pre-stress stiffness matrix, so Eq. (7.21) can be rewritten as,

$$\left[-\omega^2\mathbf{I} + (\boldsymbol{\Lambda}_{\mathbf{K}} + \mathbf{A_a})\right]\mathbf{Y} = \mathbf{0} \tag{7.22}$$

where $\mathbf{A_a}$ is the pre-stress stiffness matrix.It is noted that the pre-stress induced by aerodynamics is much less than Young's modulus, so pre-stress stiffness matrix $\mathbf{A_a}$ can be considered as small fluctuation to original stiffness matrix, namely, stiffness mistuning of the structure. Hence, it is available that the non-uniform of aerodynamics in passages can be considered as stiffness mistuning to investigate its influence on mode localization.

It is well-known that there are few works relevant to influence of aerodynamics mistuning on mode localization. Several works focus on the influences of aerodynamics mistuning of unsteady fluctuation flow induced by vibration of blades on dynamics of bladed disk [38–40]. As a conclusion, it can be drawn that studying mistuning of steady aerodynamics in passages plays important role in understanding mode or vibration localizations.

7.4 Influences of Material Mistuning on Mode Localization

7.4.1 Model of Material Mistuning of Impeller

7.4.1.1 Meshing and 3D Model

For sake of verification, the centrifugal impeller, which has been applied in engineering, will be analyzed as an example, and the results presented are verified by this impeller, as shown in Fig. 7.3.

It is well-known that centrifugal impeller is one of the most complicated cyclic structures, so the typical element, namely, Brick 20 node 186, is used to mesh and approach the system, as shown in Fig. 7.4.

7.4.1.2 Stochastic Distribution of Material Mistuning

For sake of simplification, some assumptions are given as following,

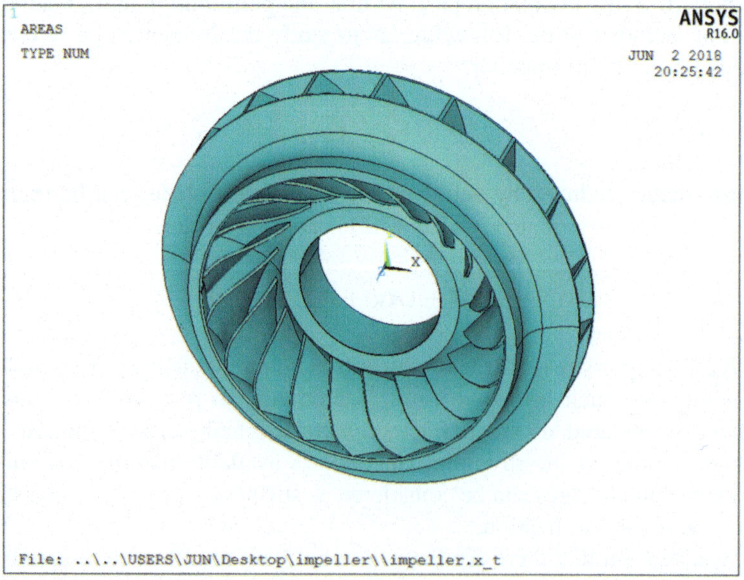

Fig. 7.3 Impeller of centrifugal compressor

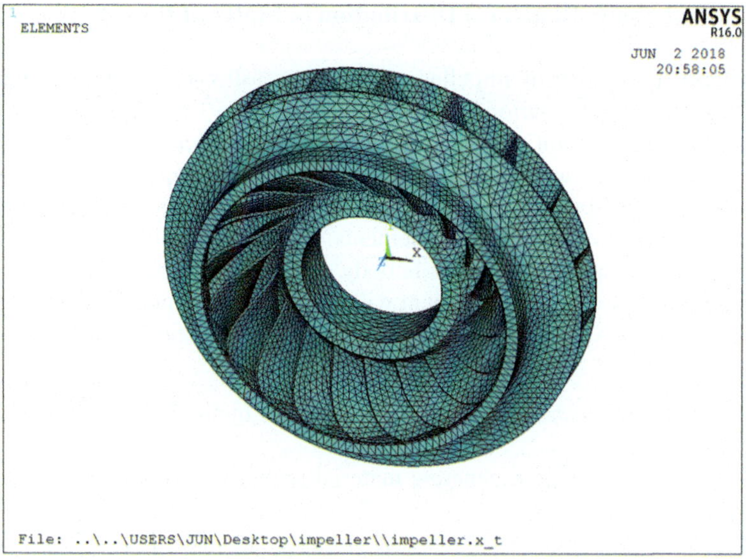

Fig. 7.4 Mesh of the Impeller

1. Only material in each element is uniform, though it is non-uniform in whole impeller. This assumption is available since the mesh used in this study is very small.
2. Material is isotropic, and general Hooke's law is satisfied in each element.

$$\sigma = E\varepsilon \tag{7.23}$$

3. There is no couplings among Young's modulus, Poisson's ratio and density, implying the distribution of Young's modulus has no influence on the distributions of Poisson's ratio and density, and vice versa.

Gaussian distribution, which is feasible in engineering, is introduced to approach the material mistuning in impeller.

The probability density function of Gaussian distribution is expressed as following,

$$f(x) = \frac{1}{\sqrt{2\pi}\sigma} e^{-\frac{(x-\mu)^2}{2\sigma^2}}, -\infty < x < \infty \tag{7.24}$$

where μ and $\sigma > 0$ are two constants.Uniform distribution is one of probability statistical distribution and is used frequently in practice, the probability density function can be expressed as following,

$$f(x) = \frac{1}{b-a}, \quad a \le x \le b \tag{7.25}$$

where a and b are two constants.

7.4.1.3 Probability Statistical Distribution of Material Properties

The material properties of impeller are listed as following, Young's modulus is 2.0×10^{11}Pa, Poisson's ratio is 0.3, and density 7.85×10^3 kg/m^3.

Under the assumption, the material properties in every element of mistuned impeller are uniform, and however the differences of material properties between adjoint elements will defined as the probability statistical distribution given above. For an impeller, the number of elements will considerable as finite element method is used to approach the whole body, and how to implement the probability statistical distribution in every element will become the key point. Following APDL module defined in ANSYS, a method for probability statistical distribution by loop statement is given as following.

1. The impeller is meshed firstly by finite element method, without consideration of material properties.
2. APDL module is used to generate material properties under normal or uniform distributions as defined.
3. Following loop statement, the material properties generated in step 2 are assigned for each element.
4. Using PlotCtrls>Style>Colors>Reverse Video defined in ANSYS, the final model with material mistuning can be obtained, as shown in Fig. 7.5.

As shown in Fig. 7.5, the blocks in colors means different material properties, and the specific property can be obtained as the function of list>properties>all materials in ANSYS is used. Figure 7.6 is the probability statistical distribution of Young's modulus as the method mentioned above is carried out, and it is clear that the distribution is the form of normal or Gaussian distribution, implying the mistuning obtained by this method could be accepted in this study.

7.4.2 Numerical Examples

In this section, the mode localization of impeller, with speed 5000 rpm and under prestress, will be studied with mistuning of material.

7.4.2.1 Mode Analysis of Tuned Impeller

At first, the tuned impeller will be analyzed, in order to make comparison. Figure 7.7 is the frequencies of tuned impeller.

From Fig. 7.7, it can be seen that the distribution of frequencies is different from the normal structures, that is, the frequencies are closed with one cluster. This is intrinsic properties of thin-walled structures, and the structure is sensitive to dynamic instabilities as initial imperfections are introduced. Figure 7.8 shows the typical mode shapes.

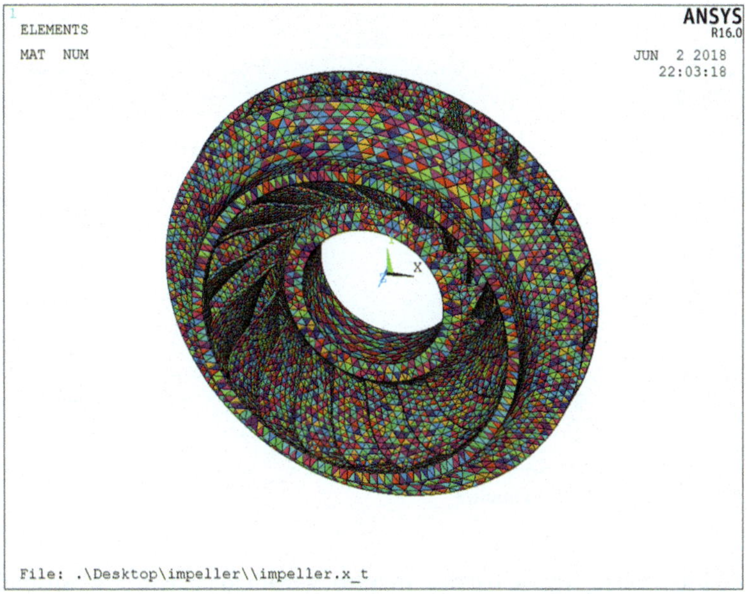

Fig. 7.5 Impeller with material mistuning

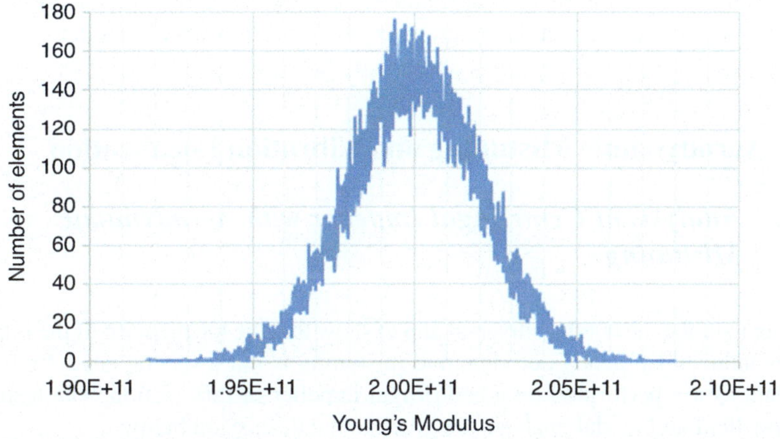

Fig. 7.6 Probability statistical distribution of Young's modulus

7.4.3 Mode Analysis of Mistuned Impeller

In the following, the mistuning of material properties, including Young's modulus, Poisson's ratio and density, will be assigned from 3 to 30% with increment 3% following Eq.(7.24). Then, the frequencies of mistuned impellers will be compared with that of tuned impellers, and the results are shown in Fig. 7.9.

Fig. 7.7 Frequencies of tuned impeller

For clarity, changing of natural frequency with increasing of mistuning are presented as shown in Fig. 7.10.

From the results, it can be concluded generally that density of material has a most significant influence on the natural frequencies of mistuned impeller, and Poisson's ratio has slight influence on the natural frequencies of mistuned impeller.

7.5 Aerodynamic Mistuning and Vibration Localization

7.5.1 Analysis of Centrifugal Impeller with Aerodynamic Mistuning

In order to make the results universal and to show that the localization of the impeller modes induced by steady aerodynamic mistuning is not a special case. Numerical simulations are performed on a centrifugal impeller in this section. The results of the flow field and modal analysis are given in the discussion below.

7.5.1.1 Numerical Simulation of the Flow Field

The centrifugal impeller in the case has 24 blades, its operating speed is 22,360 rpm, two passages are simulated. The mesh type of blades is chosen as Hex20, two blades are divided into 4950 cells and 26,304 nodes. The fluid domain is divided into 298,970 cells and 323,370 nodes. Mesh of the computational domain is shown in Fig. 7.11.

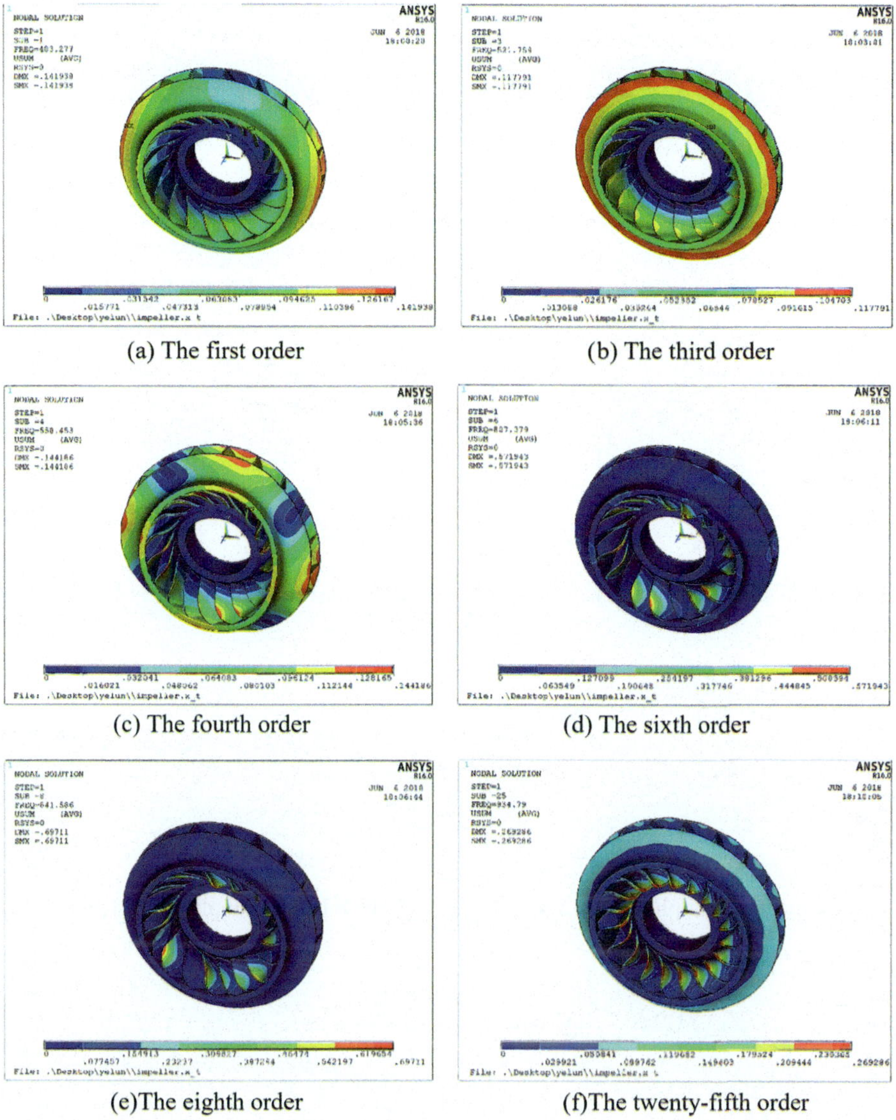

(a) The first order (b) The third order

(c) The fourth order (d) The sixth order

(e)The eighth order (f)The twenty-fifth order

Fig. 7.8 Typical modal shapes

The inlet boundary conditions of the impeller are: total pressure P_{Tin}, total temperature 293.15 K. The outlet boundary conditions are: mass flow rate (two channels) 0.334 kg/s. The expression for P_{Tin} is:

$$P_{Tin} = 101325 + 7500\cos(12\theta) \quad \text{(Pa)} \tag{7.26}$$

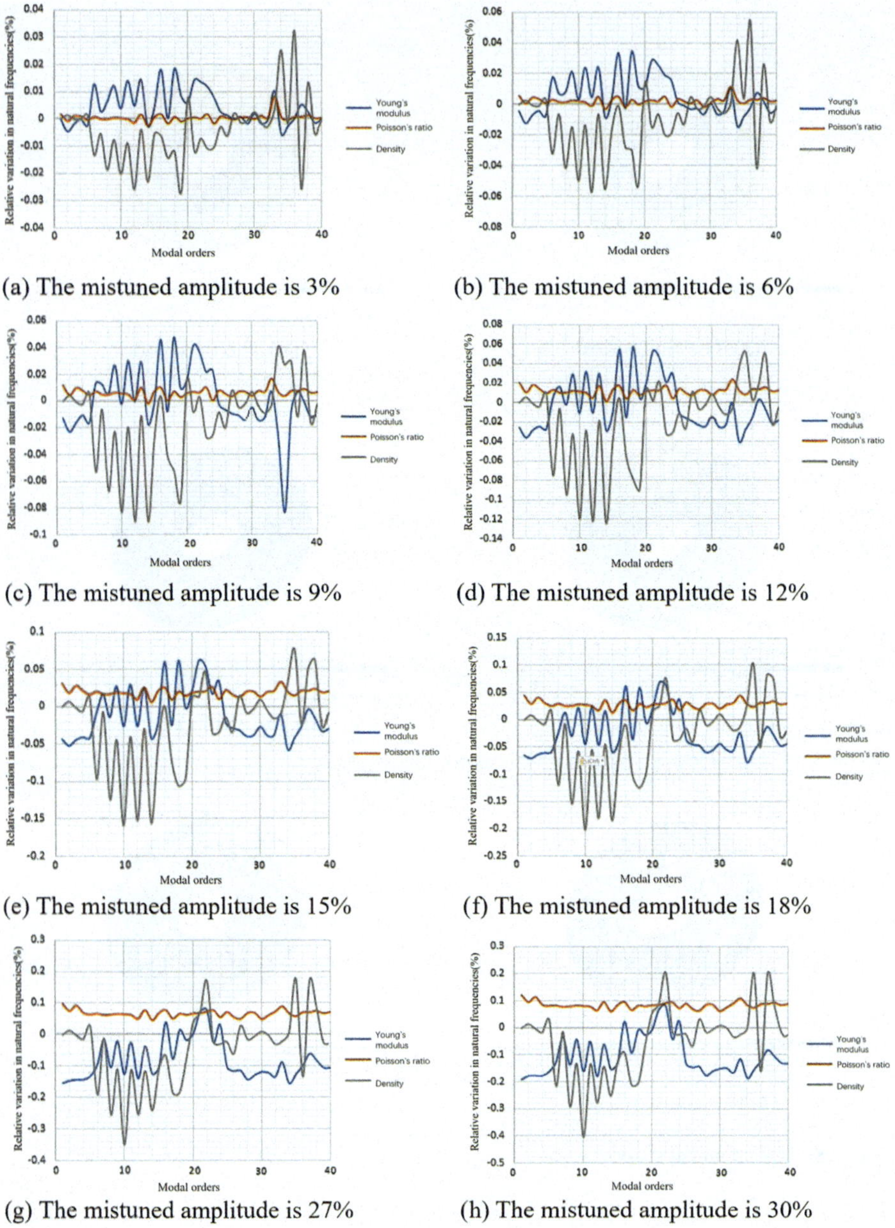

(a) The mistuned amplitude is 3%
(b) The mistuned amplitude is 6%
(c) The mistuned amplitude is 9%
(d) The mistuned amplitude is 12%
(e) The mistuned amplitude is 15%
(f) The mistuned amplitude is 18%
(g) The mistuned amplitude is 27%
(h) The mistuned amplitude is 30%

Fig. 7.9 Comparisons of frequencies of mistuned impellers

The whole field was simulated with a unidirectional fluid-structure coupling method, and the results are shown in Fig. 7.12. In this figure, (a)–(c) are the pressure distribution of the flow field at sections of different heights. Due to inlet pressure fluctuations, the pressure distributions in the two passages are different.

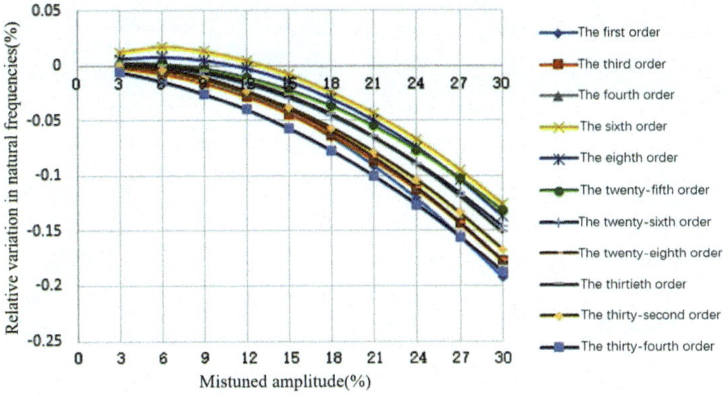

(a) Mistuning of Young's modulus

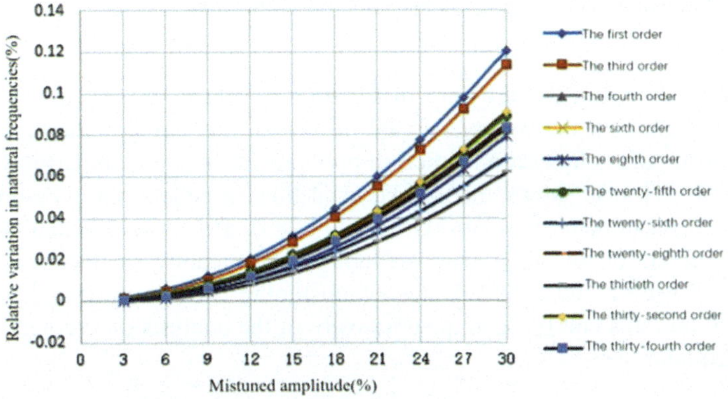

(b) Mistuning of Poisson's ratio

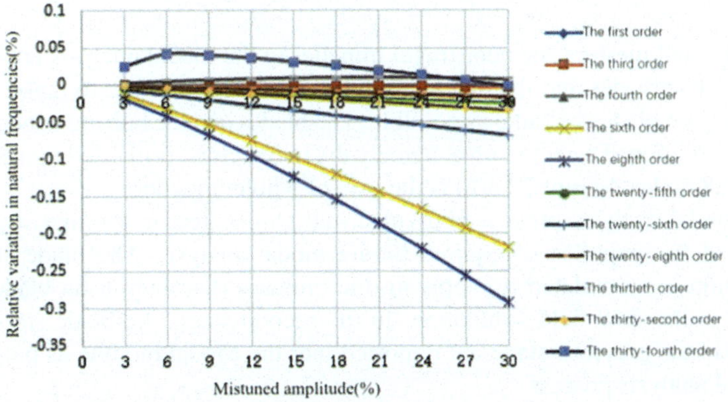

(c) Mistuning of density

Fig. 7.10 Changing of natural frequency with increasing of mistuning

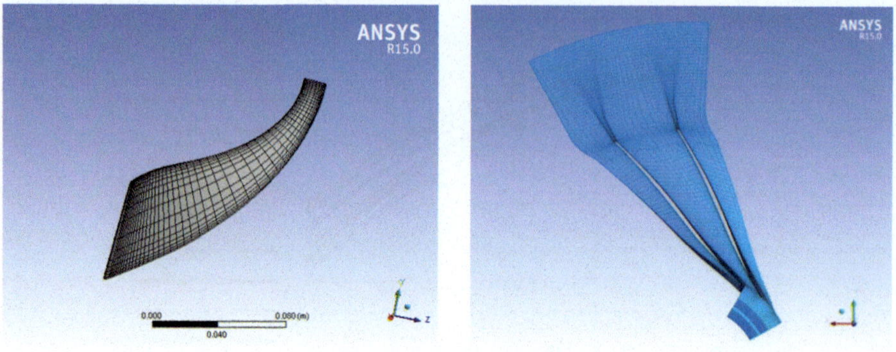

(a) Mesh of the blade (b) Mesh of the fluid domain

Fig. 7.11 Mesh of the computational domain of the centrifugal impeller

Figure 7.12(d) shows the pressure distribution on blade surface, it can be found that distributions on the two blades are different. The aerodynamic mistuning of axial flow impellers has been studied in previous literature. Comparing with these results [38, 39], it is shown that unlike axial flow impellers, the pressure-ratio of this centrifugal impeller is relatively high (4.5089). Therefore, fluctuations of inlet pressure effects on the flow field are not as significantly as those on the axial impeller. However, the actual degree of mistuning of the centrifugal impeller is not small, and this can be seen quantitatively in the analysis of average mistuning performed later.

7.5.1.2 Modal Analysis

The model and mesh of the centrifugal impeller in the modal analysis are shown in Fig. 7.13. The blades and the hub are divided separately in order to make the mesh grid of the two blades be same as possible. The blades and the hub are contacted with Bonded condition. The entire area, namely the two sectors, is divided into 3576 cells and 18,238 nodes. Figure 7.13(b) is the grid of whole impeller.

The modal analysis process is divided into two stages. In the first stage, static analysis of the impeller is required before modal analysis. The blade stress and deformation are simulated by applying the pressure distribution on blade surface obtained in the flow field simulation. In the second stage, the blade stress results obtained in the previous stage are imported into the Modal module as prestress for the modal analysis process.

The modal analysis is mainly divided into three parts. The first part is calculating the modes of the original impeller. The result is mainly used as the reference for subsequent analysis. Simulate the two-passage model mentioned above without

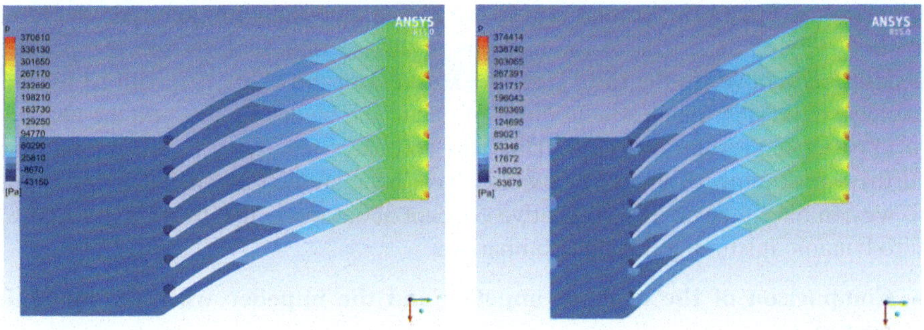

(a) Pressure contour at 20% blade height (b) Pressure contour at 50% blade height

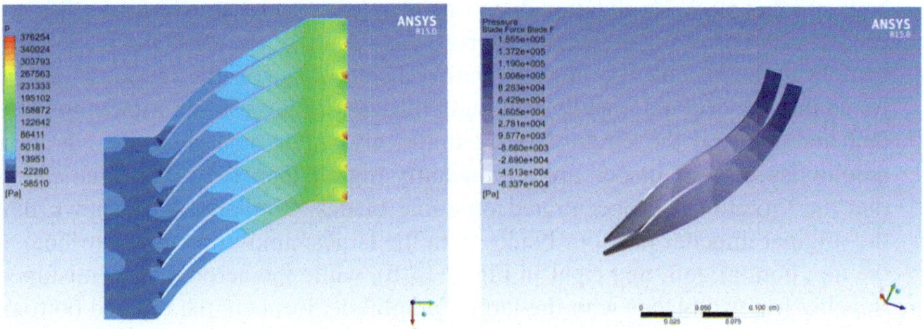

(c) Pressure contour at 80% blade height (d) Pressure contour at the blade surface

Fig. 7.12 Pressure contours of the centrifugal impeller

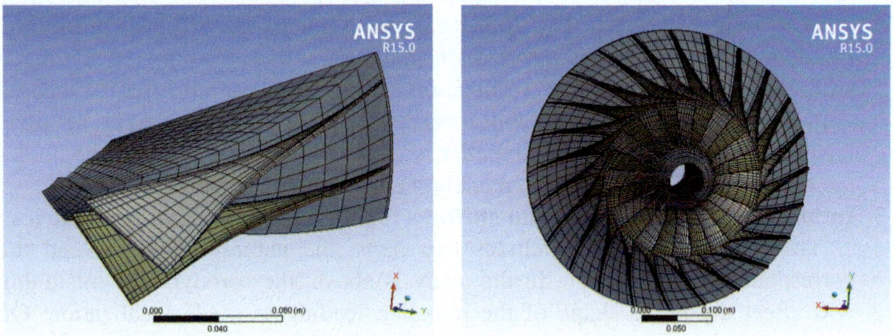

(a) Mesh of two-blade sector (b) Mesh of whole impeller

Fig. 7.13 Mesh for centrifugal impeller modal analysis

any external force. The second part is the modal analysis with aerodynamic force applied. This is the focus of the analysis in this section, and the specific process is described above. In the third part, the stiffness of the blade is mistuned by 1%, namely decreasing the elastic modulus of one of the blades by 1%, and the changes of impeller modes are analyzed. This part compares aerodynamic mistuning with stiffness mistuning. A number of works have been studied on stiffness mistuning, so we can have a basic and qualitative concept about mode localization induced by aerodynamic mistuning through comparison.

1. **Comparison of the original impeller and the impeller with aerodynamic mistuning**
   Figure 7.14 is a comparison of the first-order 4 and 5 nodal diameter-bending modes of the centrifugal impeller before and after aerodynamic mistuned. It can be found that mode localization due to steady-state aerodynamic mistuning is not obvious. By observing the warm color area on the top of the blade in Fig. 7.14(a), namely the area with the largest relative displacement in the mode, it can be found that the original impeller has eight blades with largest vibration amplitude that distribute on the circumference uniformly. After mistuned, the vibration amplitudes of four blades are significantly higher than the others, indicating that the vibration is concentrated on some blades. Similarly, it is shown that the original impeller has four blades with the largest amplitude that distribute at the top, bottom, left, and right in Fig. 7.14(b), while the aerodynamic mistuned impeller has two blades with the largest amplitude, locate at the top and bottom, meaning the mode shape changed indeed after mistuned.
2. **Comparison of aerodynamic mistuning and stiffness mistuning**
   Figure 7.15 show the comparison of the mode localization in centrifugal impeller with aerodynamic mistuning and 1% stiffness mistuning, and it can be seen that the degree of mode localization induced by aerodynamic mistuning is much smaller than the degree of localization induced by stiffness mistuning. In addition, the degree of localization induced by the mistuning of the blade stiffness is also smaller than localization of the axial flow impeller. Compared with the axial flow impeller (with a relatively small hub), the degree of mode localization in the centrifugal impeller is greatly reduced, since only the blades are mistuned and the main part, namely the hub, remains unchanged.
3. **Influences of aerodynamic and stiffness mistunings on natural frequencies**
   The concept of mode includes two parts, the natural frequency and the corresponding mode shape. In the analysis above, the aerodynamic mistuning will affect the mode shape of the impeller, leading to mode localization. On the other hand, the effects of mistuning on the first-order mode frequency of the impeller are given in Fig. 7.16. Aero is the mode frequencies curve of the aerodynamic-mistuned impeller, 1%E for the stiffness-mistuned, tuned for the original.

As aerodynamic force is the pressure on the blades, opposed to stiffening effect of the centrifugal force on the impeller, aerodynamic force reduces the stiffness of the impeller. In Fig. 7.16, a general decrease of the natural frequency of the impeller

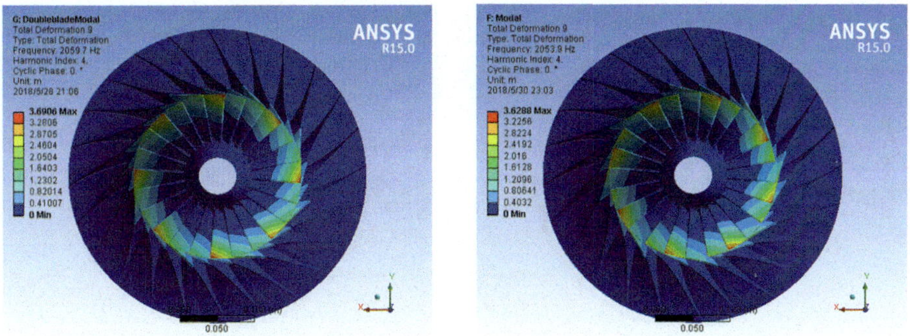

(a) Comparison of 4 nodal diameter mode of the original impeller (left) and the impeller with aerodynamic mistuning (right)

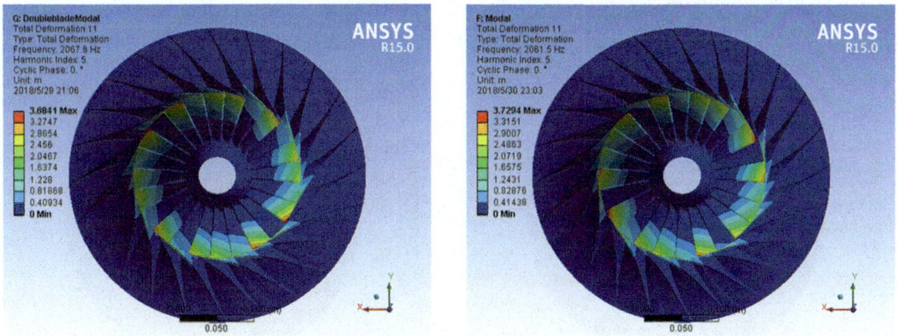

(b) Comparison of 5 nodal diameter mode of the original impeller (left) and the impeller with aerodynamic mistuning (right)

Fig. 7.14 Comparison of the centrifugal impeller modes before and after aerodynamic mistuning

is shown. In addition, as the stiffness mistuning reduces the elastic modulus by 1%, the natural frequency also decreases slightly. But in general, taking the mode natural frequency with a high nodal diameter number as an example, the natural frequency hardly changes. In addition, considering that the impeller is also stiffened by centrifugal force, it can be considered that the natural frequency of the impeller does not change much before and after mistuned.

The first-order natural frequency of the centrifugal impeller is significantly higher than the natural frequency of the axial flow impeller, indicating that the blade stiffness of the centrifugal impeller is indeed higher than that of the axial flow impeller. According to the theory of aerodynamic mistuning mentioned above, as the strength of mistuning keeps constant, the additional stiffness matrix induced by mistuning effects on the original stiffness matrix relatively. As the elements of the original stiffness matrix are mostly large, the strength of mistuning is relatively small and the effect on modes is also relatively small. As the elements in the original stiffness matrix are mostly small, the strength of the same mistuning becomes

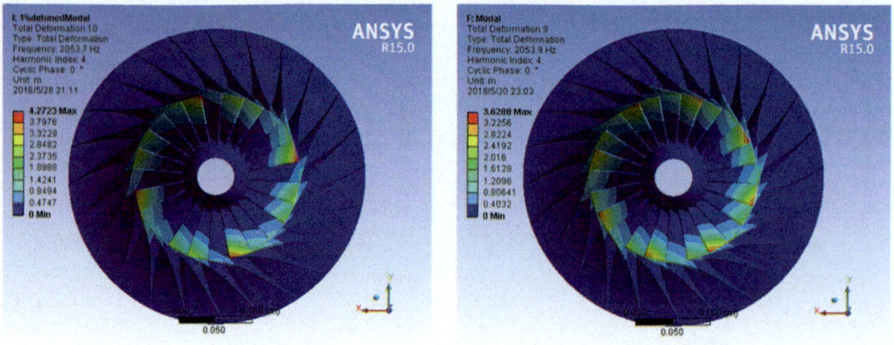

(a) Comparison of 4 nodal diameter mode of impeller with stiffness mistuning (left) and aerodynamic mistuning (right)

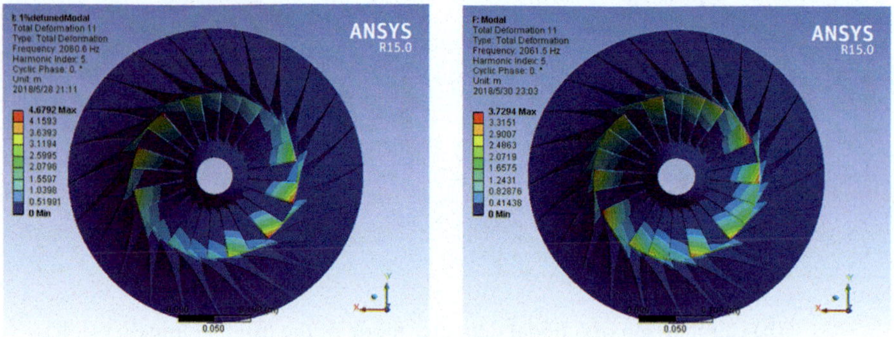

(b) Comparison of 5 nodal diameter mode of impeller with stiffness mistuning (left) and aerodynamic mistuning (right)

Fig. 7.15 Comparison of the modes of the impellers with 1% stiffness mistuning and aerodynamic mistuning

relatively large, so its effect on modes is also relatively large. To sum up, the effects of aerodynamic mistuning on mode of a centrifugal impeller are less than the effect on the axial flow impeller, this can also be seen from the comparison of natural frequencies. Whether it is aerodynamic or elastic modulus, the effect on the natural frequency of the centrifugal impeller is small.

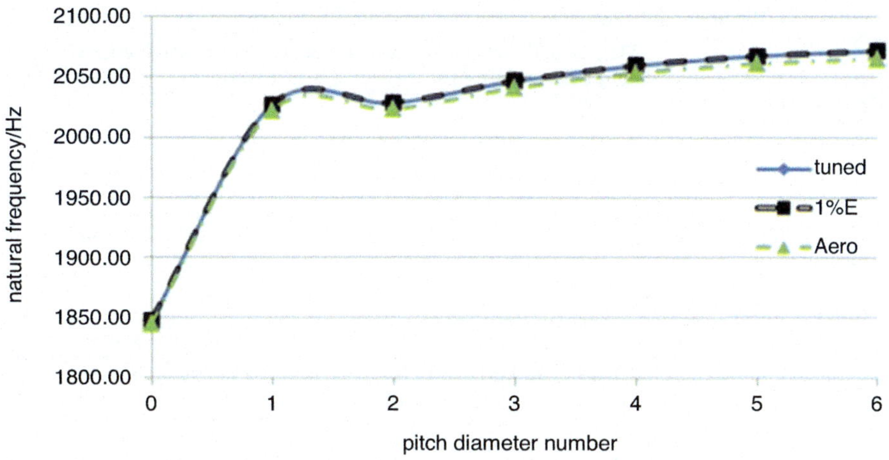

Fig. 7.16 Comparison of first-order natural frequency before and after mistuned

Table 7.1 Stress mistuning in the centrifugal impeller

Magnification of aerodynamic mistuning	Average stress (1×10^5 Pa)		Average stress mistuning (1×10^5 Pa)
	Blade 1	Blade 2	
1	62.448	81.182	18.734
21	1315.1	1706.3	391.20
27	1693.2	2196.5	503.36

7.5.2 *Relationship Between Degree of Localization and Average Aerodynamic Mistuning*

Mistuning analysis of axial flow impellers has been studied earlier than centrifugal impellers, so the contents below are based on the comparison to the results of the axial flow impeller. For centrifugal impellers, calculate the average stress mistuning in the case of original aerodynamic mistuning at first, then select an appropriate mistuning magnification to make the stress mistuning become about 30 times of the mistuning on axial flow impeller with similar aerodynamic performance. The result of the average stress mistuning of the centrifugal impeller is shown in Table 7.1.

It can be seen from Table 7.1 that, as the aerodynamic force on the centrifugal impeller is magnified to 21 times of normal one, its average stress mistuning is equivalent to 30 times of the average stress mistuning in the axial flow impeller. In order to quantitatively evaluate the degree of localization, Wang Jianjun [41] proposed the localization index of mode shape, and evaluated the localization phenomenon induced by aerodynamic mistuning. The definition of localization index is briefly introduced below.

Table 7.2 Maximum dimensionless displacements of impeller modes in cases

Item		Maximum dimensionless displacement versus nodal diameter numbers (1×10^{-5})					
		0	1	2	3	4	5
Tuned impeller		3.3921	10.0903	10.2608	9.8733	10.8873	11.5869
1% stiffness mistuning		3.4818	10.6308	10.5414	12.6395	13.2664	16.0933
aerodynamic mistuning of different magnifications	1	3.3954	9.9825	10.1750	9.9232	10.8330	11.6536
	21	3.7549	10.8986	10.0738	9.2502	12.6486	15.5841
	27	3.7449	10.9851	10.2115	9.6265	13.0392	16.3846

Firstly, the dimensionless displacements of the impeller modes before and after mistuning are required. The dimensionless displacement of mode is defined as [41],

$$u_i = \frac{\overline{u}}{\sum_{i=1}^{n} \overline{u_i}} \tag{7.27}$$

where $\overline{u}$ is the node displacement vector of the mode shape, $\overline{u_i}$ the component of $\overline{u}$ (the displacement vector value of the *i*th node), *n* the total number of nodes.For the cases in Table 7.1, Eq. (7.27) is used to calculate the maximum dimensionless displacement of the first-order or second-order modes of different nodal diameter numbers. The results are shown in Table 7.2.

Using the maximum mode displacement before and after mistuning calculated with the formula $\overline{u_t}_{\max}$, $\overline{u_m}_{\max}$, the mode displacement localization index is defined as [41],

$$R_u = \frac{\overline{u_m}_{\max} - \overline{u_t}_{\max}}{\overline{u_t}_{\max}} \tag{7.28}$$

Obviously, the localization of mode energy will increase if R_u is positive, and decrease if negative.

Take the maximum dimensionless displacement of the tuned impeller as the unit, displacement localization indices for each mode are calculated and shown in Fig. 7.17. Curse with *1%E* is the reference, which shows the mode localization indices of the centrifugal impeller with 1% stiffness mistuning, the *NAero* curve shows the mode localization indices of the impeller as aerodynamic force is magnified by N times.

First, the degree of mode localization of 0–2 nodal diameters are all below 10%, so the mode changes due to aerodynamic mistuning cannot be shown clearly. Therefore, the 3–5 nodal diameter modes are mainly analyzed. For high nodal

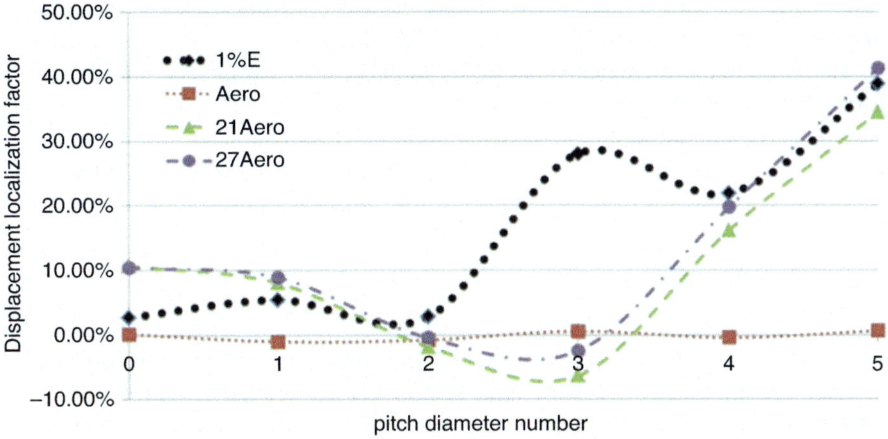

Fig. 7.17 Localization indices in cases with different dynamic force magnifications

diameter modes, the mode localization of the centrifugal impeller is significantly smaller than that of the axial flow impeller. This phenomenon has been explained in previous sections, the reason is that stiffness of the centrifugal impeller is greater than that of the axial flow impeller. Therefore, the mistuning of same strength effects more slightly on the centrifugal impeller. Observing the 4 and 5 nodal diameter modes, it can be considered that the degree of mode localization in the centrifugal impeller is equivalent to the degree of localization as the stiffness is mistuned by 1%, under 21–27 times aerodynamic force. The average stress mistuning among the blades is between (391.2~503.36) $\times$ 10^5 Pa in this condition. That means, the corresponding mean aerodynamic stress mistuning is 393.70 $\times$ 10^5 Pa as the degree of centrifugal impeller mode localization being equivalent to the case of 1% stiffness mistuned, that is basically the same as axial flow impeller.

In addition, it is necessary to pay special attention to the 3 nodal diameter mode, in which strong localization phenomenon is not produced, to the contrary, the maximum vibration value decreases. This phenomenon is also mentioned in Meng Jigang's article: mistuning with particular form may improve the original modes [28].

In summary, with the modal analysis of a typical centrifugal impeller, the following conclusions can be drawn as following,

Steady-state aerodynamic mistuning of the blade will have influence on the mode shapes of the impeller, leading to mode localization, namely making the vibration concentrate on some blades.

Mistuning just slightly reduces the natural frequency. It can be considered that the impeller will respond to same excitation, and the corresponding modes will be excited, before and after aerodynamic mistuning.

The degree of mode localization of the impeller increases with the average pressure mistuning. The effect of average pressure mistuning on the stiffness matrix is the same as Young's modulus mistuning. Severe mode localization will occur as the average pressure mistuning exceeds a certain value.

The three nodal diameter mode of the impeller selected is very sensitive to the form of aerodynamic mistuning induced by pressure fluctuations. The maximum amplitude of the three nodal diameter mode is suppressed as the aerodynamic mistuning is small, then it increases rapidly with the aerodynamic mistuning.

7.6 Conclusion

In comparison with non-cyclic structure, the impellers of turbo-machinery have distinct dynamics, such as pass- and stop-bands of frequency, and sensitivity to mistuning. In practice, the structural property of impeller, cyclic symmetry, will be destroyed slightly, because of the mistuning from manufacture, running wears, adherence of working medium, corrosion and symmetry-breaking of aerodynamics etc. These slight changes induced by mistuning in structure will lead to great changes in dynamics of the whole structure. Specifically, as an example, the pass-bands in natural frequencies of ideal impeller could become as stop-bands. Under the excitation, the vibration localization on mistuned impeller will occur with energy localization. As a result, such vibration localization will lead to fatigue failure on impeller and instability of the whole rotor system. Hence, study on the nature of mistuned impeller's modes and vibration localization of impeller and its controlling play important role in the development of turbo-machinery with higher performance, gaining fundamental understanding of vibration control, optimum design and improving safety and stability of rotor system.

At first, the influences of mistuning from structure on the dynamics of impeller are analyzed in detail, especially the changing of pass- and stop-bands of frequency. The results show that slight mistuning could induce vibration localization of impeller as the modes of impeller become closed together; Then, the finite element available for analyzing material mistuning and mathematical model for aerodynamic mistuning are given in detail, and the mistuning from material properties, such as Young's modulus, Possion's ratio, density, are considered, and probability distribution function is introduced to model the stochastic sequence of mistuning. Some important results are obtained; Further, the influences of mistuning from aerodynamics on vibration localization are studied in detail. In the study, the stress stiffening effect, which is induced by steady aerodynamic forces, is regarded as a main factor to mode localization. The results show that, more attentions should be paid to the aerodynamic mistuning of impellers working under high pressure conditions, since it may lead to rather violent mode or vibration localization easily.

In summary, the nature of vibration localization of impeller with typical mistuning is studied and hence some controlling strategies are proposed, and the results play important role in the development of turbo-machinery with higher performance.

Acknowledgments This work is supported by National Natural Science Foundation of China (No. 51775437), the Key Research and Development Program of Shaanxi Province (No. 2017ZDCXL-GY-02-02), the Key Laboratory of Compressor of China (No. SKL-YSJ201802) and the World-Class Universities (Disciplines) and the Characteristic Development Guidance Funds for the Central Universities (No. PY3A056).

References

1. Zhang, J., Y. Liu, and P. Wang. 2007. Study on the model localization in the centrifugal impeller with periodic structure. In 9th Conference on Theory and Application of Vibration, 17th–19th, October, 2007, Hang Zhou, China.
2. Meng, J., Y. Wang, and S. Yang. 2010. Research on dynamic characteristics of mistuned shrouded impeller with salt crystallization in centrifugal compressor. *Chinese Journal of Turbomachinery* 3: 7–12.
3. Ewins, D.J. 1976. Vibration modes of mistuned bladed disks[J]. *ASME Journal of Engineering for Power* 98 (7): 349–355.
4. Fabunmi, J.A. 1980. Forced vibration of a single stage axial compressor rotor. *ASME Journal of Engineering for Power* 102 (2): 322–329.
5. Li, F., Y. Wang, W. Huang, et al. 2005. Advances of vibration localization in disordered periodic structures[J]. *Advances in Mechanics* 35 (4): 498–512.
6. Hodges, C.H. 1981. Confinement of vibration by structural irregularity. *Journal of Sound & Vibration* 69 (3): 411–424.
7. Hodges, C.H., and J. Woodhouse. 1983. Vibration isolation from irregularity in a nearly periodic structure: theory and measurements. *Journal of the Acoustical Society of America* 74 (3): 894–905.
8. Wei, S.-T. and C. Pierre. 1988. Localization phenomena in mistuned assemblies with cyclic symmetry, Part I: Free vibrations. *ASME Journal of Vibration, Acoustics, Stress and Reliability in Design* 110: 429–438.
9. Wei, S.-T. and C. Pierre. 1988. Localization phenomena in mistuned assemblies with cyclic symmetry, Part II: Forced vibrations. *ASME Journal of Vibration, Acoustics, Stress and Reliability in Design* 110 (4): 439–449.
10. Wei, S.-T. and C. Pierre. 1989. Effects of dry friction damping on the occurrence of localized forced vibrations in nearly cyclic structures. *Journal of Sound & Vibration* 129 (3): 397–416.
11. Pierre, C., and D.V. Murthy. 2012. Aeroelastic modal characteristics of mistuned blade assemblies: mode localization and loss of eigenstructure. *AIAA Journal* 30 (10): 2483–2496.
12. Watson, B.C., and M.P. Kamat. 1995. Analysis of mistuned cyclic systems using mistune modes. *Applied Mathematics and Computation* 67: 61–79.
13. Happawana, G.S., O.D.I. Nwokah, A.K. Bajaj, et al. 1998. Free and forced response of mistuned linear cyclic systems: a singular perturbation approach. *Journal of Sound & Vibration* 211 (5): 761–789.
14. Sextro, W. 2001. Localization in nonlinear mistuned systems with cyclic symmetry. *Nonlinear Dynamics* 25 (1–3): 207–220.
15. Judge, J., C. Pierre, and O. Mehmed. 2001. Experimental investigation of mode localization and forced response amplitude magnification for a mistuned bladed disk. *Journal of Engineering for Gas Turbines and Power* 123: 940–950.
16. Judge, J.A., S.L. Ceccio, and C. Pierre. 2003. Traveling-wave excitation and optical measurement techniques for non-contacting investigation of bladed disk dynamics. *Shock and Vibration Digest* 35 (3): 183–190.
17. Xie, W.-C. and S.T. Ariaratnam. 1996. Vibration mode localization in disordered cyclic structures, I: Single substructure mode. *Journal of Sound and Vibration* 189 (5): 625–645.

18. Xie, W.-C. and S.T. Ariaratnam. 1996. Vibration mode localization in disordered cyclic structures, II: Multiple substructure modes. *Journal of Sound and Vibration* 189 (5): 647–660.
19. Wang, H., and W. Wang. 1990. Natural vibration analysis of turbomachinery disk with small mistuned blades[J]. *Journal of Aerospace Power* 15 (4): 357–360.
20. Yan, Y., and J. Gu. 1993. Vibration characteristics of mistuned disk-bladed. *Journal of Aerospace Power* 3: 234–240.
21. Wang, J., J. Xu, and Q. Li. 2004. Analytical models of mistuned bladed disk assembles—a review. *Turbine Technology* 46 (4): 256–259.
22. Wang, H., E. He, and S. Yu. 2005. An approximate method applicable to calculating forced response of mistuned bladed disks in medium coupling range. *Journal of Northwestern Polytechnical University* 23 (6): 716–719.
23. He, E., S. Yu, and H. Wang. 2007. Statistical analysis of the forced response of random mistuning bladed disk. *Journal of Mechanical Strength* 29 (1): 20–24.
24. Wang, H., E. He, and S. Yu. 2006. Forced response characteristics of bladed disks with disordered dry friction. *Journal of Aerospace Power* 21 (4): 711–715.
25. Dai, J., F. Li, and W. Shi. 2005. Influence of disorder on vibration characteristics of bladed disk assemblies. *China Mechanical Engineering* 13: 1158–1161.
26. Zhao, W., J. Zhang, and C. Zhou. 2012. Study on the vibration localization in the centrifugal Impeller with periodic structures. *Chinese Journal of Applied Mechanics* 29 (6): 699–704.
27. Anderson, P.W. 1958. Absence of diffusion in certain random lattices[J]. *Physical Review* 109 (5): 1492–1505.
28. Meng, J. 2010. *Dynamic characteristics of mistuned shrouded impellers of centrifugal compressors[D]*. Dalian University of Technology.
29. Wang, Y. 2006. Study on the elastic wave dispersion relation of periodic structure and vibration localization[D]. *Harbin Institute of Technology*.
30. Zhou, Y. 2007. Model localization of bladed-disk system[D]. *Harbin Institute of Technology*.
31. Yao, Z., and G. Yu. 2005. Research on vibration model localization in mistuned cyclic periodic structures. *Science Technology and Engineering* 5 (21): 1616–1622. https://doi.org/10.3969/j.issn.1671-1815.2005.21.010.
32. Wang J, Li Q. Substructure model of localized modes for mistuned bladed disks. 2003.
33. Hu, C., F. Li, J. Zou, et al. 2003. A study on localization of vibration mode in mistuned bladed disk assemblies[J]. *Proceedings of the CSEE* 23 (11): 189–194.
34. Kuang, J.H., and B.W. Huang. 1999. The effect of blade crack on mode localization in rotating bladed disks. *Journal of Sound & Vibration* 227 (1): 85–103.
35. Kuang, J.H., Huang, et al. 1999. Mode localization of a cracked blade disk. *ASME Journal of Engineering for Gas Turbines and Power* 121 (2): 335–341.
36. Huang, B.W., and J.H. Kuang. 2001. Mode localization in a rotating mistuned turbo disk with Coriolis effect. *International Journal of Mechanical Sciences* 43 (7): 1643–1660.
37. Wang, Xucheng, and Min Shao. 1996. *Fundamental theory of finite element method and numerical method*. Tsinghua University Press.
38. Ekici, K., R.E. Kielb, and K.C. Hall. 2013. The effect of aerodynamic asymmetries on turbomachinery flutter. *Journal of Fluids and Structures* 36: 1–17.
39. ———. 2010. Aerodynamic asymmetry analysis of unsteady flows in turbomachinery. *Journal of Turbomachinery* 132 (1): 011006.
40. Pierre, C., T.E. Smith, and D.V. Murthy. 1994. Localization of aeroelastic modes in mistuned high-energy turbines. *Journal of Propulsion and Power* 10 (3): 318–328.
41. Wang, J., C. Yu, J. Yao, et al. 2009. Vibratory mode localization factors of mistuned bladed disk assemblies. *Journal of Propulsion Technology* 30 (4): 457–461, 473.

Chapter 8
Fluid-Structure Interactions of a Perimeter-Reinforced Membrane Wing in Laminar Shear Flow

Xu Sun, C. Steve Suh, and Bo Yu

8.1 Introduction

With rapid developments of Micro Air Vehicles (MAVs), novel designs of wings with good aerodynamic performance at low Reynolds numbers ($Re < 10^5$) have attracted a lot of attention from the aerodynamics community. After years of exploration [1–5], two types of bio-inspired wings have been developed for MAVs to overcome the unfavorable aerodynamic conditions at low Reynolds numbers. One is the flapping wing and the other is the fixed membrane wing. Unlike the conventional fixed rigid wing, the flow-induced deformation (FID) and flow-induced vibration (FIV) of which are considered harmful, the fixed membrane wing is flexible and deformable under aerodynamic load, and FID and FIV are utilized actively to improve the aerodynamic performance of the wing [1, 2, 4].

Many investigations have been conducted on the aerodynamic performances and the dynamic behaviors of the membrane wings. Stanford et al. [4] and Albertani et al. [6] proposed two designs of the fixed membrane wing with oval configuration, namely, the batten-reinforced (BR) membrane wing and the perimeter-reinforced (PR) membrane wing, and studied their elastic deformations and aerodynamic coefficients using both wind tunnel experiment and numerical simulation. It was

X. Sun (✉)
National Engineering Laboratory for Pipeline Safety/MOE Key Laboratory of Petroleum Engineering, China University of Petroleum-Beijing, Beijing, P. R. China
e-mail: xsun@cup.edu.cn

C. S. Suh
Nonlinear Engineering and Control Lab, Mechanical Engineering Department, Texas A&M University, College Station, TX, USA

B. Yu
School of Mechanical Engineering, Beijing Institute of Petrochemical Technology, Beijing, P. R. China

J. Zhang (ed.), *Dynamics and Fault Diagnosis of Nonlinear Rotors and Impellers*, Nonlinear Systems and Complexity 34, https://doi.org/10.1007/978-3-030-94301-1_8

found that the structural deformation of a BR membrane wing can improve the stall, while the flow-induced camber of a PR membrane wing can increase both lift and longitudinal static stability. Johnston et al. [7] and Attar et al. [8, 9] proposed a rectangular BR membrane wing model and studied its dynamic responses in a variety of flow and structural parameter combinations. At small angles of attack, the onset of flutter and post-flutter limit cycle oscillation (LCO) above an inflow velocity were observed, and the flutter velocity was found to be affected significantly by the membrane pre-strain and the number of structure battens. At high angles of attack, LCOs of the membrane led to large oscillations of the wing. Rojratsirikul et al. [10–12] proposed a benchmark PR membrane wing model and conducted experiments on the unsteady fluid-membrane interactions. They found that the flow-induced camber helped the wing to attain a larger lift, and the FIV excited the roll-up and reattachment of the separated shear layer, which postponed the stall angle. Gordnier [13], Visbal et al. [14] and Gordnier and Attar [15] established the mechanical and mathematical models of the PR membrane wing proposed by Rojratsirikul et al. [10–12], and carried out high-fidelity fluid-structure interaction (FSI) simulations. Their numerical results confirmed the findings reported by Rojratsirikul et al. [10–12]. Arbós-Torrent et al. [16] and Sun and Zhang [17] studied the leading- and trailing-edge effects on the aeromechanics of the PR membrane wing. The work of Arbós-Torrent et al. [16] showed that the geometry and size of the leading- and trailing-edge supports have significant influences on the mean camber, membrane vibration and lift of the wing. Sun and Zhang [17] indicated that slight reinforcement at the leading edge does not affect much the lift of the wing. Bleischwitz et al. [18–20] investigated aerodynamic performance of a PR membrane wing in ground-effect by wind tunnel experiments. It was found that the ground-effect can help the membrane wing attain higher lift at low to moderate angles of attack. Sun et al. [21, 22] and Tiomkin and Raveh [23] studied numerically the stability of the PR membrane wing in laminar flows. The angle of attack, the membrane mass-ratio and the Reynolds number were found to influence greatly the dynamic response of the membrane wing.

In previous studies, the aerodynamic performance and dynamic response of the fixed membrane wing in uniform flow have been investigated extensively. However, few works focus on the effect of the shear inflow, which usually encounters in the out-door flying of MAVs. In the present paper, aerodynamics of a PR membrane wing in laminar shear flow is studied using FSI simulation. The membrane wing has the same configuration with that proposed by Rojratsirikul et al. [10–12], and the mechanical model and mathematical formulation reported by Gordnier [13] are employed to describe the unsteady fluid-membrane interaction in laminar flow. The well-validated FSI solver based on the finite element method (FEM) [24] is introduced to numerical simulation. To study the effect of the shear flow, the Reynolds number, membrane mass-ratio, non-dimensional rigidity and pre-strain are fixed at $Re = 2500$, $\rho_S = 589$, $Eh' = 50$ and $\delta_0 = 0$, while the angle of attack and velocity gradient of the free stream vary in $0^o \sim 20^o$ and $k = 0 \sim 0.06$, respectively. The numerical results obtained may help people understand in depth the performance of the PR membrane wing in the out-door environment.

8.2 Mechanical Model and Mathematical Formulation

Figure 8.1 shows the mechanical model of a fixed PR membrane wing in shear flow. As shown in the figure, the benchmark membrane wing model proposed by Rojratsirkul et al. [10, 11] is used here, which is formed by gluing a rectangular membrane sheet to two rigid mounts at the leading and trailing edges. For this wing model, the membrane sheet has large aspect ratio and is not confined at the root and tip. As a result, both of flow field and membrane deformation exhibit 2D characteristics at low Reynolds numbers. As seen in Fig. 8.1, the chord length, membrane length and membrane thickness are indicated by c, L and h, respectively. To investigate the FSI of this membrane wing model in shear flow, the free stream is assumed to be non-uniform and have a constant velocity gradient, which can be expressed as $u_\infty = u_0 + ky$, where u_∞ is the inflow velocity, u_0 the inflow velocity at $y = 0$ and k the velocity gradient. At $k = 0$, the mechanical model presented in Fig. 8.1 degenerates to the model of a fixed PR membrane wing in a uniform flow, which has been investigated extensively in previous studies [13, 21–23].

Considering the Reynolds number with respect to u_0 and c is fixed at $Re = 2500$ and the angle of attack is not too high, the flow is assumed to be two-dimensional (2D), incompressible, viscous and laminar. Moreover, the effect of the air gravity is ignored. Based on these assumptions, the incompressible Navier-Stokes (NS) Equations are taken as the governing equations of the flow, the non-dimensional form of governing equations can be written as

$$\frac{\partial u_i}{\partial x_i} = 0, \tag{8.1}$$

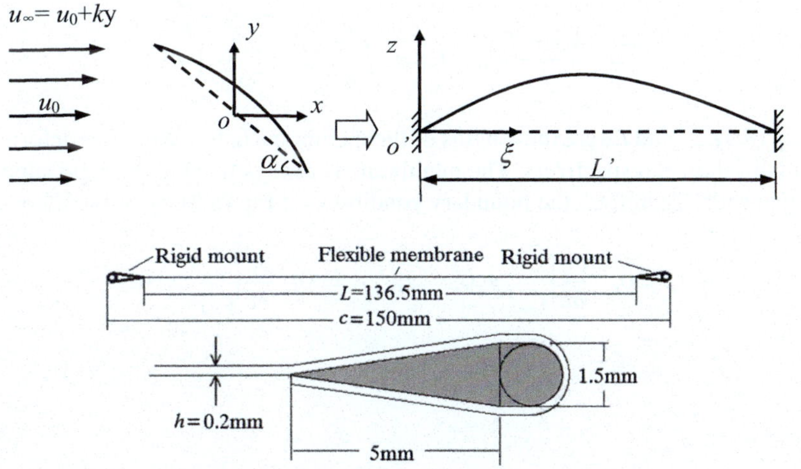

Fig. 8.1 Schematic of the shear flow past a fixed PR membrane wing

$$\frac{\partial u_i}{\partial t} + u_j \frac{\partial u_i}{\partial x_j} = -\frac{\partial p}{\partial x_i} + \frac{1}{\mathrm{Re}} \frac{\partial^2 u_i}{\partial x_j \partial x_j}, \tag{8.2}$$

where u_i are velocity components normalized by u_0, x_i are coordinate components normalized by c, t the time normalized by c/u_0, p the pressure normalized by $\rho_F u_0^2$ and $Re = u_0 c/\upsilon$, with ρ_F and υ being the density and kinematic viscosity of the fluid, respectively.As seen in Fig. 8.1, a local coordinate $\xi o'z$ is introduced to describe the membrane vibration. The coordinate origin o' locates at the leading point of the membrane, the ξ axis is along the wing chord and the z axis is perpendicular to the wing chord. The displacement of the membrane is restricted in z direction and governed by

$$\rho_S h' \frac{\partial^2 z}{\partial t^2} + \rho_S C_d \frac{\partial z}{\partial t} - T \frac{\partial^2 z}{\partial \xi^2} \left[1 + \left(\frac{\partial z}{\partial \xi} \right)^2 \right]^{-\frac{3}{2}} = \Delta p, \tag{8.3}$$

where ρ_S is the membrane density normalized by ρ_F, h' the membrane thickness normalized by c, z the membrane displacement normalized by c, C_d the structural damping normalized by u_0, T the tension on the membrane normalized by $\rho_F u_0^2/c$, ξ the local coordinate normalized by c, and Δp the pressure difference between the lower and upper surfaces normalized by $\rho_F u_0^2$. The tension T is uniform along the membrane and can be obtained by

$$T = Eh' \left(\delta_0 + \overline{\delta} \right), \tag{8.4}$$

where E is the elastic modulus normalized by $\rho_F u_0^2$, δ_0 is the pre-strain and the strain $\overline{\delta}$ can be written as

$$\overline{\delta} = \frac{L_S - L'}{L'}, \quad L_S = \int_0^{L'} \sqrt{1 + \left(\frac{\partial z}{\partial \xi} \right)^2} d\xi. \tag{8.5}$$

In Eq. (8.5), L' and L_S are the lengths of the membrane before and after deformation normalized by c, respectively. The membrane is always fixed to the rigid support at the two ends. Therefore, the boundary conditions of Eq. (8.3) can be written as

$$z|_{\xi=0} = 0, \quad \left. \frac{\partial z}{\partial t} \right|_{\xi=0} = 0; \quad z|_{\xi=L'} = 0, \quad \left. \frac{\partial z}{\partial t} \right|_{\xi=L'} = 0. \tag{8.6}$$

8.3 Numerical Method

8.3.1 *Flow Solver*

Incorporating a modified characteristic-based split (CBS) Scheme [25, 26] with the dual-time stepping (DTS) method [27] and the spring analogy method [28], an implicit solver for the 2D incompressible flows with moving boundaries was developed and some results are published in several of our previous papers. The solver is of excellent accuracy and stability in FIV simulations of a PR membrane wing [17, 21, 22, 24], FIV modeling of a flexible bottom in a lid driven cavity [29], forced oscillation of a circular cylinder in uniform flow [26, 30], FIV calculations of a flexible plate behind a circular cylinder [30] and of two staggered circular cylinders [31].

Following the idea of our flow solver, a derivative term of u_i with respect to the pseudo time τ is added first to Eq. (8.2), namely

$$\frac{\partial u_i}{\partial \tau} + \frac{\partial u_i}{\partial t} + u_j \frac{\partial u_i}{\partial x_j} = -\frac{\partial p}{\partial x_i} + \frac{1}{\mathrm{Re}} \frac{\partial^2 u_i}{\partial x_j \partial x_j}. \tag{8.7}$$

Discretizing Eq. (8.7) in τ using the characteristic-Galerkin method [32] yields

$$\begin{aligned} u_i^{m+1} - u_i^m &= -\Delta\tau \left[u_j^m \frac{\partial u_i^m}{\partial x_j} - \frac{1}{\mathrm{Re}} \frac{\partial^2 u_i^m}{\partial x_j \partial x_j} + \frac{\partial p^{m+\theta}}{\partial x_i} \right] \\ &+ \frac{\Delta\tau^2}{2} u_k^m \frac{\partial}{\partial x_k} \left[u_j \frac{\partial u_i}{\partial x_j} + \frac{\partial p}{\partial x_i} \right]^m - \Delta\tau \frac{3u_i^m - 4u_i^n + u_i^{n-1}}{2\Delta t}, \end{aligned} \tag{8.8}$$

where $\Delta\tau$ is the pseudo time step normalized by c/u_0, Δt is the real time step normalized by c/u_0, $\theta \in [0, 1]$ is a control parameter, and the superscripts m and n indicate the pseudo and real time steps, respectively. In Eq. (8.8), the last term is a discretized form of $\partial u_i/\partial t$ in Eq. (8.7).Subsequently, Eq. (8.8) is solved using the three-step procedure below following the CBS algorithm [32],

$$\text{Step 1}: u_i^* = u_i^m - \Delta\tau \left[u_j \frac{\partial u_i}{\partial x_j} - \frac{1}{\mathrm{Re}} \frac{\partial^2 u_i}{\partial x_j \partial x_j} \right]^m + \frac{\Delta\tau^2}{2} u_k^m \frac{\partial}{\partial x_k} \left[u_j \frac{\partial u_i}{\partial x_j} \right]^m, \tag{8.9}$$

$$\text{Step 2}: \theta \frac{\partial}{\partial x_i} \left(\frac{\partial p^{m+1}}{\partial x_i} \right) = \frac{1}{\Delta\tau} \frac{\partial}{\partial x_i} \left[u_i^* - \Delta\tau (1 - \theta) \frac{\partial p^m}{\partial x_i} \right], \tag{8.10}$$

$$\text{Step } 3: u_i^{m+1} = u_i^* - \Delta\tau \frac{\partial p^{m+\theta}}{\partial x_i} + \frac{\Delta\tau^2}{2} u_k^m \frac{\partial}{\partial x_k}\left(\frac{\partial p}{\partial x_i}\right)^m - \Delta\tau \frac{3u_i^m - 4u_i^n + u_i^{n-1}}{2\Delta t} \tag{8.11}$$

where u_i^* are the intermediate velocity components normalized by u_0. In each real time interval $[t^n, t^{n+1}]$, taking $u_i^0 = u_i^n$ and $p^0 = p^n$ as the initial conditions and solving Eqs. (8.9)–(8.11) iteratively, unknowns u_i^{n+1} and p^{n+1} at t^{n+1} can be approached when $u_i^{m+1} \approx u_i^m$ and $p^{m+1} \approx p^m$. With a dynamic mesh being employed to address the moving boundary induced by the FIVs of the flexible membrane, it is noted that, u_i^*, u_i^m, u_i^{m+1}, u_i^{n+1}, p^m, p^{m+1} and p^{n+1} in Eqs. (8.9)–(8.11) are nodal values of the new mesh at t^{n+1}, while u_i^n, p^n and u_i^{n-1} are not nodal values due to mesh movement. In our flow solver, u_i^n, p^n and u_i^{n-1} are obtained by Taylor expansion using the nodal values of old meshes at t^{n-1} and t^n.

Finally, an unstructured triangular grid with linear shape functions for both the velocity and pressure is utilized to discretize the flow domain, followed by solving Eqs. (8.9)–(8.11) using the standard Galerkin FEM. More details about the spatial discretization can be found in Sun and Zhang [24] and Zienkiewicz et al. [32].

8.3.2 Structure Solver

The standard Galerkin FEM is used to solve Eq. (8.3). For spatial discretization, the membrane is divided equally into *NM* parts. On each element, the Hermite polynomial function is employed to approach the displacement and velocity of the membrane. Following the Galerkin FEM, the spatially discretized form of Eq. (8.3) can be obtained as

$$\mathrm{M}\ddot{\mathrm{X}} + \mathrm{C}\dot{\mathrm{X}} + \mathrm{K}(\mathrm{X})\,\mathrm{X} = \mathrm{Q} \tag{8.12}$$

where

$$\begin{cases} \mathrm{X} = \left[z_1, (\partial z/\partial\xi)_1, z_2, (\partial z/\partial\xi)_2 \cdots z_{(n_M+1)}, (\partial z/\partial\xi)_{(n_M+1)}\right]^{\mathrm{T}}, \\ \mathrm{M} = \rho_S h \int_0^{L'} \mathrm{N}^{\mathrm{T}}\mathrm{N} d\xi, \quad \mathrm{C} = \rho_S C_d \int_0^{L'} \mathrm{N}^{\mathrm{T}}\mathrm{N} d\xi, \\ \mathrm{K}(\mathrm{X}) = \int_0^{L'} \frac{\partial}{\partial\xi} \left\{ \mathrm{N}^{\mathrm{T}} T \left[1 + \left(\frac{\partial \mathrm{N}\cdot\mathrm{X}}{\partial\xi}\right)^2\right]^{-\frac{3}{2}} \right\} \frac{\partial \mathrm{N}}{\partial\xi} d\xi, \\ \mathrm{Q} = \int_0^{L'} \mathrm{N}^{\mathrm{T}} \Delta p d\xi, \quad \mathbf{N} = \left[N_1, N_2, \cdots, N_{2(n_M+1)}\right]. \end{cases} \tag{8.13}$$

In Eq. (8.13), **N** is a vector constructed by the Hermite polynomial function at each grid node. Discretizing Eq. (8.12) in time using the generalized-α method [33], we have

$$\mathrm{M}\ddot{\mathrm{x}}_{n+1-\alpha_m} + \mathrm{C}\dot{\mathrm{x}}_{n+1-\alpha_f} + \mathrm{K}\left(\mathrm{x}_{n+1-\alpha_f}\right)\mathrm{x}_{n+1-\alpha_f} = \mathrm{Q}\left(t_{n+1-\alpha_f}\right) \tag{8.14}$$

where

$$\begin{cases} \ddot{\mathrm{x}}_{n+1-\alpha_m} = (1-\alpha_m)\ddot{\mathrm{x}}_{n+1} + \alpha_m\ddot{\mathrm{x}}_n, \quad \dot{\mathrm{x}}_{n+1-\alpha_f} = \left(1-\alpha_f\right)\dot{\mathrm{x}}_{n+1} + \alpha_f\dot{\mathrm{x}}_n, \\ \mathrm{x}_{n+1-\alpha_f} = \left(1-\alpha_f\right)\mathrm{x}_{n+1} + \alpha_f\mathrm{x}_n, \dot{\mathrm{x}}_{n+1} = \dot{\mathrm{x}}_n + \Delta t\left[(1-\gamma)\ddot{\mathrm{x}}_n + \gamma\ddot{\mathrm{x}}_{n+1}\right], \\ \mathrm{x}_{n+1} = \mathrm{x}_n + \Delta t\dot{\mathrm{x}}_n + \Delta t^2\left[\left(\frac{1}{2}-\beta\right)\ddot{\mathrm{x}}_n + \beta\ddot{\mathrm{x}}_{n+1}\right] \\ \mathrm{Q}\left(t_{n+1-\alpha_f}\right) = \left(1-\alpha_f\right)\mathrm{Q}\left(t_{n+1}\right) + \alpha_f\mathrm{Q}\left(t_n\right), \end{cases} \tag{8.15}$$

The parameters in Eq. (8.15) are defined as

$$\alpha_m = \frac{2\varsigma - 1}{\varsigma + 1}, \quad \beta = \frac{1}{4}\left(1 - \alpha_m + \alpha_f\right)^2, \alpha_f = \frac{\varsigma}{\varsigma + 1}, \gamma = \frac{1}{2} - \alpha_m + \alpha_f, \tag{8.16}$$

where $\varsigma \in [0, 1]$ is one control parameter. Substituting Eq. (8.15) into Eq. (8.14) yields,

$$\begin{aligned} &\left[\tfrac{(1-\alpha_m)}{\beta\Delta t^2}\mathrm{M} + \tfrac{(1-\alpha_f)\gamma}{\beta\Delta t}\mathrm{C} + \left(1-\alpha_f\right)\mathrm{K}\left(\mathrm{x}_{n+1-\alpha_f}\right)\right]\mathrm{x}_{n+1} + \alpha_f\mathrm{K}\left(\mathrm{x}_{n+1-\alpha_f}\right)\mathrm{x}_n \\ &= \mathrm{M}\left[\tfrac{(1-\alpha_m)}{\beta\Delta t^2}\mathrm{x}_n + \tfrac{(1-\alpha_m)}{\beta\Delta t}\dot{\mathrm{x}}_n + \tfrac{(1-\alpha_m-2\beta)}{2\beta}\ddot{\mathrm{x}}_n\right] \\ &+ \mathrm{C}\left[\tfrac{(1-\alpha_f)\gamma}{\beta\Delta t}\mathrm{x}_n + \left(\tfrac{\gamma}{\beta} - \tfrac{\gamma\alpha_f}{\beta} - 1\right)\dot{\mathrm{x}}_n + \Delta t\left(1-\alpha_f\right)\left(\tfrac{\gamma}{2\beta} - 1\right)\ddot{\mathrm{x}}_n\right] \\ &+ \mathrm{Q}\left(t_{n+1-\alpha_f}\right). \end{aligned} \tag{8.17}$$

Equation (8.17) is a group of nonlinear algebra equations on x_{n+1}, which is solved by the Newton-Raphson method. Once x_{n+1} is obtained, the velocity vector at t_{n+1}, namely $\dot{\mathrm{x}}_{n+1}$, can be computed by the relationships in Eq. (8.15).

8.3.3 FSI Solution Procedure

The flow and structure solvers are coupled following the loosely-coupled partitioned approach. At each real time step, the FSI solution procedure is as follows:

1. *Compute the fluid load* Δp *on the membrane using the flow variables at* t^n*;*
2. *Compute the displacement and velocity of the membrane at* t^{n+1} *by Eqs. (8.12)–(8.17);*
3. *Use the displacement obtained in Step 2 to compute the new mesh of the flow domain at* t^{n+1} *by the segment spring analogy method;*
4. *Compute the flow unknowns at* t^{n+1} *on the new mesh:*

(a) *Compute* u_i^n, p^n, *and* u_i^{n-1} *by Taylor expansion using the nodal values of old meshes at* t^{n-1} *and* t^n*;*
(b) *Compute* u_i^* *by Eq. (8.9);*
(c) *Compute* p^{m+1} *by Eq. (8.10);*
(d) *Compute* u_i^{m+1} *by Eq. (8.11);*
(e) *Set* $u_i^m = u_i^{m+1}$, $p^m = p^{m+1}$ *and* $u_i^m = u_i^{m+1}$. *Return to step (b) until the convergence criterion is achieved;*

5. *Set* $u_i^n = u_i^{n+1}$, $p^n = p^{n+1}$ *and* $u_i^{n-1} = u_i^n$ *and loop over to the next real time step.*

At Step (e), the Aitken- Δ^2 method is utilized to accelerate the iteration process with the following convergence criterion

$$\begin{cases} \sqrt{\sum_{J=1}^{NP}\left[\left(u_1^{m+1}\right)_J - \left(u_1^m\right)_J\right]^2} \Big/ \sqrt{\sum_{J=1}^{NP}\left(u_1^{m+1}\right)_J^2} < 10^{-6} \\ \sqrt{\sum_{J=1}^{NP}\left(p_J^{m+1} - p_J^m\right)^2} \Big/ \sqrt{\sum_{J=1}^{NP}\left(p_J^{m+1}\right)^2} < 10^{-4} \end{cases} \quad (8.18)$$

where *NP* is the total number of grid nodes in the flow domain.

8.4 Numerical Results

8.4.1 Computational Model and Code Verification

Figure 8.2 shows the solution domain and boundary conditions of a PR membrane wing in shear flow. As seen in the figure, the inlet and the two side boundaries locate at 8*c* from the membrane center, where the free stream velocity governed by $u_\infty = u_0 + ky$ is imposed. The outlet is placed at 16*c* downstream, where the pressure condition $p = 0$ is used. On the membrane wing, we use the no-slip boundary conditions. On the rigid supports, the flow velocity is zero. On the flexible membrane, the fluid and membrane have same displacement, velocity and stress on the interface.

To examine the accuracy of the FSI scheme proposed in Sect. 8.3, the aerodynamic performance of the PR membrane wing in uniform inflow ($k = 0$) is computed firstly. The flow and structure parameters are set as $Re = 2500$, $Eh' = 50$, $\rho_S h' = 0.589$, $C_d = 0$ and $\alpha = 8°$. Table 8.1 presents the time-averaged lift coefficients ($\overline{C}_L$) and drag coefficients ($\overline{C}_D$) computed from three meshes of different grid densities. For Mesh_2 of medium grid density, three time-step are performed to test the effect of time step selection. As seen in the table, the numerical results computed from different cases are very close. Figure 8.3 shows the time-averaged deflection and pressure distribution of the membrane computed from

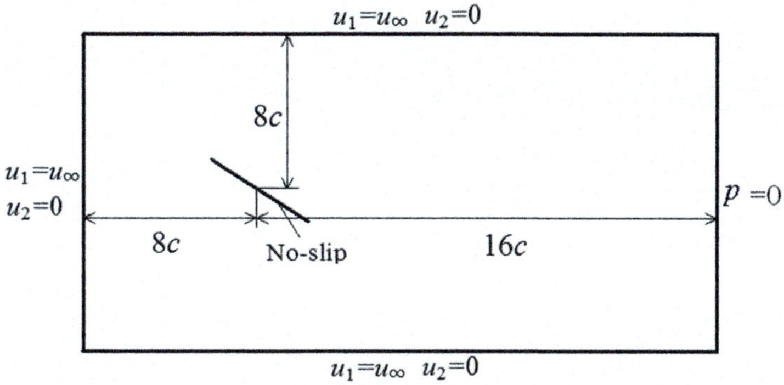

Fig. 8.2 A schematic of the computational model

Table 8.1 Mean lift and drag coefficients of a fixed PR membrane wing at $Re = 2500$ and $\alpha = 8°$

	NM	*NP*	*NE*	Δt	$\overline{C}_L$	$\overline{C}_D$
Mesh_1	60	5699	11,138	0.01	0.998	0.112
				0.02	0.993	0.113
Mesh_2	80	7581	14,851	0.01	0.998	0.113
				0.005	0.995	0.113
Mesh_3	100	10,103	19,850	0.01	1.0	0.113

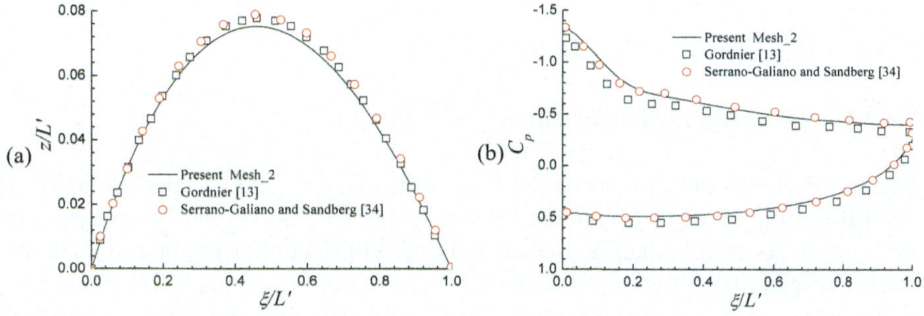

Fig. 8.3 Time-averaged (**a**) deflection and (**b**) pressure coefficient of the membrane at $\alpha = 8°$

Mesh_2 and $\Delta t = 0.01$. It can be found that our results agree very well with those reported by Gordnier [13] and Serrano-Galiano and Sandberg [34]. Figure 8.4 shows the initial and deformed Mesh_2. As seen in the figure, the body-fitted grid is introduced near the wing and the mesh is refined in the wake, to capture the boundary layer, flow separation and vortex structures in these regions. Comparing Fig. 8.4(b) with Fig. 8.4(a), it can be found that the quality of the computational mesh is not decreased with the large deformation of the flexible membrane. Therefore, Mesh_2 is used for further simulation at $\alpha = 8°$. At other angles of attack, the computational meshes are generated using the same grid density as Mesh_2.

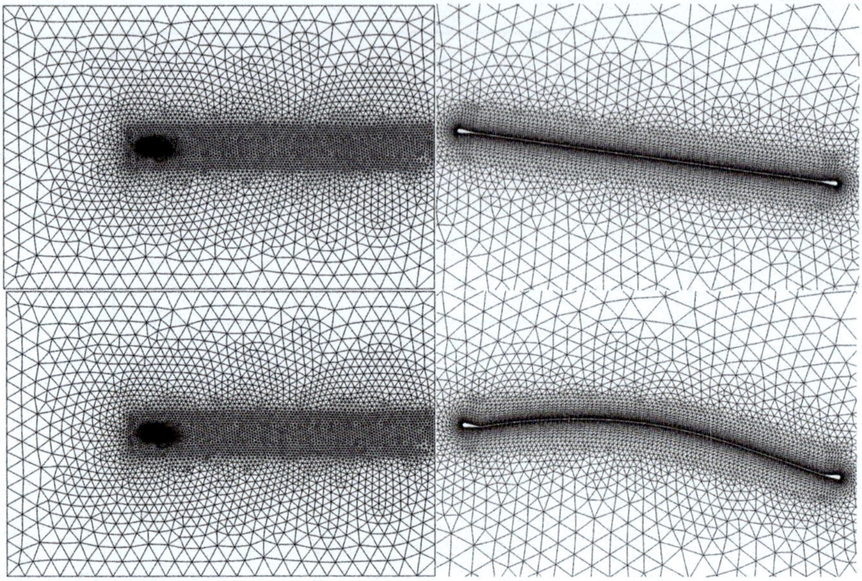

Fig. 8.4 Computational mesh (Mesh_2) at $\alpha = 8°$: (**a**) initial mesh; (**b**) deformed mesh

8.4.2 *Membrane Wing in Shear Flow*

8.4.2.1 Aerodynamic Performance

Figure 8.5 shows the time-averaged lift coefficients ($\overline{C}_L$), drag coefficients ($\overline{C}_D$) and lift-to-drag ratios ($\overline{L/D}$) of the PR membrane wing in laminar shear flow with $Re = 2500$. As seen in the Fig. 8.5(a), in the case of the uniform inflow ($k = 0$), the time-averaged lift coefficient of the membrane wing increases gradually from $\overline{C}_L = 0.366$ at $\alpha = 1°$ to $\overline{C}_L = 1.71$ at $\alpha = 20°$. As the velocity gradient increases from $k = 0$ to $k = 0.02$, 0.04 and 0.06, the time-averaged lift coefficients at $\alpha = 2° \sim 20°$ are not changed much. At $\alpha = 1°$, however, it drops to a negative value of $\overline{C}_L = -0.149$, which means that the PR membrane wing flying at very small angles of attack has a risk of losing the lift in shear flows. As seen in Fig. 8.5(b), the time-averaged drag of the membrane wing also increases gradually with the angle of attack. At most angles of attack considered, the velocity gradient of the inflow has negligible effect on the drag. At high angles of attack ($\alpha = 19° \sim 20°$), the membrane wing in shear flow has a slightly smaller drag than that in uniform inflow. As shown in Fig. 8.5(c), with increase of the angle of attack, the lift-to-drag ratio of the PR membrane wing in uniform flow increases first in $\alpha = 1° \sim 7°$ and then decreases in $\alpha = 7° \sim 20°$, with a maximum value of $\overline{L/D} = 8.828$ at $\alpha = 7°$. For cases with shear inflow ($k = 0.02$, 0.04 and 0.06), the membrane wing has a negative time-averaged lift-to-drag ratio ($\overline{L/D} = -2.327$) at $\alpha = 1°$; at other angles of attack, the lift-to-drag ratios are very close to those in the case of the uniform inflow. Compared

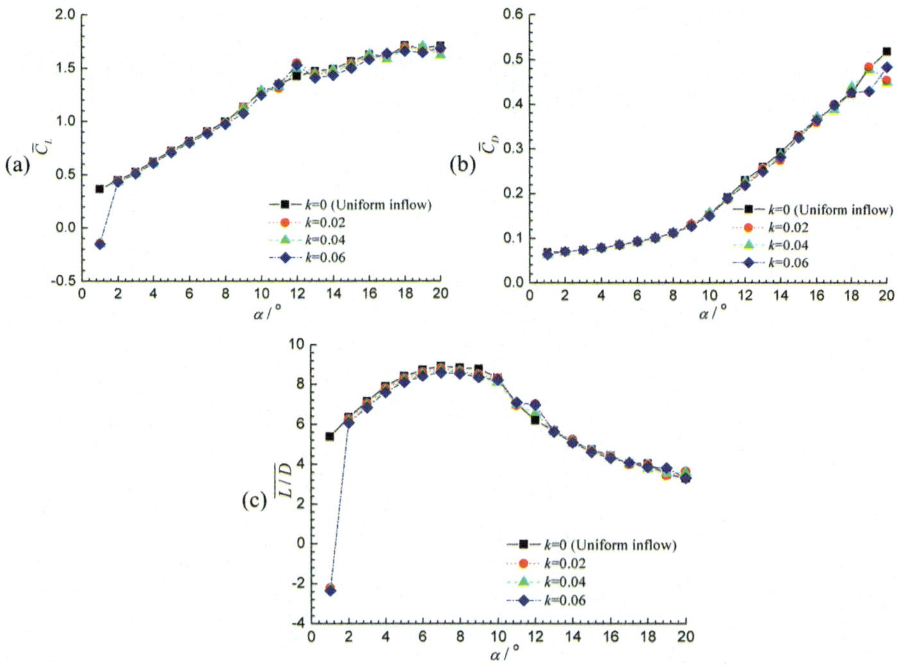

Fig. 8.5 Aerodynamic loads of a PR membrane wing in shear flow at $Re = 2500$: (**a**) lift coefficient; (**b**) drag coefficient; (**c**) lift-to-drag ratio

with those at moderate and high angles of attack, the aerodynamic loads of the PR membrane wing at very low angles of attack are more sensitive to the shear flow.

8.4.2.2 Structure Response

Figure 8.6 is the bifurcation diagram of FIV at the membrane center, with respect to the angle of attack. As shown in Fig. 8.6(a), at $k = 0$ (uniform inflow) the FIV is so weak in $\alpha = 1°$ ~$7°$ that the membrane wing can be taken as in the static equilibrium state. As the angle of attack increases slightly from $\alpha = 7°$ to $\alpha = 8°$, period-1 LCO with apparent vibrating amplitude appears, and however, it does not change much in $\alpha = 8°$ ~$10°$. At $\alpha = 11°$ ~$20°$, the chaotic response of the membrane wing is observed. When the membrane wing is put into shear flows with $k = 0.02$, 0.04 and 0.06, its response changes significantly at some angles of attacks. At $\alpha = 1°$, the deformation of the membrane center jumps from 0.056 to a negative value, which changes slightly from −0.0449 to −0.0455 when k increases from 0.02 to 0.06. At $\alpha = 8°$, the FIV amplitude of the membrane center reduces a lot at $k = 0.02$ and 0.04. It seems that the velocity gradient of the inflow is prone to delay the Hopf bifurcation of the FIV near $\alpha = 8°$. At $\alpha = 9°$ and/or $\alpha = 10°$, the flow-induced Period-1 LCO is amplified in the shear inflow with $k = 0.02$ and 0.04. At $\alpha = 12°$,

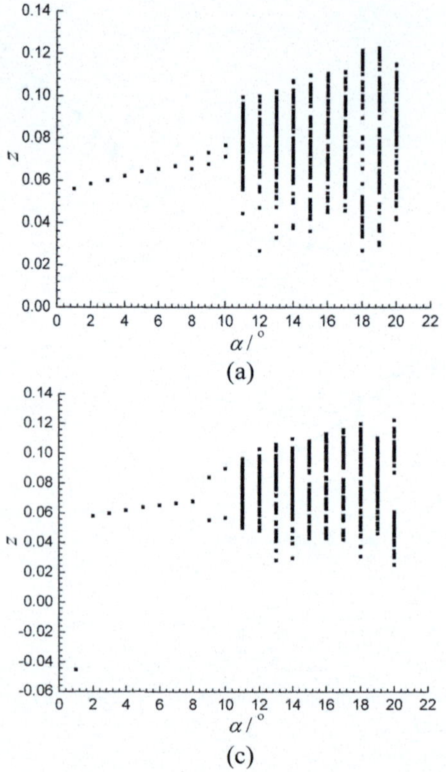

Fig. 8.6 Bifurcation diagram of FIV at the membrane centre with respect to the angle of attack: (**a**) $k = 0$ (uniform flow); (**b**) $k = 0.02$; (**c**) $k = 0.04$; (**d**) $k = 0.06$

the chaotic LCO response in the uniform inflow is stabilized significantly at $k = 0.02$ and 0.06.

To show the effect of the shear flow on the response of the whole membrane wing, Fig. 8.7 shows the time-averaged membrane deflections at $\alpha = 1°, 5°, 8°, 10°, 11°$ and $12°$. As seen in Fig. 8.7(a), at $\alpha = 1°$ the membrane wing deflects upward and forms a positive camber when it locates in a uniform inflow. In shear flows with $k = 0.02$, 0.04 and 0.06, however, the membrane deflects downward, resulting in the negative lift as shown in Fig. 8.5(a). As seen in Fig. 8.7(b) and (c), at $\alpha = 5°$ and $8°$ the membrane wing deflects upwards in both uniform and shear inflows, and the deflection increases with the angle of attack. At $\alpha = 1°, 5°$ and $8°$, the time-averaged deflection does not change much at $k = 0.02$, 0.04 and 0.06. At $\alpha = 10°$ and $11°$, the deflection of the membrane wing further increases and the effect of shear flow becomes noticeable. As seen in Fig. 8.7(d) and (e), the time-averaged deformations at $\alpha = 10°$ and $11°$ drops slightly in the shear flow. At $\alpha = 12°$, however, a different trend is observed. As shown in Fig. 8.7(f), the camber of the membrane wing amplifies in shear flows.

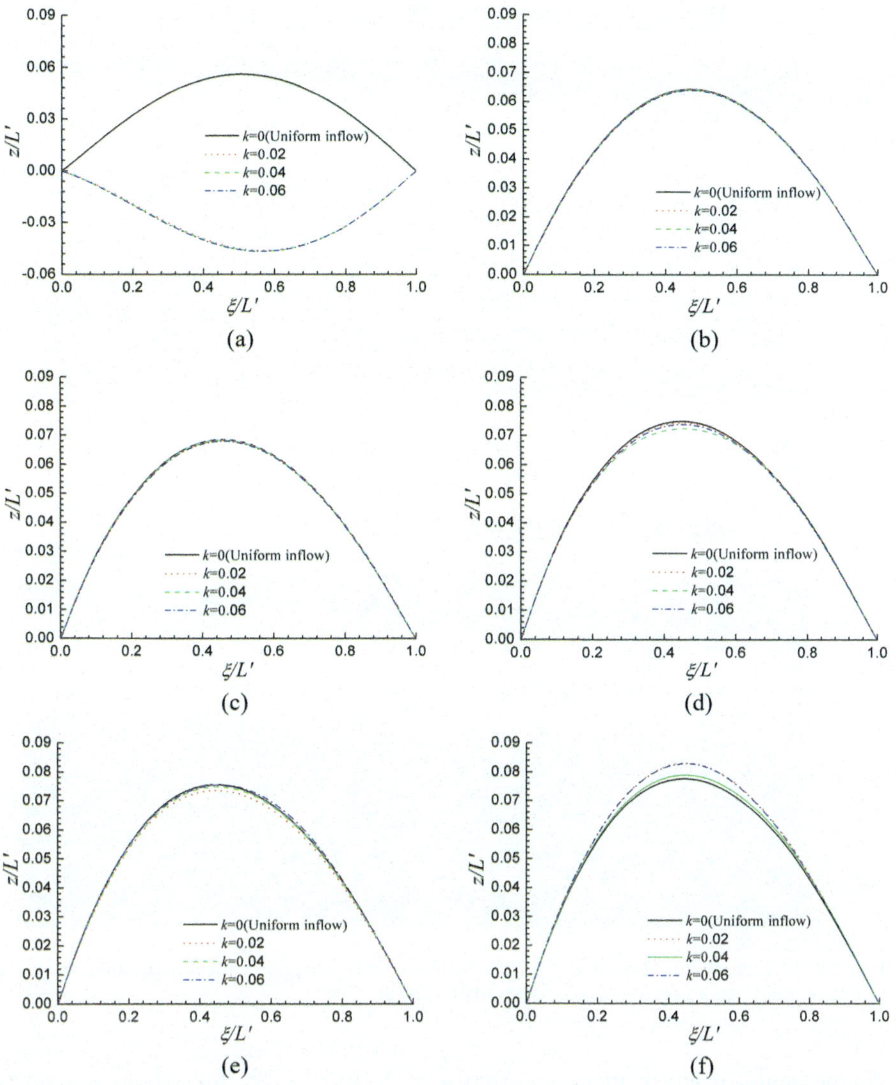

Fig. 8.7 Time-averaged membrane deflections at: (**a**) $\alpha = 1°$; (**b**) $\alpha = 5°$; (**c**) $\alpha = 8°$; (**d**) $\alpha = 10°$; (**e**) $\alpha = 11°$; (**f**) $\alpha = 12°$

Figures 8.8, 8.9, 8.10, 8.11, 8.12, and 8.13 display the evolution of the membrane oscillation relative to its time-averaged deflection position. As shown in Fig. 8.8, the membrane wing is in the stationary state at $\alpha = 1°$ in both uniform and shear inflows and there is no oscillation on the membrane. As shown in Fig. 8.9, at $\alpha = 5°$ the membrane wing exhibits the fourth-mode response with amplitude in an order of 2×10^{-4}, and the shear flow shows little effect on the dynamic response.

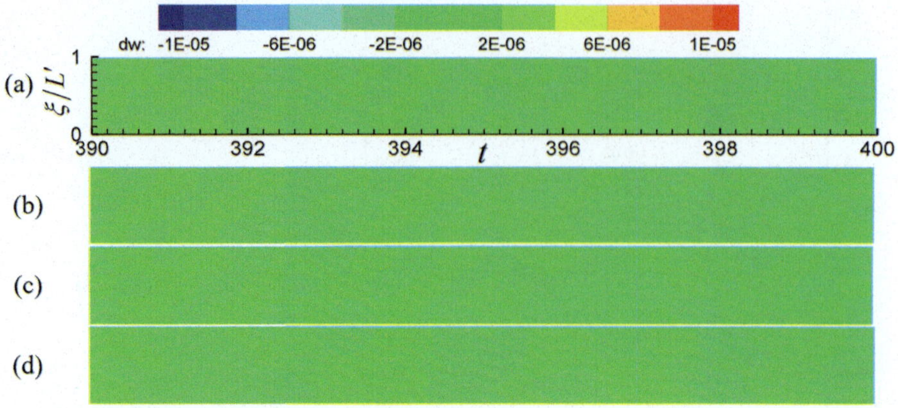

Fig. 8.8 Membrane vibration at $\alpha = 1°$ relative to the time-averaged deflection position: (**a**) $k = 0$ (uniform flow); (**b**) $k = 0.02$; (**c**) $k = 0.04$; (**d**) $k = 0.06$

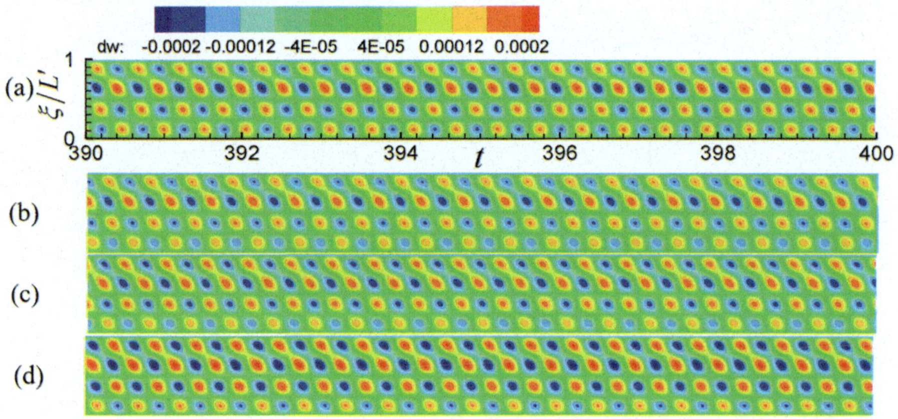

Fig. 8.9 Membrane vibration at $\alpha = 5°$ relative to the time-averaged deflection position: (**a**) $k = 0$ (uniform flow); (**b**) $k = 0.02$; (**c**) $k = 0.04$; (**d**) $k = 0.06$

As the angle of attack increases from $\alpha = 5°$ to $\alpha = 8°$, the vibration state in uniform flow turns from fourth-mode to third-mode and the vibrating amplitude increases to 2×10^{-3}, as seen in Fig. 8.10(a). In this case, the shear flow shows significant influences on the membrane vibration. As seen in Fig. 8.10(b) and (c), the third-mode response is disturbed by the shear flow and the vibrating amplitude is suppressed greatly at $k = 0.02$ and $k = 0.04$. At $\alpha = 10°$, the membrane wing in the uniform flow experiences the second-mode response, as seen in Fig. 8.11(a). At $k = 0.04$, the shear inflow amplifies the vibrating amplitude, as seen in Fig. 8.11(c). At $k = 0.02$ and $k = 0.06$, however, the effect of shear flow is negligible, as seen in Fig. 8.11(b) and (d). When the angle of attack further increases to $\alpha = 11°$, the membrane vibration becomes chaotic, as seen in Fig. 8.12(a). With increase of the

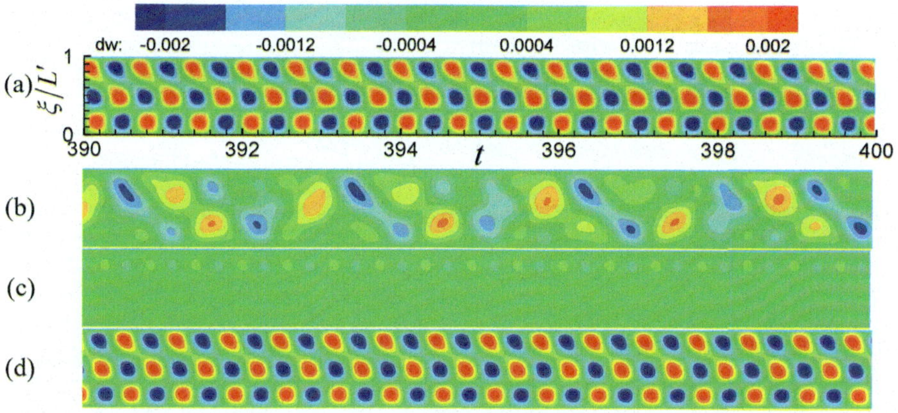

Fig. 8.10 Membrane vibration at $\alpha = 8°$ relative to the time-averaged deflection position: (**a**) $k = 0$ (uniform flow); (**b**) $k = 0.02$; (**c**) $k = 0.04$; (**d**) $k = 0.06$

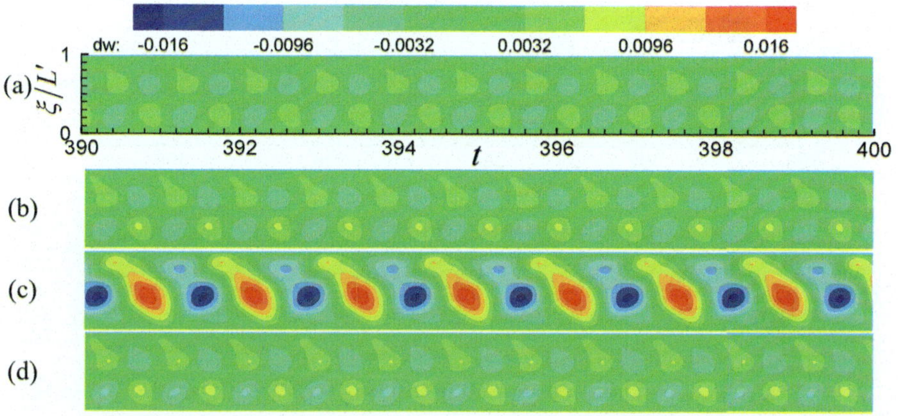

Fig. 8.11 Membrane vibration at $\alpha = 10°$ relative to the time-averaged deflection position: (**a**) $k = 0$ (uniform flow); (**b**) $k = 0.02$; (**c**) $k = 0.04$; (**d**) $k = 0.06$

velocity gradient in the free stream, the irregular response of the membrane wing is stabilized, as seen in Fig. 8.12(b)–(d). At $\alpha = 12°$, the chaotic vibration of the membrane wing in the uniform flow becomes more apparent, although its amplitude does not change much, as seen in Fig. 8.13(a). At $k = 0.02$ and $k = 0.06$, the shear inflow significantly suppresses the membrane vibration and makes it more regular. Compared with its effect on the time-averaged deflection of the membrane wing as shown in Fig. 8.7, it is evident that the shear inflow shows larger influence on the FIV of the membrane wing.

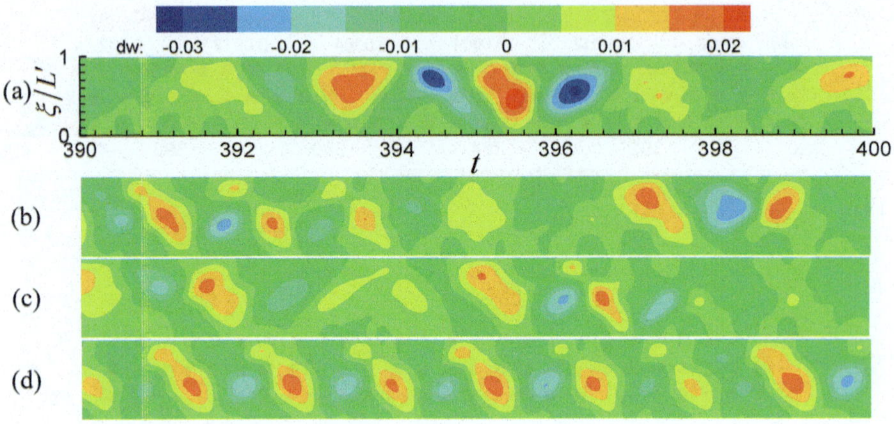

Fig. 8.12 Membrane oscillation of the membrane wing at $\alpha = 11°$ relative to the time-averaged deflection position: (**a**) $k = 0$ (uniform flow); (**b**) $k = 0.02$; (**c**) $k = 0.04$; (**d**) $k = 0.06$

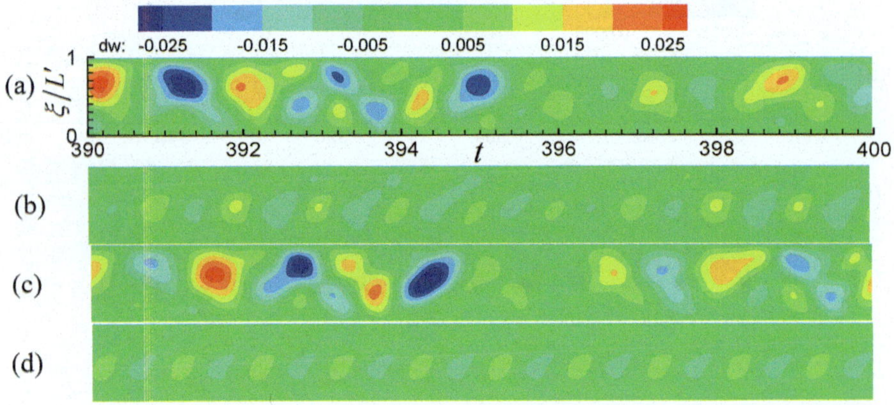

Fig. 8.13 Membrane oscillation of the membrane wing at $\alpha = 12°$ relative to the time-averaged deflection position: (**a**) $k = 0$ (uniform flow); (**b**) $k = 0.02$; (**c**) $k = 0.04$; (**d**) $k = 0.06$

8.4.2.3 Flow Field

To understand in depth the above responses of the PR membrane wing in shear flows, the flow field around the membrane wing is further analyzed. Figures 8.14, 8.15, 8.16, 8.17, 8.18, and 8.19 present the instantaneous streamlines at $\alpha = 1°$, $5°$, $8°$, $10°$, $11°$, $12°$ and $k = 0$ (uniform flow), $k = 0.02$, $k = 0.04$ and $k = 0.06$. At each combination of α and k, the streamlines at four time instants in one period, which is obtained from the dominant vibrating frequency of the membrane center, are presented to show the evolution of the vortex structure around the membrane wing. At the first time instant, the membrane center has the minimum deflection. At the second time instant, the membrane center is moving upward through the mean

(a) (b) (c) (d)

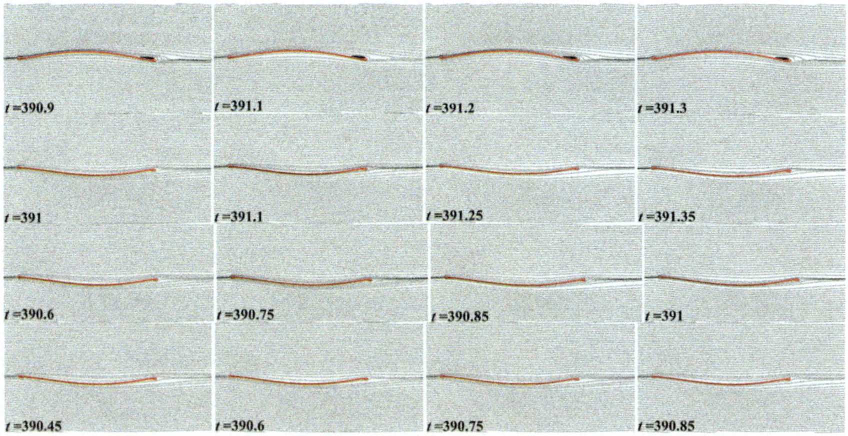

Fig. 8.14 Instantaneous streamlines around the PR membrane wing at $\alpha = 1°$: (**a**) $k = 0$ (uniform flow); (**b**) $k = 0.02$; (**c**) $k = 0.04$; (**d**) $k = 0.06$

(a) (b) (c) (d)

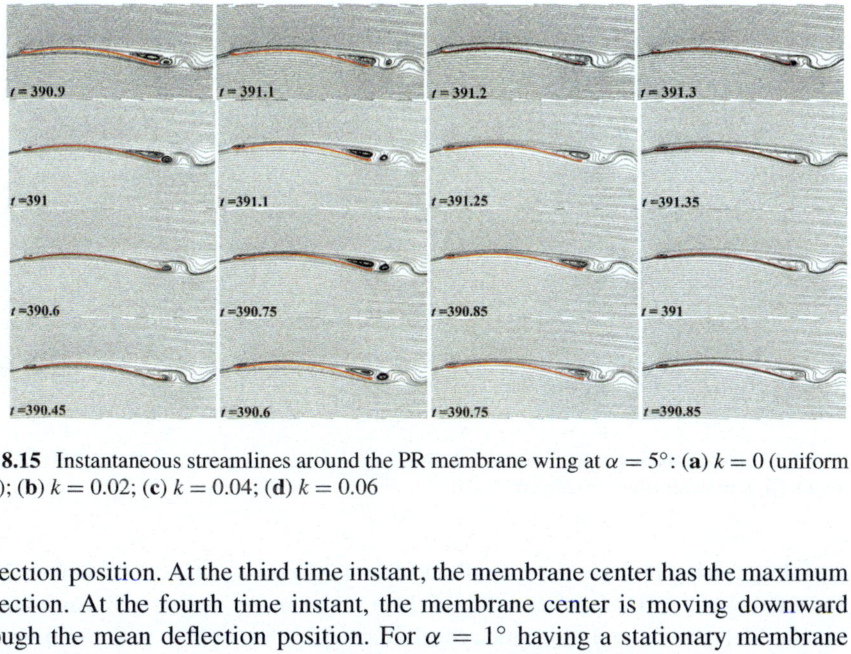

Fig. 8.15 Instantaneous streamlines around the PR membrane wing at $\alpha = 5°$: (**a**) $k = 0$ (uniform flow); (**b**) $k = 0.02$; (**c**) $k = 0.04$; (**d**) $k = 0.06$

deflection position. At the third time instant, the membrane center has the maximum deflection. At the fourth time instant, the membrane center is moving downward through the mean deflection position. For $\alpha = 1°$ having a stationary membrane response, the time instants are selected as same as those at $\alpha = 5°$.

As seen in Fig. 8.14(a), the membrane wing deflects upward and forms a positive camber at $\alpha = 1°$ and $k = 0$ (uniform flow). In this case, the fluid flows smoothly around the membrane wing, with a small separation bubble near the trailing edge. The flow is stationary and not changes with time, resulting in a stationary response of the membrane wing as seen in Fig. 8.8(a). As the inflow has a small velocity gradient of $k = 0.02$, the flow pattern varies significantly. As seen in Fig. 8.14(b), the

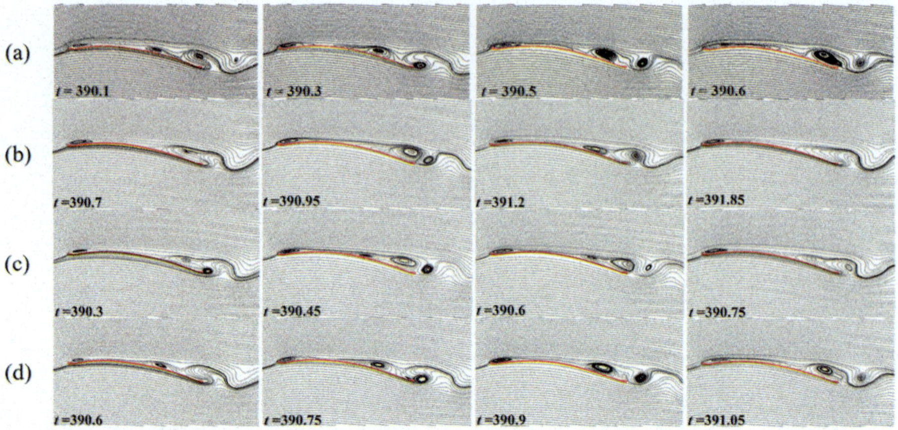

Fig. 8.16 Instantaneous streamlines around the PR membrane wing at $\alpha = 8°$: (**a**) $k = 0$ (uniform flow); (**b**) $k = 0.02$; (**c**) $k = 0.04$; (**d**) $k = 0.06$

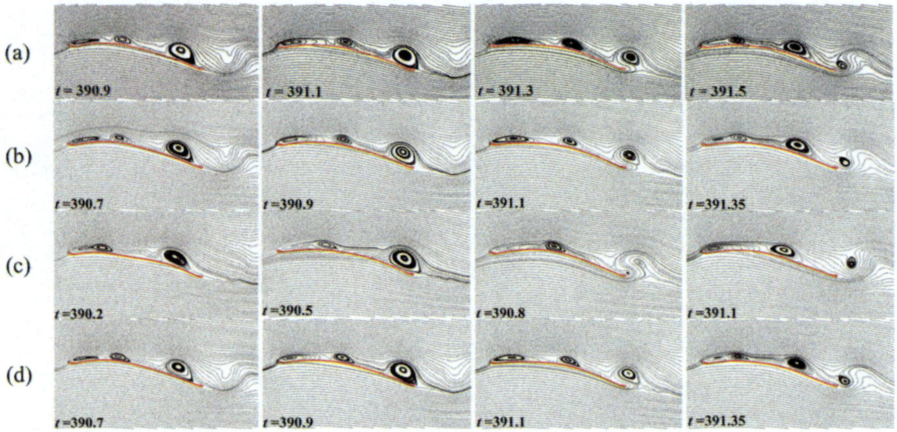

Fig. 8.17 Instantaneous streamlines around the PR membrane wing at $\alpha = 10°$: (**a**) $k = 0$ (uniform flow); (**b**) $k = 0.02$; (**c**) $k = 0.04$; (**d**) $k = 0.06$

membrane wing deflects downward and forms a negative camber. Compared with the camber at $k = 0$ as shown in Fig. 8.14(a), the camber at $k = 0.02$ has a slightly smaller deflection amplitude. As a result, the separation bubble at the trailing edge disappears in Fig. 8.14(b). Moreover, the flow remains steady at $k = 0.02$. As the velocity gradient of the free stream increases from $k = 0.02$ to $k = 0.04$ and 0.06, the flow field is almost unchanged, as seen in Fig. 8.14(c) and (d), leading to very close membrane responses as shown in Figs. 8.7(a) and 8.8.

At $\alpha = 5°$, the deflection of the membrane wing in the uniform flow increases, which generates a larger vortex at the trailing edge. In this case, the flow becomes unsteady and the trailing-edge vortex sheds periodically into the wake, as seen in

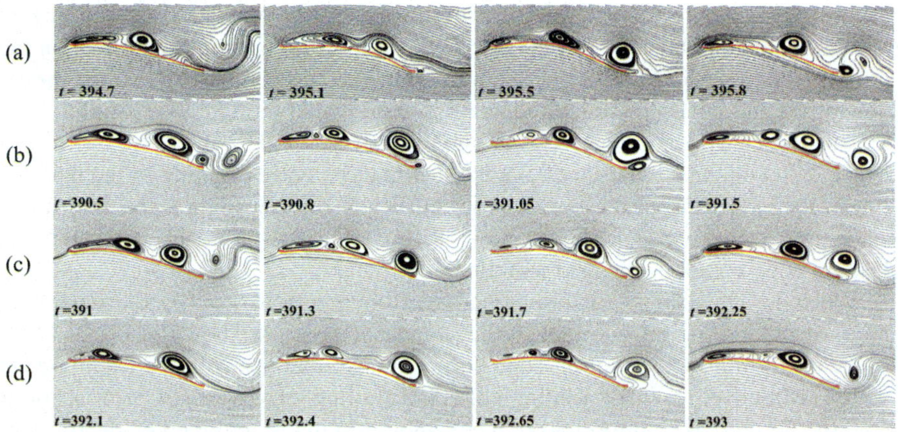

Fig. 8.18 Instantaneous streamlines around the membrane wing at $\alpha = 11°$: (**a**) $k = 0$ (uniform flow); (**b**) $k = 0.02$; (**c**) $k = 0.04$; (**d**) $k = 0.06$

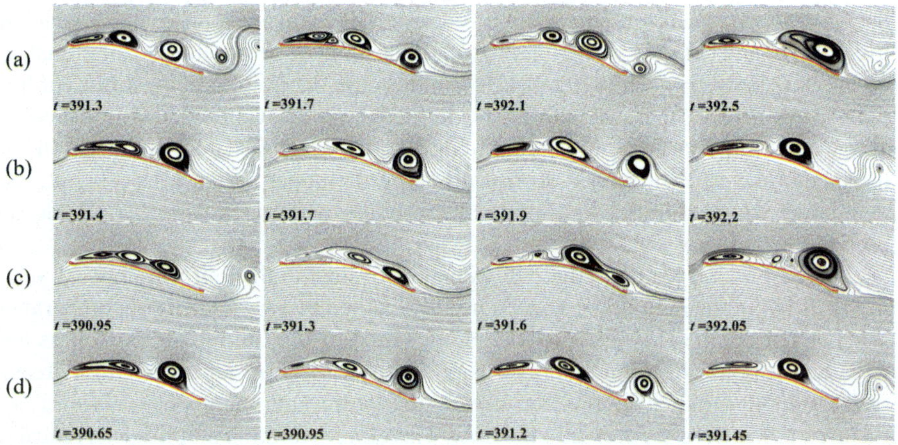

Fig. 8.19 Instantaneous streamlines around the membrane wing at $\alpha = 12°$: (**a**) $k = 0$ (uniform flow); (**b**) $k = 0.02$; (**c**) $k = 0.04$; (**d**) $k = 0.06$

Fig. 8.15(a). Perturbed by the vortex shedding process, the membrane wing exhibits the Forth-order standing wave response as shown in Fig. 8.9(a). As the membrane wing is put into the shear flows with $k = 0.02$, 0.04 and 0.06, the scale of the separation bubble at the leading edge increases slightly, while the variation of the vortex shedding process of the trailing edge vortex is negligible. Therefore, the time-averaged deflection and the dynamic response of the membrane wing in uniform and shear flows are very close, as shown in Figs. 8.7(b) and 8.9.

At $\alpha = 8°$, the scale and strength of the trailing-edge vortex further increase in the uniform flow, and its formation position moves upwind, as shown in Fig. 8.16(a). As more part of the membrane wing is perturbed by the trailing-edge vortex, the third-mode standing wave response with larger amplitude appears as seen in Fig. 8.10(a). At $k = 0.02$ and $k = 0.04$, the shear flow shows suppressing effect on the trailing-edge vortex. As seen in Fig. 8.16(b) and (c), the trailing-edge vortex moves downward and its scale reduces apparently. As a result, the vibrating amplitude of the membrane wing at $\alpha = 8°$ reduces significantly at $k = 0.02$ and $k = 0.04$, as seen in Fig. 8.10(b) and (c). At $k = 0.06$, the flow pattern corresponding to the uniform flow (see Fig. 8.10(a)) is recovered, as seen in Fig. 8.10(d), leading to the third-mode standing wave response as shown in Fig. 8.10(d). In all shear flows considered, the leading-edge vortex is not shed with time.

When the angles of attack further increase from $\alpha = 8°$ to $\alpha = 10°$, the generation point of the trailing-edge vortex further moves upward. As shown in Fig. 8.17(a), in the uniform flow the trailing-edge vortex almost interacts with the leading-edge vortex. Moreover, both of leading-edge and trailing-edge vortices grow a lot at $\alpha = 10°$. In this case, almost half of the membrane wing is perturbed by the trailing-edge vortex, which generates the second-order standing wave response as shown in Fig. 8.11(a). At $k = 0.02$ and $k = 0.06$, the scale of the leading- and trailing-edge vortices decreases slightly, but their pattern is not changed much, as seen in Fig. 8.17(b) and (d). Therefore, the second-order standing wave response also appears in Fig. 8.11(b) and (d). At $k = 0.04$, however, the shear flow shows significant influence on the flow field. As seen in Fig. 8.17(c), in this case the trailing-edge vortex further moves upward and interacts with the leading-edge vortex. Compared with the cases at $k = 0$, 0.02 and 0.06, the trailing-edge vortex imposes more interruption on the membrane when it moves downward along the upper surface. As a result, the membrane wing exhibits a first-mode like response with much larger amplitude, as seen in Fig. 8.11(c).

At $\alpha = 11°$, the leading-edge vortex begins to shed in the uniform flow. As seen in Fig. 8.18(a), the membrane wing is usually excited by two vortices of comparable scale and strength. This leads to the complication of the fluid loads as well as the dynamic response of the membrane wing, as shown in Fig. 8.12(a). With increase of the velocity gradient in the free stream, the vortices on the upper surface of the membrane wing are suppressed gradually. Therefore, vibration amplitude is reduced and the chaotic response becomes more regular when k increases from 0 to 0.06, as seen in Fig. 8.12.

At $\alpha = 12°$, secondary vortices appear during the vortex shedding process, making the flow pattern as well as the dynamic response of the membrane wing more complex, as seen in Figs. 8.13(a) and 8.19(a). In this case, the fluid-membrane system seems very sensitive to the velocity gradient of the free stream. As seen in Fig. 8.19(b)–(d), flow pattern quite different from that in the case of the uniform inflow can be observed at $k = 0.02$, 0.04 and 0.06. Similar with that recorded at $\alpha = 8°$, 10° and 11°, the shear flow shows a suppressing effect on the vortex shedding process. Therefore, the chaotic response of the membrane wing is significantly suppressed at $k = 0.02$ and 0.06, as seen in Fig. 8.13(b) and (d).

8.5 Conclusions

Aerodynamic performances and dynamic responses of a PR membrane wing in shear laminar flow are investigated in detail, using FSI simulation. The followings summarize the many observations made in this Chapter,

1. By combining the modified CBS scheme, dual-time step method, segment spring analogy technique, Galerkin finite element method, generalized-α method and loosely-coupled partitioned method, the proposed FSI solution procedure demonstrates excellent accuracy and stability in computing the aeromechanics of the PR membrane wing in two-dimensional laminar flow.
2. At very small angle of attack, the flow is steady and the PR membrane wing stay at an equilibrium position. In the uniform flow condition, the membrane wing has an upward camber, which generates positive lift and lift-to-drag ratio. In shear flows, however, the membrane wing deflects downward and generates negative lift and lift-to-drag ratio, which is dangerous to the membrane-wing-based MAVs.
3. With increase of the angle of attack, the trailing-edge vortex begins to shed periodically at moderate angle of attack, leading to standing wave response of the membrane wing. At high angles of attack, the leading-edge vortex sheds and moves downward along the upper surface of the wing, leading to chaotic response of the membrane wing. At moderate and high angles of attack, the shear flow shows little effect on the lift and lift-to-drag ratio of the PR membrane wing. At high angles of attack, the membrane wing undergoes smaller drag in shear flows. In most cases considered, the shear flow is beneficial to decreasing the scale and strength of the leading-edge and trailing-edge vortices, which further suppresses the FIV of the PR membrane wing.

Acknowledgments This work is supported by the National Key R&D Program of China (No. 2016YFE0204200) and the National Natural Science Foundation of China (No. 51506224). The authors gratefully acknowledge the support of these funding.

References

1. Lian, Y., W. Shyy, D. Viieru, and B. Zhang. 2003. Membrane wing aerodynamics for micro air vehicles. *Progress in Aerospace Sciences* 39: 425–465.
2. Shyy, W., P. Ifju, and D. Viieru. 2005. Membrane wing-based micro air vehicles. *Applied Mechanics Reviews* 58: 203–301.
3. Ansari, S.A., R. Zbikowski, and K. Knowles. 2006. Aerodynamic modeling of insect-like flapping flight for micro air vehicles. *Progress in Aerospace Sciences* 42: 129–172.
4. Stanford, B., P. Ifju, R. Albertani, and W. Shyy. 2008. Fixed membrane wings for micro air vehicles: experimental characterisation, numerical modelling and tailoring. *Progress in Aerospace Sciences* 46: 258–294.
5. Orlowski, C.T., and A.R. Girard. 2012. Dynamics, stability, and control analyses of flapping wing micro-air vehicles. *Progress in Aerospace Sciences* 51: 18–30.

6. Albertani, R., B. Stanford, J.P. Hubner, and P.G. Ifju. 2007. Aerodynamic coefficients and deformation measurements on flexible micro air vehicle wings. *Experimental Mechanics* 47: 625–635.
7. Johnston, J.W., W. Romber, P.J. Attar, and R. Parthasarathy. 2010. Experimental characterization of limit cycle oscillations in membrane wing Micro Air Vehicles. *Journal of Aircraft* 47: 1300–11308.
8. Attar, P.J., R.E. Gordnier, J.W. Johnston, W.A. Romberg, and R.N. Parthasarathy. 2011. Aeroelastic analysis of membrane micro air vehicles-Part I: flutter and limit cycle analysis for fixed-wing configurations. *Journal of Vibration and Acoustics* 133: 021008.
9. Attar, P.J., J.M. Brian, W.A. Romberg, J.W. Johnston, and R.N. Parthasarathy. 2012. Experimental characterization of aerodynamic behavior of membrane wings in low-Reynolds-number flow. *AIAA Journal* 50: 1525–1537.
10. Rojratsirikul, P., Z. Wang, and I. Gursul. 2009. Unsteady fluid–structure interactions of membrane airfoils at low Reynolds numbers. *Experiments in Fluids* 46: 859–872.
11. Rojratsirikul P., Z. Wang and I. Gursul. 2010. Effect of pre-strain and excess length on unsteady fluid-structure interactions of membrane airfoils. *Journal of Fluids and Structures* 18: 359–376.
12. Rojratsirikul, P., M. Genc, Z. Wang, and I. Gursul. 2011. Flow-induced vibrations of low aspect ratio rectangular membrane wings. *Journal of Fluids and Structures* 19: 1296–1309.
13. Gordnier, R.E. 2009. High fidelity computational simulations of a membrane wing airfoil. *Journal of Fluids and Structures* 25: 897–917.
14. Visbal, M.R., R.E. Gordnier, and M.C. Galbraith. 2009. High-fidelity simulations of moving and flexible airfoils at low Reynolds numbers. *Experiments in Fluids* 46: 903–922.
15. Gordnier, R.E., and P.J. Attar. 2014. Impact of flexibility on the aerodynamics of an aspect ratio two membrane wing. *Journal of Fluids and Structures* 45: 138–152.
16. Arbós-Torrent, S., B. Ganapathisubramani, and R. Palacios. 2013. Leading- and trailing-edge effects on the aeromechanics of membrane aerofoils. *Journal of Fluids and Structures* 38: 107–118.
17. Sun, X., and J.Z. Zhang. 2017. Effect of the reinforced leading or trailing edge on the aerodynamic performance of a perimeter-reinforced membrane wing. *Journal of Fluids and Structures* 68: 90–112.
18. Bleischwitz, R. 2016. R.de Kat, and B. Ganapathisubramani, Aeromechanics of membrane and rigid wings in and out of ground-effect at moderate Reynolds numbers. *Journal of Fluids and Structures* 62: 318–331.
19. Bleischwitz, R., R. de Kat, and B. Ganapathisubramani. 2017. On the fluid-structure interaction of flexible membrane wings for MAVs in and out of ground-effect. *Journal of Fluids and Structures* 70: 214–234.
20. Bleischwitz R., R. de Kat and B. Ganapathisubramani. 2018. Near-wake characteristics of rigid and membrane wings in ground effect. *Journal of Fluids and Structures* 80: 199–216.
21. Sun, X., X.L. Ren, and J.Z. Zhang. 2017. Nonlinear dynamic responses of a perimeter-reinforced membrane wing in laminar flows. *Nonlinear Dynamics* 88: 749–776.
22. Sun, X., S.Z. Wang, J.Z. Zhang, and Z.H. Ye. 2018. Bifurcations of vortex-induced vibrations of a fixed membrane wing at Re$\leq$1000. *Nonlinear Dynamics* 91: 2097–2112.
23. Tiomkin, S., and D.E. Raveh. 2019. On membrane-wing stability in laminar flow, Journal of Fluids and Structures. *Journal of Fluids and Structures* 91: 102694.
24. Sun, X., and J.Z. Zhang. 2016. Chap. 3: Finite-element analysis of nonlinear fluid-membrane interactions using a modified characteristic-based split (CBS) scheme. In *Complex motions and chaos*, ed. V. Afraimovich, J.A. Tenreiro Machado, and J.Z. Zhang, 75. Switzerland: Springer.
25. Sun, X., J.Z. Zhang, and G.H. Mei. 2012. An Improved characteristic-based split (CBS) scheme for compressible and incompressible moving boundary flows. *International Journal of Aerospace and Lightweight Structures* 2: 281–297.
26. Sun, X., J.Z. Zhang, and X.L. Ren. 2012. Characteristic-based split (CBS) finite element method for incompressible viscous flow with moving boundaries. *Engineering Applications of Computational Fluid Mechanics* 6: 461–474.

27. Jameson, J. 1991. Time dependent calculations using multigrid with application to unsteady flows past airfoils and wings. *AIAA paper*: 91–1596.
28. Blom, F.J. 2000. Considerations on the spring analogy. *International Journal for Numerical Methods in Fluids* 32: 647–668.
29. Sun, X., W.X. Li, and Z.H. Ye. 2017. Flow-induced vibration of flexible bottom wall in a lid-driven cavity, Journal of Vibration Testing and System. *Dynamics (Pembroke, Ont.)* 1: 333–341.
30. Sun, X., Z.H. Ye, J.J. Li, K. Wen, and H. Tian. 2019. Forced convection heat transfer from a circular cylinder with a flexible fin. *International Journal of Heat and Mass Transfer* 128: 319–334.
31. Ye, Z.H., X. Sun, and J.Z. Zhang. 2019. Flow-induced vibrations of two staggered circular cylinders at low Reynolds number, Journal of Vibration Testing and System. *Dynamics (Pembroke, Ont.)* 3: 39–54.
32. Zienkiewicz, O.C., R.L. Taylor, and P. Nithiarasu. 2013. *The finite element method for fluid dynamics*. 7th ed. Elsevier: Butterworth-Heinemann.
33. Chung, J., and G. Hulbert. 1993. A time integration algorithm for structural dynamics with improved numerical dissipation: the generalized-α method. *Journal of Applied Mechanics-Transactions of the ASME* 60: 371–375.
34. Serrano-Galiano, S., and R.D. Sandberg. 2015. Direct numerical simulations of membrane wings at low Reynolds number. In 53rd AIAA Aerospace Sciences Meeting, 5–9 January 2015, Kissimmee, Florida, USA (AIAA Paper 2015–1300).

Chapter 9
Periodic Motions and Bifurcations in a Double Pendulum

Chuan Guo and Albert C. J. Luo

9.1 Introduction

Linearized equations of the double pendulum system have been used in many decades to present the two-degree-of-freedom system. Recently, numerical method has been used to demonstrate the complex motions in the double pendulum. However, so far, one did not give a comprehensive study on complex motions in such a double pendulum. This is because no any efficient mathematical tools can be used. In 2018, Luo and Guo [1] initially studied period-1 motions to chaos in a periodically forced, damped, double-pendulum through the implicit mapping method. The implicit mapping method is based on the discretization of the differential equations of nonlinear dynamical systems to generate specific implicit mappings. From the discrete mapping, the mapping structures of periodic motions are used to determine the discrete nodes on the periodic motions.

A brief literature survey on investigation of the double-pendulum is presented. In 1966, Herrmann et al. [2] examined the linear stability of the inverted double-pendulum with a partially follower load. In 1967, Roorda and Nemat-Nasser [3] used an energy method to analyze a double-pendulum system with a tangential load. In 1988, Jin and Matsuzaki [4] used the normal form theory to investigate the codimension-2 bifurcations of the partially follower-loaded double pendulum. The further bifurcation analysis of such a double pendulum system can be found

C. Guo · A. C. J. Luo (✉)
Department of Mechanical and Mechatronics Engineering, Southern Illinois University Edwardsville, Edwardsville, IL, USA
e-mail: aluo@siue.edu

J. Zhang (ed.), *Dynamics and Fault Diagnosis of Nonlinear Rotors and Impellers*, Nonlinear Systems and Complexity 34, https://doi.org/10.1007/978-3-030-94301-1_9

in Jin and Matsuzaki [5]. In 1995, Thomsen [6] studied chaotic motions of a elastically restrained double pendulum with non-conservative follower-type loading and linear damping. In 2006, de Paula et al. [7] experimentally investigated chaos in al nonlinear pendulum.

The aforementioned studies were initially researches on nonlinear dynamics of the double-pendulum systems. The results did not show the inherent characteristics of nonlinear dynamics of the double-pendulum. Thus, in 2015, Luo [8] developed the discrete implicit mapping method for periodic motions in nonlinear dynamical systems. This approach can be easily applied to non-polynomial nonlinear systems. In addition, the comprehensive discussion of the implicit mapping method can be referenced to Luo [9]. In 2017, Guo and Luo [10] used the discrete implicit mapping method to investigate nonlinear dynamics of a periodically forced, damped pendulum system. The complicated bifurcation trees of period-1 motions to chaos in the periodically forced pendulum were presented, and the bifurcation scenario with excitation frequency possessed the dual-spiral structures of the stable and unstable periodic motions. In 2018, Luo and Guo [1] presented a bifurcation tree of period-1 motion to chaos in the double-pendulum system.

In this chapter, period-1 to period-4 motions and an independent branch of period-3 motions of a double pendulum system will be studied through the discrete mapping method. The discrete mapping will be developed first. Using such discrete mapping, periodic motions existing in such a double pendulum will be determined. Numerical illustrations of the periodic motions will be presented.

9.2 Mechanical Model

Consider a tangentially loaded double pendulum as shown in Fig. 9.1. There are two particles connected with two massless bars. The lengths of the massless bars and the masses of the two particles can be different. Two following tangential driven forces are applied on each particle. The double pendulum system is described as

$$\begin{aligned}
&(m_1+m_2)\,l_1\ddot{x}_1+m_2l_2\ddot{x}_2\cos(x_1-x_2)+m_2l_2\dot{x}_2^2\sin(x_1-x_2)\\
&+(m_1+m_2)\,g\sin x_1+(\delta_1+\delta_2)\,l_1\dot{x}_1+\delta_2l_2\dot{x}_2\cos(x_1-x_2)\\
&=-Q_1\cos\Omega t-Q_2\cos\Omega t\cos(x_1-x_2)\,,\\
&m_2l_2\ddot{x}_2+m_2l_1\ddot{x}_1\cos(x_1-x_2)-m_2l_1\dot{x}_1^2\sin(x_1-x_2)\\
&+m_2g\sin x_2+\delta_2l_2\dot{x}_2+\delta_2l_1\dot{x}_1\cos(x_1-x_2)\\
&=-Q_2\cos\Omega t
\end{aligned}\tag{9.1}$$

where

m_1, m_2 are the masses of the two particles,
l_1, l_2 are the lengths of the two bars,
δ_1, δ_2 are the damping coefficients of the two particles

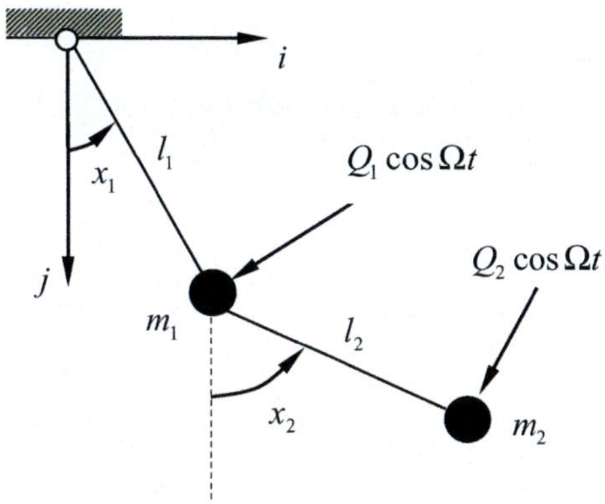

Fig. 9.1 Physical model of a tangentially loaded double pendulum system

$x_1 = x_1(t)$ and $x_2 = x_2(t)$ represent the angular displacements of the first and second masses in the pendulum.

The state space form of Eq. (9.1) is,

$$\begin{aligned}
&\dot{x}_1 = y_1, \quad \dot{x}_2 = y_2, \\
&\dot{y}_1 = -\tfrac{1}{(m_1+m_2)l_1}\big[(\delta_1 + \delta_2)\, l_1 y_1 + \delta_2 l_2 y_2 \cos(x_1 - x_2) + (m_1 + m_2)\, g \sin x_1 \\
&\qquad + m_2 l_2 \dot{y}_2 \cos(x_1 - x_2) + m_2 l_2 y_2^2 \sin(x_1 - x_2) \\
&\qquad + Q_1 \cos \Omega t + Q_2 \cos \Omega t \cos(x_1 - x_2) \big], \\
&\dot{y}_2 = \tfrac{1}{m_2 l_2}\big[- \delta_2 l_2 y_2 - \delta_2 l_1 y_1 \cos(x_1 - x_2) - Q_2 \cos \Omega t \\
&\qquad - m_2 l_1 \dot{y}_1 \cos(x_1 - x_2) + m_2 l_1 y_1^2 \sin(x_1 - x_2) - m_2 g \sin x_2 \big].
\end{aligned} \tag{9.2}$$

Let $\mathbf{x} = (x_1, x_2)^{\mathrm{T}}$ and $\mathbf{y} = (y_1, y_2)^{\mathrm{T}}$. At discrete time $t_k = kh$ ($k = 0, 1, 2, \ldots$), $\mathbf{x}_k = (x_{1k}, x_{2k})^{\mathrm{T}}$ and $\mathbf{y}_k = (y_{1k}, y_{2k})^{\mathrm{T}}$. For the time interval $t \in [t_{k-1}, t_k]$ ($k = 1, 2, \ldots$), the differential equation in Eq. (9.2) are discretized by the midpoint scheme to form an implicit mapP_k:

$$P_k : (\mathbf{x}_{k-1}, \mathbf{y}_{k-1}) \to (\mathbf{x}_k, \mathbf{y}_k) \Rightarrow (\mathbf{x}_k, \mathbf{y}_k) = P_k\,(\mathbf{x}_{k-1}, \mathbf{y}_{k-1})\,. \tag{9.3}$$

The corresponding implicit relations for the implicit map (9.3) are

$$
\begin{aligned}
x_{1k} &= x_{1(k-1)} + \tfrac{1}{2}h(y_{1(k-1)} + y_{1k}),\\
x_{2k} &= x_{2(k-1)} + \tfrac{1}{2}h(y_{2(k-1)} + y_{2k}),\\
y_{1k} &= y_{1(k-1)} - \tfrac{h}{(m_1+m_2)l_1}\{\tfrac{1}{2}(\delta_1+\delta_2)l_1(y_{1(k-1)} + y_{1k})\\
&\quad + \delta_2 l_2 \tfrac{1}{2}\left(y_{2(k-1)} + y_{2k}\right)\cos[\tfrac{1}{2}(x_{1(k-1)} + x_{1k}) - \tfrac{1}{2}(x_{2(k-1)} + x_{2k})]\\
&\quad + m_2 l_2 \tfrac{y_{2k} - y_{2(k-1)}}{h}\cos[\tfrac{1}{2}(x_{1(k-1)} + x_{1k}) - \tfrac{1}{2}(x_{2(k-1)} + x_{2k})]\\
&\quad + m_2 l_2 [\tfrac{1}{2}(y_{2(k-1)} + y_{2k})]^2 \sin[\tfrac{1}{2}(x_{1(k-1)} + x_{1k}) - \tfrac{1}{2}(x_{2(k-1)} + x_{2k})]\\
&\quad + (m_1+m_2)g\sin[\tfrac{1}{2}(x_{1(k-1)} + x_{1k})] + Q_1\cos\Omega(t_{k-1} + \tfrac{1}{2}h)\\
&\quad + Q_2\cos\Omega(t_{k-1} + \tfrac{1}{2}h)\cos[\tfrac{1}{2}(x_{1(k-1)} + x_{1k}) - \tfrac{1}{2}(x_{2(k-1)} + x_{2k})]\},\\
y_{2k} &= y_{2(k-1)} - \tfrac{h}{m_2 l_2}\{\delta_2 l_2 \tfrac{1}{2}(y_{2(k-1)} + y_{2k})\\
&\quad + \delta_2 l_1 \tfrac{1}{2}(y_{1(k-1)} + y_{1k})\cos[\tfrac{1}{2}(x_{1(k-1)} + x_{1k}) - \tfrac{1}{2}(x_{2(k-1)} + x_{2k})]\\
&\quad + m_2 l_1 \tfrac{y_{1k} - y_{1(k-1)}}{h}\cos[\tfrac{1}{2}(x_{1(k-1)} + x_{1k}) - \tfrac{1}{2}(x_{2(k-1)} + x_{2k})]\\
&\quad - m_2 l_1 [\tfrac{1}{2}(y_{1(k-1)} + y_{1k})]^2 \sin[\tfrac{1}{2}(x_{1(k-1)} + x_{1k}) - \tfrac{1}{2}(x_{2(k-1)} + x_{2k})]\\
&\quad + m_2 g\sin[\tfrac{1}{2}(x_{2(k-1)} + x_{2k})] + Q_2\cos\Omega(t_{k-1} + \tfrac{1}{2}h)\}.
\end{aligned}
\tag{9.4}
$$

9.3 Periodic Motions and Stability

Assuming a periodic motion exists in the tangentially loaded double pendulum. The periodic motion can be discretized into mN points. A discrete mapping structure of a period-m motion is constructed as

$$
\begin{aligned}
&P = \underbrace{P_{mN} \circ P_{mN-1} \circ \cdots \circ P_2 \circ P_1}_{N-\text{actions}} : (\mathbf{x}_0, \mathbf{y}_0) \to (\mathbf{x}_{mN}, \mathbf{y}_{mN}),\\
&(\mathbf{x}_{mN}, \mathbf{y}_{mN}) = P(\mathbf{x}_0, \mathbf{y}_0).
\end{aligned}
\tag{9.5}
$$

with

$$
P_k : (\mathbf{x}_{k-1}, \mathbf{y}_k) \to (\mathbf{x}_k, \mathbf{y}_k)\ (k = 1, 2, \cdots, mN). \tag{9.6}
$$

The set of discrete points $\mathbf{x}_k^*$ ($k = 1, 2, \cdots, mN$) on the periodic motion can be obtained by

$$
\begin{aligned}
&\mathbf{g}_k\left(\mathbf{x}_{k-1}^*, \mathbf{x}_k^*; \mathbf{y}_{k-1}^*, \mathbf{y}_k^*, \mathbf{p}\right) = \mathbf{0}, (k = 1, 2, \cdots, mN),\\
&\mathbf{x}_0^* = \mathbf{x}_{mN}^* + [2k_1\pi, 2k_2\pi] \text{ and } \mathbf{y}_0^* = \mathbf{y}_{mN}^*, (k_1, k_2 \in \mathbf{Z}).
\end{aligned}
\tag{9.7}
$$

From Eq.(9.4), the algebraic equations for periodic motion can be obtained. i.e., for P_k, $\mathbf{g}_k = (g_{k1}, g_{k2}, g_{k3}, g_{k4})^{\mathrm{T}} = \mathbf{0}$ gives

$$
\begin{aligned}
g_{k1} &= x_{1k} - x_{1(k-1)} - \tfrac{1}{2}h(y_{1(k-1)} + y_{1k}) \\
&= 0, \\
g_{k2} &= x_{2k} - x_{2(k-1)} - \tfrac{1}{2}h(y_{2(k-1)} + y_{2k}) \\
&= 0, \\
g_{k3} &= y_{1k} - y_{1(k-1)} + \tfrac{h}{(m_1+m_2)l_1}\{\tfrac{1}{2}(\delta_1 + \delta_2)\, l_1(y_{1(k-1)} + y_{1k}) \\
&\quad + \tfrac{1}{2}\delta_2 l_2(y_{2(k-1)} + y_{2k})\cos[\tfrac{1}{2}(x_{1(k-1)} + x_{1k}) - \tfrac{1}{2}(x_{2(k-1)} + x_{2k})] \\
&\quad + \tfrac{1}{h}m_2 l_2(y_{2k} - y_{2(k-1)})\cos[\tfrac{1}{2}(x_{1(k-1)} + x_{1k}) - \tfrac{1}{2}(x_{2(k-1)} + x_{2k})] \\
&\quad + m_2 l_2(\tfrac{1}{2}(y_{2(k-1)} + y_{2k}))^2 \sin[\tfrac{1}{2}(x_{1(k-1)} + x_{1k}) - \tfrac{1}{2}(x_{2(k-1)} + x_{2k})] \\
&\quad + (m_1 + m_2)\, g \sin[\tfrac{1}{2}(x_{1(k-1)} + x_{1k})] + Q_1 \cos\Omega(t_{k-1} + \tfrac{1}{2}h) \\
&\quad + Q_2 \cos\Omega(t_{k-1} + \tfrac{1}{2}h)\cos[\tfrac{1}{2}(x_{1(k-1)} + x_{1k}) - \tfrac{1}{2}(x_{2(k-1)} + x_{2k})]\} \\
&= 0, \\
g_{k4} &= y_{2k} - y_{2(k-1)} + \tfrac{h}{m_2 l_2}\{\tfrac{1}{2}\delta_2 l_2(y_{2(k-1)} + y_{2k}) \\
&\quad + \tfrac{1}{2}\delta_2 l_1(y_{1(k-1)} + y_{1k})\cos[\tfrac{1}{2}(x_{1(k-1)} + x_{1k}) - \tfrac{1}{2}(x_{2(k-1)} + x_{2k})] \\
&\quad + \tfrac{1}{h}m_2 l_1(y_{1k} - y_{1(k-1)})\cos[\tfrac{1}{2}(x_{1(k-1)} + x_{1k}) - \tfrac{1}{2}(x_{2(k-1)} + x_{2k})] \\
&\quad - m_2 l_1[\tfrac{1}{2}(y_{1(k-1)} + y_{1k})]^2 \sin[\tfrac{1}{2}(x_{1(k-1)} + x_{1k}) - \tfrac{1}{2}(x_{2(k-1)} + x_{2k})] \\
&\quad + m_2 g \sin[\tfrac{1}{2}(x_{2(k-1)} + x_{2k})] + Q_2 \cos\Omega(t_{k-1} + \tfrac{1}{2}h)\} \\
&= 0. \\
&\quad (k = 1, 2, \cdots, mN)
\end{aligned}
\tag{9.8}
$$

From Eqs. (9.7)–(9.8), discrete nodes of periodic motions in the double pendulum can be determined by $4(mN + 1)$ equations. If the discrete nodes $(\mathbf{x}_k^*, \mathbf{y}_k^*)$ $(k = 1, 2, \cdots, mN)$ of the period-m motion is obtained, the corresponding stability of the periodic motion can be discussed by the eigenvalue analysis of the implicit mappings of the periodic motion. Consider the vicinity of $\mathbf{x}_k^*$ and $\mathbf{y}_k^*$. With $\mathbf{x}_k = \mathbf{x}_k^* + \Delta\mathbf{x}_k$ and $\mathbf{y}_k = \mathbf{y}_k^* + \Delta\mathbf{y}_k$, the linearized equation of $\mathbf{g}_k = \mathbf{0}$ is

$$
\sum_{j=k-1}^{k} \frac{\partial \mathbf{g}_k}{\partial \mathbf{x}_j}\Delta\mathbf{x}_j + \frac{\partial \mathbf{g}_k}{\partial \mathbf{y}_j}\Delta\mathbf{y}_j = \mathbf{0}, \text{ with } j = k-1, k; \ (k = 1, 2, \cdots, mN). \tag{9.9}
$$

Define a vector **z** as

$$\begin{aligned}\mathbf{z}_k &= (\mathbf{x}_k, \mathbf{y}_k)^{\mathrm{T}},\\ \mathbf{z}_{k-1} &= (\mathbf{x}_{k-1}, \mathbf{y}_{k-1})^{\mathrm{T}},\\ \Delta\mathbf{z}_k &= (\Delta\mathbf{x}_k, \Delta\mathbf{y}_k)^{\mathrm{T}},\\ \Delta\mathbf{z}_{k-1} &= (\Delta\mathbf{x}_{k-1}, \Delta\mathbf{y}_{k-1})^{\mathrm{T}}.\end{aligned} \tag{9.10}$$

The resultant Jacobian matrices of the periodic motion in the periodic forced, dumped double pendulum are

$$DP = DP_{mN(mN-1)\cdots 1} = \left[\frac{\partial \mathbf{z}_{mN}}{\partial \mathbf{z}_0}\right]_{(\mathbf{z}_0^*, \mathbf{z}_1^*, \ \cdots, \mathbf{z}_{mN}^*)} \tag{9.11}$$

and

$$\begin{aligned}&\Delta\mathbf{z}_{mN} = \mathbf{A}\Delta\mathbf{z}_0 = \mathbf{A}_{mN}\mathbf{A}_{mN-1}\cdots\mathbf{A}_1\Delta\mathbf{z}_0,\\ &\Delta\mathbf{z}_k = \mathbf{A}_k\Delta\mathbf{z}_{k-1}, \quad \mathbf{A}_k \equiv \left[\frac{\partial \mathbf{z}_k}{\partial \mathbf{z}_{k-1}}\right]_{(\mathbf{z}_{k-1}^*, \mathbf{z}_k^*)}\end{aligned} \tag{9.12}$$

with

$$\mathbf{A}_k = -\left[\frac{\partial \mathbf{g}_k}{\partial \mathbf{z}_k}\right]^{-1}\left[\frac{\partial \mathbf{g}_k}{\partial \mathbf{z}_{k-1}}\right]. \tag{9.13}$$

The eigenvalues of *DP* matrices for the periodic motions in the tangentially loaded double pendulum are computed by

$$| \, DP - \lambda\mathbf{I}_{4\times 4} \, | = 0. \tag{9.14}$$

1. If the magnitudes of all eigenvalues of *DP* are less than one (i.e., $|\lambda_i| < 1$, $i = 1$, 2, 3, 4), the approximate periodic solution is stable.
2. If the magnitude of at least one eigenvalue of *DP* is greater than one (i.e., $|\lambda_i| > 1$, $i \in \{1, 2, 3, 4\}$), the approximate periodic solution is unstable.
3. The boundaries between stable and unstable periodic flow with higher order singularity give bifurcation and stability conditions with higher order singularity.

The bifurcation conditions of periodic motion are presented as follow [10]

1. If one eigenvalue is equal to 1 and the other eigenvalues are within the unit circle, the saddle-node bifurcation (SN) occurs.
2. If one eigenvalue is equal to -1 and the other eigenvalues are within the unit circle, the period-doubling bifurcation (PD) occurs.
3. If the magnitude of a pair of complex eigenvalues is equal to 1, and the other eigenvalues are within the unit circle, Neimark bifurcation (NB) occurs.

From the node points $\mathbf{z}_j^{(m)} = \left(\mathbf{x}_j^{(m)}, \mathbf{y}_j^{(m)}\right)^{\mathrm{T}}$ ($j = 0, 1, 2, \ldots, mN$) of period-*m* motion, the corresponding analytical expression of the period-*m* motion is

$$\mathbf{z}^{(m)}(t) = \mathbf{a}_0^{(m)} + \sum_{k=1}^{mN/2} \mathbf{b}_{k/m} \cos\left(\frac{k}{m}\Omega t\right) + \mathbf{c}_{j/m} \sin\left(\frac{k}{m}\Omega t\right) \tag{9.15}$$

where

$$\begin{aligned}
\mathbf{a}_0^{(m)} &= \frac{1}{mN} \sum_{j=0}^{mN} \mathbf{z}_j^{(m)}, \\
\mathbf{b}_{k/m} &= \frac{2}{mN} \sum_{j=0}^{mN} \mathbf{z}_j^{(m)} \cos\left(\frac{k}{m}\frac{2\pi j}{N}\right), \\
\mathbf{c}_{k/m} &= \frac{2}{mN} \sum_{j=0}^{mN} \mathbf{z}_j^{(m)} \sin\left(\frac{k}{m}\frac{2\pi j}{N}\right), \\
&(k = 1, 2, \ldots mN/2).
\end{aligned} \tag{9.16}$$

The harmonic amplitudes and phases for a period-m motion of a periodically forced, dumped double pendulum can be expressed for $(i = 1, 2, 3, 4)$ as

$$\begin{aligned}
A_{i,k/m} &= \sqrt{\left(b_{i,k/m}\right)^2 + \left(c_{i,k/m}\right)^2}, \\
\varphi_{i,k/m} &= \arctan \frac{c_{i,k/m}}{b_{i,k/m}}.
\end{aligned} \tag{9.17}$$

Then the periodic solution of the periodic forced, dumped double pendulum system can be rewritten as

$$\begin{Bmatrix} x_1^{(m)} \\ x_2^{(m)} \\ y_1^{(m)} \\ y_2^{(m)} \end{Bmatrix} \approx \begin{Bmatrix} a_{1,0}^{(m)} \\ a_{2,0}^{(m)} \\ a_{3,0}^{(m)} \\ a_{4,0}^{(m)} \end{Bmatrix} + \sum_{k=1}^{mN/2} \begin{Bmatrix} A_{1,k/m} \cos\left(\frac{k}{m}\Omega t - \varphi_{1,k/m}\right) \\ A_{2,k/m} \cos\left(\frac{k}{m}\Omega t - \varphi_{2,k/m}\right) \\ A_{3,k/m} \cos\left(\frac{k}{m}\Omega t - \varphi_{3,k/m}\right) \\ A_{4,k/m} \cos\left(\frac{k}{m}\Omega t - \varphi_{4,k/m}\right) \end{Bmatrix} \tag{9.18}$$

Based on the Fourier series expression of the periodic motions, the frequency-amplitude characteristics of period-m motion can be obtained.

9.4 Period-1 to Period-4 Motions

The bifurcation diagram of period-1 to period-4 motions in a tangentially loaded double pendulum will be presented in Fig. 9.2. The corresponding stability analysis of the periodic motions is carried out through eigenvalue analysis. Without loss of generality, a set of arbitrary parameters is chosen as

$$\begin{aligned}
&m_1 = 1.0, \ \ m_2 = 0.2, \ \ l_1 = 1.0, \ \ l_2 = 1.0, \\
&d_1 = 0.8, d_2 = 0.5, Q_1 = 0, Q = 10.
\end{aligned} \tag{9.19}$$

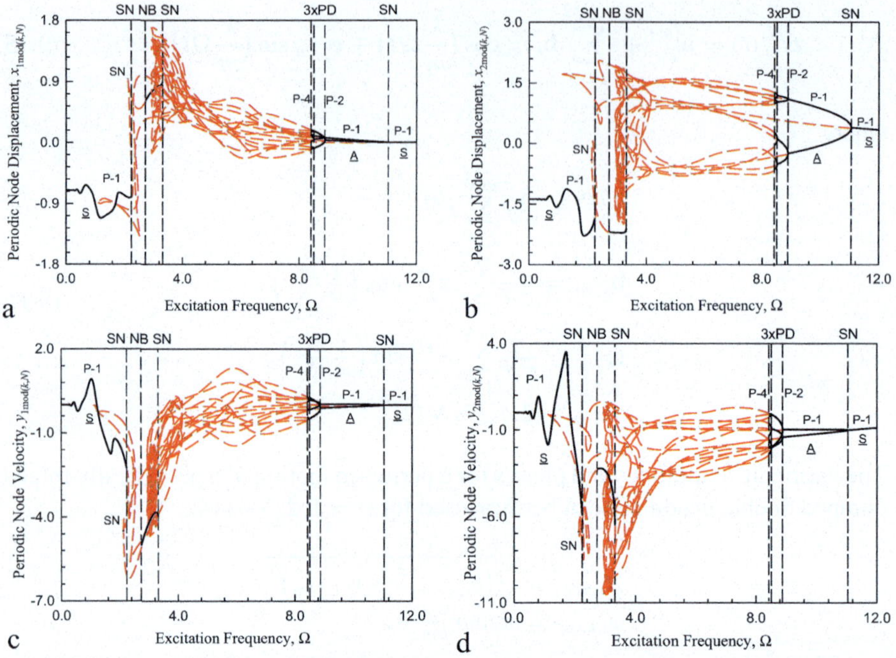

Fig. 9.2 Discrete nodes of periodic motions (mod$(k, N) = 0$) varying with excitation frequency: (**a**) Displacement $x_{1\,\mathrm{mod}\,(k,N)}$; (**b**) Displacement $x_{2\,\mathrm{mod}\,(k,N)}$; (**c**) Velocity $y_{1\,\mathrm{mod}\,(k,N)}$ and (**d**) Velocity $y_{2\,\mathrm{mod}\,(k,N)}$. Parameters ($m_1 = 1.0$, $m_2 = 0.2$, $l_1 = 1.0$, $l_2 = 1.0$, $d_1 = 0.8$, $d_2 = 0.5$, $Q_1 = 0$, $Q_2 = 10$)

The angular displacements ($x_{1\,\mathrm{mod}\,(k,N)}$, $x_{2\,\mathrm{mod}\,(k,N)}$) and the angular velocities ($y_{1\,\mathrm{mod}\,(k,N)}$, $y_{2\,\mathrm{mod}\,(k,N)}$) of the periodic nodes for $\Omega \in (0, 12)$ are depicted in Fig. 9.2(a) and (b), respectively. The index of the plotted periodic nodes satisfies mod$(k, N) = 0$. The periodic motions for $\Omega > 12$ are simple stable period-1 motions. The stable and unstable periodic motions are presented by solid and dash curves, respectively. The acronyms 'SN', 'PD' and 'NB' are for saddle-node, period-doubling, and Neimark bifurcations, respectively.

The stable and unstable symmetric period-1 motions existed in $\Omega \in (0, 12)$. The asymmetric period-1 motion occurs at $\Omega = 11.05$ and only exists in the frequency range $\Omega \in (2.93, 11.05)$. There are a pair of asymmetric period-1 motions, the stable asymmetric period-1 motions exist in $\Omega \in (8.89, 11.05)$ while the unstable asymmetric period-1 motions exist in $\Omega \in (2.93, 8.89)$. At $\Omega = 8.89$, there is a period-doubling bifurcation. The asymmetric period-2 motions appear from the first period-doubling bifurcation and exist in $\Omega \in (3.6451, 8.89)$. The stable and unstable period-2 motions apart at $\Omega = 8.52$, which is another period-doubling bifurcation from period-2 to period-4 motion. The asymmetric period-4 motions occur at $\Omega = 8.52$ and merge back to the unstable period-2 motion at $\Omega = 4.1363$.

In order to verify the semi-analytical solutions of periodic motions, numerical simulations of periodic motions are completed by the midpoint integration method. The initial conditions for numerical solutions are obtained from the semi-analytical solutions. The phase trajectories and harmonic amplitudes of periodic motions will be presented. In Figs. 9.3–9.5, circular symbols and solid curves represented analytical and numerical solutions, respectively. The acronym "I.C." is the initial condition denoted as a circular symbol.

A pair of stable asymmetric period-1 motions is presented for $\Omega = 9$ in Fig. 9.3. The initial conditions for one branch (black) are $(x_{1(0)}, y_{1(0)}) \approx (0.0653, -0.1039)$ and $(x_{2(0)}, y_{2(0)}) \approx (1.0735, -0.9808)$. The initial conditions for another branch (red) are $(x_{1(0)}, y_{1(0)}) \approx (0.0308, 0.0296)$ and $(x_{2(0)}, y_{2(0)}) \approx (-0.2278, -1.3959)$. The trajectories of the two paired stable asymmetric period-1 motions for the first and second pendulums are presented in Fig. 9.3(a)-(d), respectively. The period-1 motion is one closed cycle. The harmonic amplitudes and phases for the displacement of the first pendulum is presented in Fig. 9.3(e) and (f). The constant terms are $A_{1,0} = a^B_{1,0} = -a^R_{1,0} = 9.9034 \times 10^{-3}$. The main harmonic amplitudes for the first pendulum are $A_{1,1} \approx 0.0484$, and $A_{1,2} \approx 8.3421 \times 10^{-3}$. The other amplitudes for the first pendulum are $A_{1,k} \in (10^{-10}, 10^{-3})$ $(j = 3, 4, \cdots, 12)$ with $A_{1,12} = 3.7249 \times 10^{-11}$. The two harmonic phases for red and black asymmetric motions are $\varphi^B_{1,k} = \text{mod}(\varphi^R_{1,k} + (k+1)\pi, 2\pi)$. The harmonic amplitudes and phases for the displacement of the second pendulum is presented in Fig. 9.3(g) and (h). The constant terms are $A_{2,0} = a^B_{2,0} = -a^R_{2,0} = 0.6741$.The main harmonic amplitudes for the second pendulum are $A_{2,1} \approx 0.4435$, and $A_{2,2} \approx 0.0264$. The other amplitudes for the first pendulum are $A_{2,k} \in (10^{-10}, 10^{-3})$ $(j = 3, 4, \cdots, 12)$ with $A_{2,12} = 8.4828 \times 10^{-11}$. The two harmonic phases for red and black asymmetric motions also satisfy $\varphi^B_{2,k} = \text{mod}(\varphi^R_{2,k} + (k+1)\pi, 2\pi)$. The paired asymmetric period-1 motions need about 12 harmonic terms in the finite Fourier series for an approximate analytical expression with accuracy of $\varepsilon = 10^{-11}$.

A pair of stable asymmetric period-2 motions is presented for $\Omega = 8.7$ in Fig. 9.4. The initial conditions for one branch (black) are $(x_{1(0)}, y_{1(0)}) \approx (0.0545, -0.3711)$ and $(x_{2(0)}, y_{2(0)}) \approx (1.1584, -0.3164)$. The initial conditions for another branch (red) are $(x_{1(0)}, y_{1(0)}) \approx (0.1261, -0.1020)$ and $(x_{2(0)}, y_{2(0)}) \approx (-0.4500, -1.0399)$. The trajectories of the first and second pendulums or the two period-2 motion presented in Fig. 9.4(a)-(d), respectively. The period-1 motion has two closed cycles in phase plane. The harmonic amplitudes and phases for the displacement of the first pendulum is presented in Fig. 9.4(e) and (f). The constant terms are $A_{1,0/2} = a^{(2)B}_{1,0} = -a^{(2)R}_{1,0} = 8.9628 \times 10^{-3}$. The main harmonic amplitudes for the first pendulum are $A_{1,1/2} \approx 0.08416$, $A_{1,1} \approx 0.05439$, and $A_{1,3/2} \approx 8.0570 \times 10^{-3}$. The other amplitudes for the first pendulum are $A_{1,k/2} \in (10^{-10}, 10^{-3})$ $(j = 4, 5, \cdots, 30)$ with $A_{1,15} = 2.7028 \times 10^{-12}$. The two harmonic phases are $\varphi^B_{1,k/2} = \text{mod}(\varphi^R_{1,k/2} + (k/2+1)\pi, 2\pi)$ for the red and black asymmetric period-2 motions. The harmonic amplitudes and phases for the displacement of the second pendulum is presented in Fig. 9.4(g) and (h). The constant terms are $A_{2,0} = a^{(2)B}_{2,0} = -a^{(2)R}_{2,0} = 0.6862$. The main harmonic amplitudes for the second pendulum are $A_{2,1/2} \approx 0.2235$,

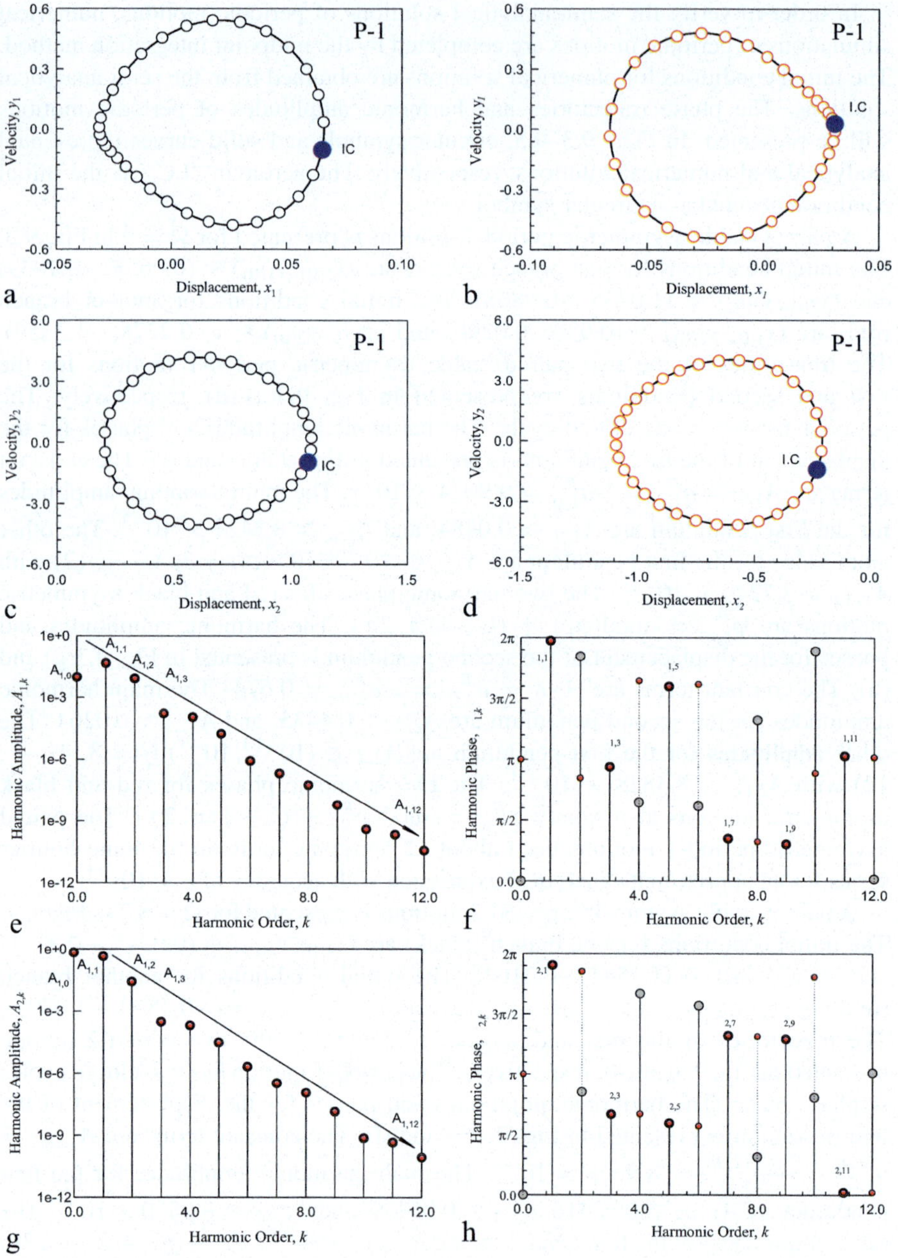

Fig. 9.3 A pair of stable asymmetric period-1 motions ($\Omega = 9$): (**a, b**) trajectory of (x_1, y_1), (**c, d**) trajectory of (x_2, y_2), (**e**) harmonic amplitude $A_{1,k}$, (**f**) harmonic amplitude $A_{2,k}$, (**g**) harmonic phase $\varphi_{1,k}$, (**h**) harmonic phase $\varphi_{2,k}$

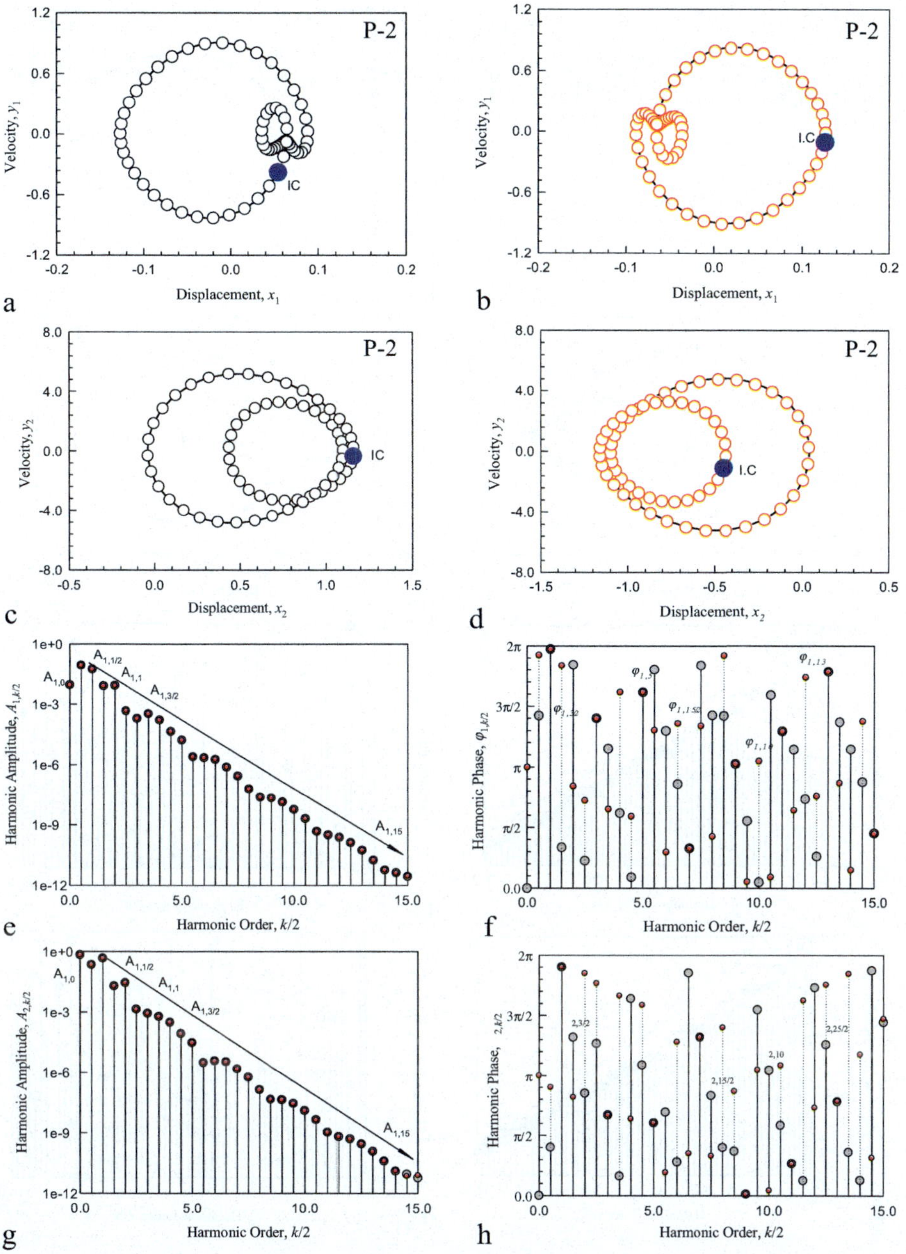

Fig. 9.4 A pair of stable asymmetric period-2 motions ($\Omega = 8.7$): (**a**, **b**) trajectory of (x_1, y_1), (**c**, **d**) trajectory of (x_2, y_2), (**e**) harmonic amplitude $A_{1,\,k/2}$, (**f**) harmonic phase $\varphi_{1,\,k/2}$, (**g**) harmonic amplitude $A_{2,\,k/2}$, (**h**) harmonic phase $\varphi_{2,\,k/2}$

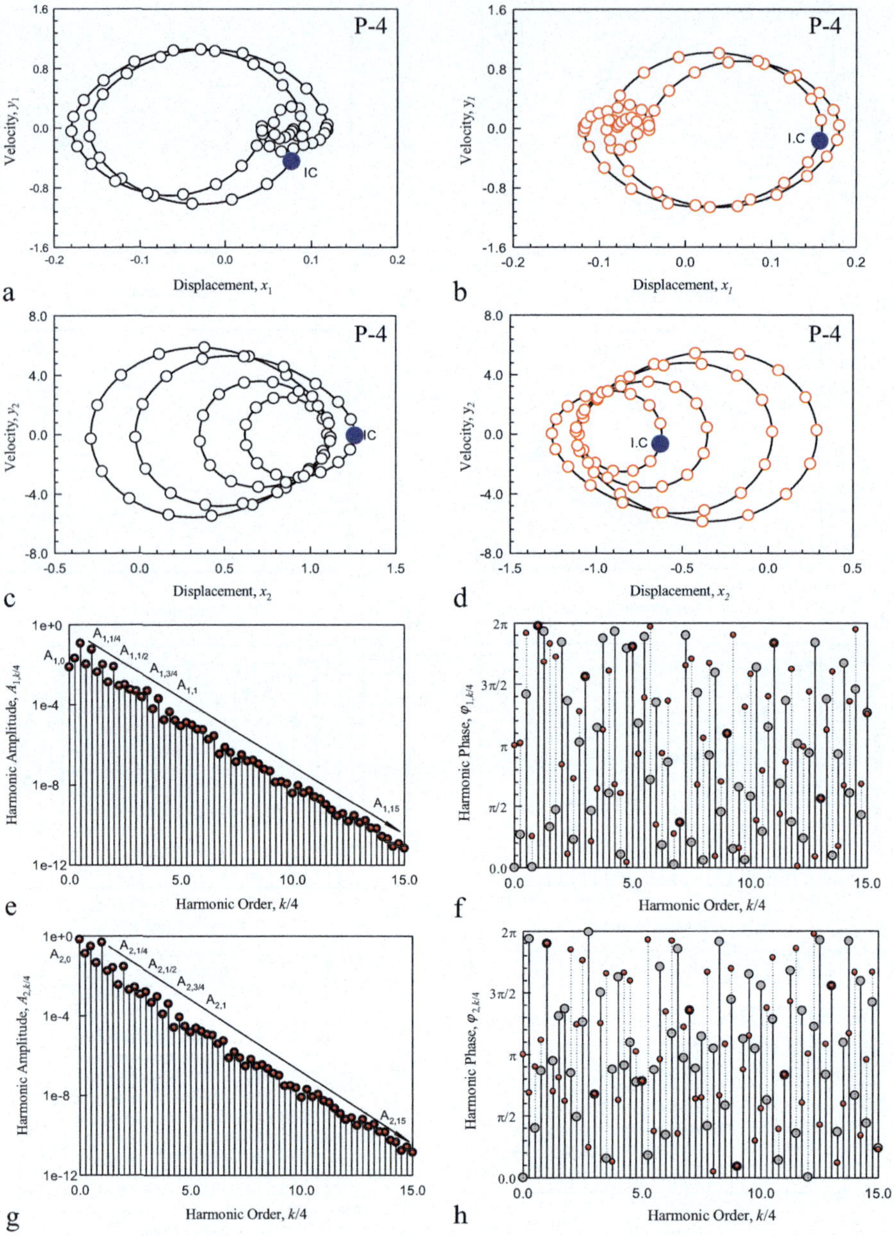

Fig. 9.5 A pair of stable asymmetric period-4 motions ($\Omega = 8.45$): (**a**, **b**) trajectory of (x_1, y_1), (**c**, **d**) trajectory of (x_2, y_2), (**e**) harmonic amplitude $A_{1,\,k/4}$, (**f**) harmonic phase $\varphi_{1,\,k/4}$, (**g**) harmonic amplitude $A_{2,\,k/4}$, (**h**) harmonic phase $\varphi_{2,\,k/4}$

$A_{2,1} \approx 0.4624$, $A_{2,3/2} \approx 0.0195$, $A_{2,2} \approx 0.0280$, and $A_{2,5/2} \approx 1.3726 \times 10^{-3}$. The other amplitudes for the first pendulum are $A_{2,k/2} \in (10^{-12}, 10^{-3})$ $(j = 6, 7, \cdots, 30)$ with $A_{2,15} = 5.7584 \times 10^{-12}$. The two harmonic phases for the red and black asymmetric period-2 motions also satisfy $\varphi^B_{2,k/2} = \text{mod}(\varphi^R_{2,k/2} + (k/2+1)\pi, 2\pi)$. The paired asymmetric period-2 motions need about 30 harmonic terms in the finite Fourier series for an approximate analytical expression with accuracy of $\varepsilon = 10^{-12}$.

A pair of stable asymmetric period-4 motions is presented for $\Omega = 8.45$ in Fig. 9.5. The initial conditions for one branch (black) are $(x_{1(0)}, y_{1(0)}) \approx (0.0767, -0.4447)$ and $(x_{2(0)}, y_{2(0)}) \approx (1.2587, -0.0592)$. The initial conditions for another branch (red) are $(x_{1(0)}, y_{1(0)}) \approx (0.1570, -0.1682)$ and $(x_{2(0)}, y_{2(0)}) \approx (-0.6274, -0.6562)$. The trajectories of the first and second pendulum for such a stable asymmetric period-4 motion are presented in Fig. 9.5(a)-(d), respectively. The phase trajectories are from the doubled responses of period-2 motion. The phase trajectories for the first pendulum have four cycles including two larges and two small cycles. For the second pendulum, phase trajectories have four large cycles. The harmonic amplitudes and phases for the first pendulum are presented in Fig. 9.5(e) and (f). The constant terms are $A_{1,0/4} = a^{(4)B}_{1,0} = -a^{(4)R}_{1,0} = 7.8905 \times 10^{-3}$. The main harmonic amplitudes for the first pendulum are $A_{1,1/4} \approx 0.0215$, $A_{1,1/2} \approx 0.1197$, $A_{1,3/4} \approx 0.0108$ $A_{1,1} \approx 0.0586$, $A_{1,5/4} \approx 4.3777 \times 10^{-3}$, $A_{1,3/2} \approx 0.0105$. $A_{1,7/4} \approx 1.3254 \times 10^{-3}$, and $A_{1,2} \approx 8.2503 \times 10^{-3}$. The other amplitudes for the first pendulum are $A_{1,k/4} \in (10^{-12}, 10^{-3})$ $(j = 9, 10, \cdots, 60)$ with $A_{1,15} = 6.7008 \times 10^{-12}$. The two harmonic phases for the red and black asymmetric period-4 motions are $\varphi^B_{1,k/4} = \text{mod}(\varphi^R_{1,k/4} + (k/4+1)\pi, 2\pi)$. The harmonic amplitudes and phases for the displacement of the second pendulum is presented in Fig. 9.5(g) and (h). The constant terms are $A_{2,0} = a^{(4)B}_{2,0} = -a^{(4)R}_{2,0} \approx 0.6776$. The main harmonic amplitudes for the second pendulum are $A_{2,1/4} \approx 0.1301$, $A_{2,1/2} \approx 0.3039$, $A_{2,3/4} \approx 0.0447$, $A_{2,1} \approx 0.4842$, $A_{2,5/4} \approx 0.0172$, $A_{2,3/2} \approx 0.0253$, $A_{2,7/4} \approx 3.4936 \times 10^{-3}$, $A_{2,2} \approx 0.0286$, $A_{2,9/4} \approx 1.9699 \times 10^{-3}$, and $A_{2,5/2} \approx 2.6240 \times 10^{-3}$. The other amplitudes for the second pendulum are $A_{2,k/4} \in (10^{-12}, 10^{-3})$ $(j = 10, 11, \cdots, 60)$ with $A_{2,15} = 1.41376 \times 10^{-11}$. The two harmonic phases for the red and black asymmetric period-4 motions also satisfy $\varphi^B_{2,k/4} = \text{mod}(\varphi^R_{2,k/4} + (k/4+1)\pi, 2\pi)$. The paired asymmetric period-4 motions need about 60 harmonic terms in the finite Fourier series for an approximate analytical expression with accuracy of $\varepsilon = 10^{-12}$.

9.5 Period-3 Motion

The bifurcation diagram of the period-3 motions in a tangentially loaded double pendulum system is presented in Fig. 9.6. The corresponding stability analysis of the period-3 motions is carried out through eigenvalue analysis. The parameters are chosen as

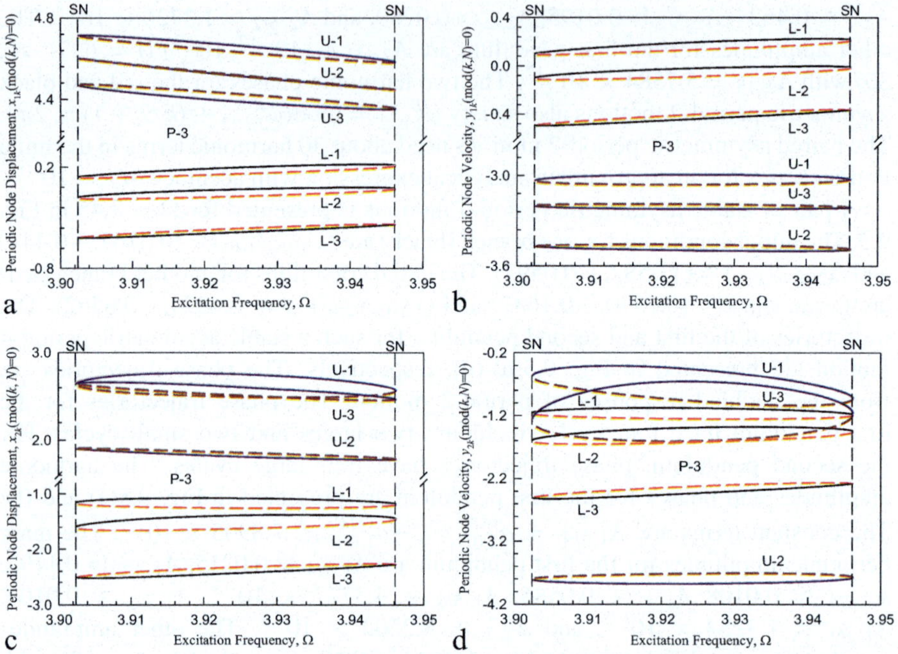

Fig. 9.6 Discrete nodes of period-3 motions (mod$(k, N) = 0$) varying with excitation frequency: (**a**) displacement $x_{1\,mod\,(k,N)}$ (**b**) velocity $y_{1\,mod\,(k,N)}$, (**c**) displacement $x_{2\,mod\,(k,N)}$ and (**d**) velocity $y_{2\,mod\,(k,N)}$. U-1, U-2, and U-3 are for three points ($t = 0, 1T, 2T$) of the upper-period-3 motion. L-1, L-2 and L-3 are for three points ($t = 0, 1T, 2T$) of the lower-period-3 motion. ($m_1 = 1.0, m_2 = 0.1$, $l_1 = 1.0,\ \ l_2 = 1.0,\ \ d_1 = 1.5, d_2 = 0.8, Q_1 = 50, Q_2 = 0$)

$$\begin{aligned} &m_1 = 1.0,\ \ m_2 = 0.1,\ \ l_1 = 1.0,\ \ l_2 = 1.0,\\ &d_1 = 1.5, d_2 = 0.8, Q_1 = 50.0, Q_2 = 0. \end{aligned} \tag{9.20}$$

The angular displacement $x_{1\,mod\,(k,N)}$, $x_{2\,mod\,(k,N)}$ and the angular velocity $y_{1\,mod\,(k,N)}$, $y_{2\,mod\,(k,N)}$ are depicted in Fig. 9.6(a)–(d), respectively. The black solid lines represent the stable period-3 motions, from $\Omega = 3.9024$ to $\Omega = 3.9461$. The red dash lines represent the unstable period-3 motions. At $\Omega = 3.9024$ and $\Omega = 3.9461$, there are saddle node bifurcations at where the period-3 motions vanish. The acronyms 'SN' is for saddle node bifurcation.

The numerical simulations of the asymmetric period-3 motions are also completed for verification with semi-analytical results in such a periodically forced double pendulum. The numerical simulations are carried out by the middle-point scheme. The initial conditions are from analytical predictions.

A pair of asymmetric stable period-3 motions is presented in Fig. 9.7 for $\Omega = 3.931$. The initial conditions of the upper period-3 motions are $(x_{10}, y_{10}) \approx (4.3581, -3.2150)$ and $(x_{20}, y_{20}) \approx (2.5314, -0.6836)$. The initial conditions of the lower period-3 motions are $(x_{1,0}, x_{2,0}) = (-1.0694\text{e}-3, -4.6479)$,

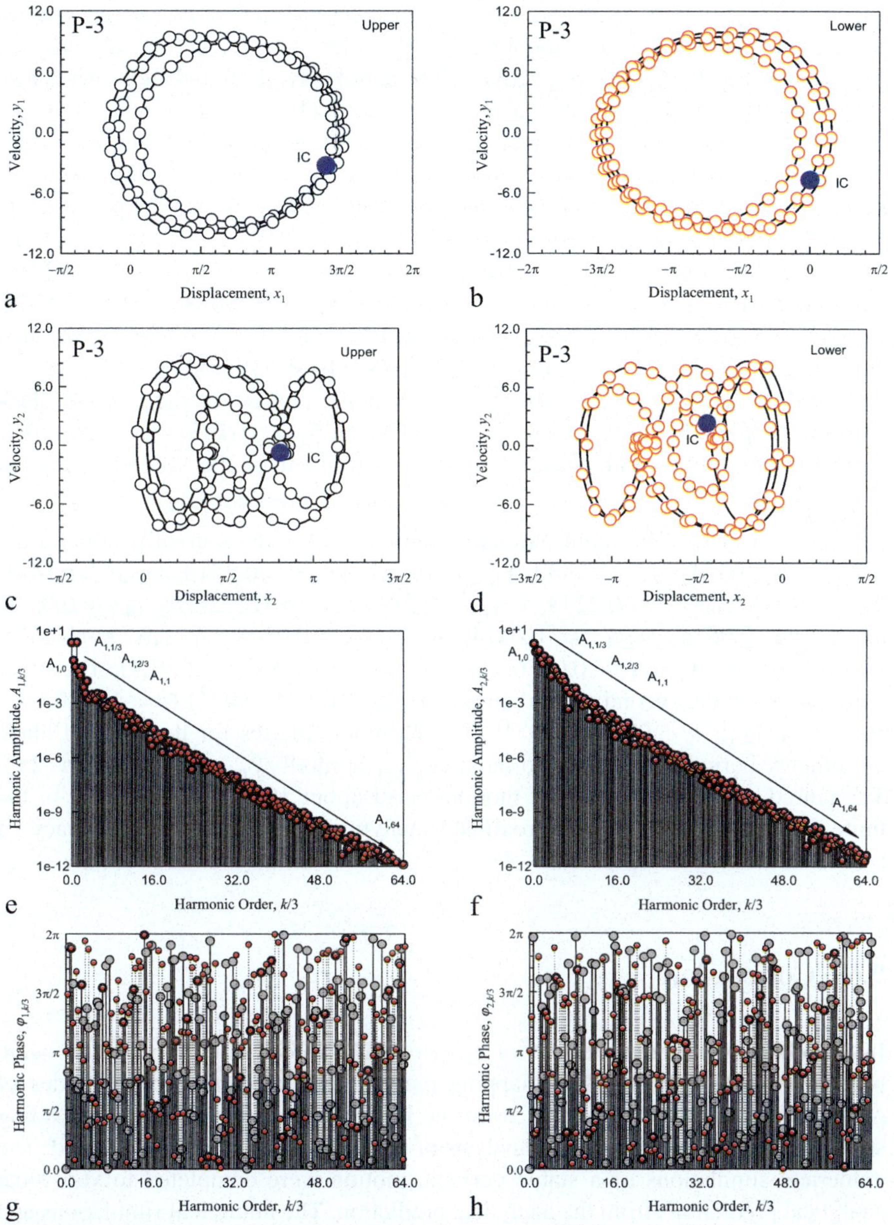

Fig. 9.7 Stable asymmetric period-3 motion ($\Omega = 3.94$): (**a**, **c**) trajectories of (x_1, y_1), (**b**, **d**) trajectories of (x_2, y_2), (**e**) harmonic amplitude $A_{1,\,k/3}$, (**f**) harmonic phase $\varphi_{1,\,k/3}$, (**g**) harmonic amplitude $A_{2,\,k/3}$, (**h**) harmonic phase $\varphi_{2,\,k/3}$

and $(y_{1,0}, y_{2,0}) = (-1.3872, 2.4155)$. The phase trajectories of the paired asymmetric stable period-3 motions for the first and second pendulums are presented in Fig. 9.7(a)-(d), respectively. The circular symbols and slide curves also are for the numerical and analytical results, respectively. The blue circular symbols show the initial conditions. The phase trajectories for the first pendulum have three cycles. For the second pendulum, phase trajectories have two large cycles with five knots. The harmonic amplitudes and phases for the first pendulum are presented in Fig. 9.7(e) and (f). The constant terms are $A_{1,0/3} = a_{1,0}^{(3)B} = -a_{1,0}^{(3)R} = 2.3200$. The main harmonic amplitudes for the first pendulum are $A_{1,1/3} \approx 0.2361$, $A_{1,2/3} \approx 0.1025$, $A_{1,1} \approx 2.3969$, $A_{1,4/3} \approx 0.0559$, $A_{1,5/3} \approx 0.0338$, $A_{1,2} \approx 0.0972$, $A_{1,7/3} \approx 7.5725 \times 10^{-3}$, $A_{1,8/3} \approx 3.4362 \times 10^{-3}$, and $A_{1,3} \approx 0.0165$. The other amplitudes for the first pendulum are $A_{1,k/3} \in (10^{-12}, 10^{-3})$ $(j = 10, 11, \cdots, 192)$ with $A_{1,64} \approx 1.2412 \times 10^{-12}$. The harmonic phases for the red and black asymmetric period-3 motions are $\varphi_{1,k/3}^{B} = \text{mod}(\varphi_{1,k/3}^{R} + (k/3 + 1)\pi, 2\pi)$. The harmonic amplitudes and phases for the displacement of the second pendulum is presented in Fig. 9.7(g) and (h). The constant terms are $A_{2,0} = a_{2,0}^{(3)B} = -a_{2,0}^{(3)R} \approx 1.8010$. The main harmonic amplitudes for the second pendulum are $A_{2,1/3} \approx 0.6503$, $A_{2,2/3} \approx 0.7482$, $A_{2,1} \approx 0.7762$, $A_{2,4/3} \approx 0.2510$, $A_{2,5/3} \approx 0.1441$, $A_{2,2} \approx 0.3757$, $A_{2,7/3} \approx 0.2243$, $A_{2,8/3} \approx 0.2125$, $A_{2,3} \approx 0.2224$, $A_{2,10/3} \approx 0.0278$, $A_{2,11/3} \approx 0.0674$, $A_{2,4} \approx 0.0874$, $A_{2,13/3} \approx 0.0568$, $A_{2,14/3} \approx 0.0335$, $A_{2,5} \approx 0.0266$, $A_{2,16/3} \approx 0.0170$, $A_{2,17/3} \approx 0.0254$, $A_{2,6} \approx 0.0150$. The other amplitudes for the second pendulum are $A_{2,k/2} \in (10^{-12}, 10^{-3})$ $(j = 19, 20, \cdots, 192)$ with $A_{2,64} = 3.9756 \times 10^{-12}$. The harmonic phases for the red and black asymmetric period-3 motions also satisfy $\varphi_{2,k/3}^{B} = \text{mod}(\varphi_{2,k/3}^{R} + (k/3 + 1)\pi, 2\pi)$. The paired asymmetric period-3 motions need about 192 harmonic terms in the finite Fourier series for an approximate analytical expression with accuracy of $\varepsilon = 10^{-12}$.

9.6 Conclusion

In this chapter, periodic motions of a periodically forced double pendulum were predicted through the implicit mapping method. The discrete periodic nodes of periodic motions were computed from the implicit mappings. The corresponding stability analysis and bifurcation analysis of periodic motions were performed. The numerical simulations of a stable periodic motion were completed to verify the analytical predictions from the analytical prediction. The numerical simulations and analytical predictions matched very well. In such a study, the double-pendulum was not expanded through the Taylor series. For such a system, the traditional perturbation analysis cannot obtain appropriate solutions of periodic motions.

Appendix

$$\frac{\partial \mathbf{g}_k}{\partial \mathbf{z}_{k-1}} = \begin{bmatrix} \frac{\partial g_{k1}}{\partial x_{1(k-1)}} & \frac{\partial g_{k1}}{\partial x_{2(k-1)}} & \frac{\partial g_{k1}}{\partial y_{1(k-1)}} & \frac{\partial g_{k1}}{\partial y_{2(k-1)}} \\ \frac{\partial g_{k2}}{\partial x_{1(k-1)}} & \frac{\partial g_{k2}}{\partial x_{2(k-1)}} & \frac{\partial g_{k2}}{\partial y_{1(k-1)}} & \frac{\partial g_{k2}}{\partial y_{2(k-1)}} \\ \frac{\partial g_{k3}}{\partial x_{1(k-1)}} & \frac{\partial g_{k3}}{\partial x_{2(k-1)}} & \frac{\partial g_{k3}}{\partial y_{1(k-1)}} & \frac{\partial g_{k3}}{\partial y_{2(k-1)}} \\ \frac{\partial g_{k4}}{\partial x_{1(k-1)}} & \frac{\partial g_{k4}}{\partial x_{2(k-1)}} & \frac{\partial g_{k4}}{\partial y_{1(k-1)}} & \frac{\partial g_{k4}}{\partial y_{2(k-1)}} \end{bmatrix}, \tag{9.21}$$

where

$$\begin{aligned} &\frac{\partial g_{k1}}{\partial x_{1(k-1)}} = -1, \frac{\partial g_{k1}}{\partial x_{2(k-1)}} = 0, \frac{\partial g_{k1}}{\partial y_{1(k-1)}} = -\frac{h}{2}, \frac{\partial g_{k1}}{\partial y_{2(k-1)}} = 0, \\ &\frac{\partial g_{k2}}{\partial x_{1(k-1)}} = 0, \frac{\partial g_{k2}}{\partial x_{2(k-1)}} = -1, \frac{\partial g_{k2}}{\partial y_{1(k-1)}} = 0, \frac{\partial g_{k2}}{\partial y_{2(k-1)}} = -\frac{h}{2}, \end{aligned} \tag{9.22}$$

$$\begin{aligned} \frac{\partial g_{k3}}{\partial x_{1(k-1)}} = {} & \frac{h}{2\,(m_1+m_2)\,l_1}\{(m_1+m_2)g\cos[\frac{1}{2}(x_{1(k-1)}+x_{1k})] \\ & - Q_2\cos\Omega(t_{k-1}+\frac{1}{2}h)\sin[\frac{1}{2}(x_{1(k-1)}+x_{1k})-\frac{1}{2}(x_{2(k-1)}+x_{2k})] \\ & - \frac{1}{h}m_2l_2(y_{2k}-y_{2(k-1)})\sin[\frac{1}{2}(x_{1(k-1)}+x_{1k})-\frac{1}{2}(x_{2(k-1)}+x_{2k})] \\ & + m_2l_2[\frac{1}{2}(y_{2(k-1)}+y_{2k})]^2\cos[\frac{1}{2}(x_{1(k-1)}+x_{1k})-\frac{1}{2}(x_{2(k-1)}+x_{2k})] \\ & - \frac{1}{2}\delta_2l_2(y_{2(k-1)}+y_{2k})\sin[\frac{1}{2}(x_{1(k-1)}+x_{1k})-\frac{1}{2}(x_{2(k-1)}+x_{2k})]\}, \end{aligned} \tag{9.23}$$

$$\begin{aligned} \frac{\partial g_{k3}}{\partial x_{2(k-1)}} = {} & \frac{h}{2\,(m_1+m_2)\,l_1}\{\frac{1}{2}\delta_2l_2(y_{2(k-1)}+y_{2k})\sin[\frac{1}{2}(x_{1(k-1)}+x_{1k}) \\ & - \frac{1}{2}(x_{2(k-1)}+x_{2k})] + Q_2\cos\Omega(t_{k-1}+\frac{1}{2}h)\sin[\frac{1}{2}(x_{1(k-1)}+x_{1k}) \\ & - \frac{1}{2}(x_{2(k-1)}+x_{2k})] + \frac{1}{h}m_2l_2(y_{2k}-y_{2(k-1)})\sin[\frac{1}{2}(x_{1(k-1)}+x_{1k}) \\ & - \frac{1}{2}(x_{2(k-1)}+x_{2k})] - m_2l_2[\frac{1}{2}(y_{2(k-1)}+y_{2k})]^2\cos[\frac{1}{2}(x_{1(k-1)}+x_{1k}) \\ & - \frac{1}{2}(x_{2(k-1)}+x_{2k})]\}, \end{aligned} \tag{9.24}$$

$$\frac{\partial g_{k3}}{\partial y_{1(k-1)}} = -1 + \frac{h\,(\delta_1 + \delta_2)}{2\,(m_1 + m_2)}, \tag{9.25}$$

$$\begin{aligned}
\frac{\partial g_{k3}}{\partial y_{2(k-1)}} = {} & \frac{h}{2\,(m_1 + m_2)\,l_1}\{\delta_2 l_2 \cos[\frac{1}{2}(x_{1(k-1)} + x_{1k}) - \frac{1}{2}(x_{2(k-1)} + x_{2k})] \\
& - \frac{2}{h} m_2 l_2 \cos[\frac{1}{2}(x_{1(k-1)} + x_{1k}) - \frac{1}{2}(x_{2(k-1)} + x_{2k})] \\
& + m_2 l_2 (y_{2(k-1)} + y_{2k}) \sin[\frac{1}{2}(x_{1(k-1)} + x_{1k}) - \frac{1}{2}(x_{2(k-1)} + x_{2k})]\},
\end{aligned} \tag{9.26}$$

$$\begin{aligned}
\frac{\partial g_{k4}}{\partial x_{1(k-1)}} = {} & \frac{h}{2m_2 l_2}\{-\frac{1}{2}\delta_2 l_1 (y_{1(k-1)} + y_{1k}) \sin[\frac{1}{2}(x_{1(k-1)} + x_{1k}) \\
& - \frac{1}{2}(x_{2(k-1)} + x_{2k})] - \frac{1}{h} m_2 l_1 (y_{1k} - y_{1(k-1)}) \sin[\frac{1}{2}(x_{1(k-1)} + x_{1k}) \\
& - \frac{1}{2}(x_{2(k-1)} + x_{2k})] - m_2 l_1 [\frac{1}{2}(y_{1(k-1)} + y_{1k})]^2 \cos[\frac{1}{2}(x_{1(k-1)} + x_{1k}) \\
& - \frac{1}{2}(x_{2(k-1)} + x_{2k})]\},
\end{aligned} \tag{9.27}$$

$$\begin{aligned}
\frac{\partial g_{k4}}{\partial x_{2(k-1)}} = {} & \frac{h}{2m_2 l_2}\{m_2 g \cos[\frac{1}{2}(x_{2(k-1)} + x_{2k})] \\
& + \frac{1}{h} m_2 l_1 (y_{1k} - y_{1(k-1)}) \sin[\frac{1}{2}(x_{1(k-1)} + x_{1k}) - \frac{1}{2}(x_{2(k-1)} + x_{2k})] \\
& + m_2 l_1 [\frac{1}{2}(y_{1(k-1)} + y_{1k})]^2 \cos[\frac{1}{2}(x_{1(k-1)} + x_{1k}) - \frac{1}{2}(x_{2(k-1)} + x_{2k})] \\
& + \frac{1}{2}\delta_2 l_1 (y_{1(k-1)} + y_{1k}) \sin[\frac{1}{2}(x_{1(k-1)} + x_{1k}) - \frac{1}{2}(x_{2(k-1)} + x_{2k})]\},
\end{aligned} \tag{9.28}$$

$$\begin{aligned}
\frac{\partial g_{k4}}{\partial y_{1(k-1)}} = {} & \frac{h}{2m_2 l_2}\{\delta_2 l_1 \cos[\frac{1}{2}(x_{1(k-1)} + x_{1k}) - \frac{1}{2}(x_{2(k-1)} + x_{2k})] \\
& - \frac{2}{h} m_2 l_1 \cos[\frac{1}{2}(x_{1(k-1)} + x_{1k}) - \frac{1}{2}(x_{2(k-1)} + x_{2k})] \\
& - m_2 l_1 (y_{1(k-1)} + y_{1k}) \sin[\frac{1}{2}(x_{1(k-1)} + x_{1k}) - \frac{1}{2}(x_{2(k-1)} + x_{2k})]\},
\end{aligned} \tag{9.29}$$

$$\frac{\partial \mathbf{g}_k}{\partial \mathbf{z}_k} = \begin{bmatrix} \frac{\partial g_{k1}}{\partial x_{1k}} & \frac{\partial g_{k1}}{\partial x_{2k}} & \frac{\partial g_{k1}}{\partial y_{1k}} & \frac{\partial g_{k1}}{\partial y_{2k}} \\ \frac{\partial g_{k2}}{\partial x_{1k}} & \frac{\partial g_{k2}}{\partial x_{2k}} & \frac{\partial g_{k2}}{\partial y_{1k}} & \frac{\partial g_{k2}}{\partial y_{2k}} \\ \frac{\partial g_{k3}}{\partial x_{1k}} & \frac{\partial g_{k3}}{\partial x_{2k}} & \frac{\partial g_{k3}}{\partial y_{1k}} & \frac{\partial g_{k3}}{\partial y_{2k}} \\ \frac{\partial g_{k4}}{\partial x_{1k}} & \frac{\partial g_{k4}}{\partial x_{2k}} & \frac{\partial g_{k4}}{\partial y_{1k}} & \frac{\partial g_{k4}}{\partial y_{2k}} \end{bmatrix}, \tag{9.30}$$

where

$$\begin{aligned} &\frac{\partial g_{k1}}{\partial x_{1k}} = 1, \frac{\partial g_{k1}}{\partial x_{2k}} = 0, \frac{\partial g_{k1}}{\partial y_{1k}} = -\frac{h}{2}, \frac{\partial g_{k1}}{\partial y_{2k}} = 0, \\ &\frac{\partial g_{k2}}{\partial x_{1k}} = 0, \frac{\partial g_{k2}}{\partial x_{2k}} = 1, \frac{\partial g_{k2}}{\partial y_{1k}} = 0, \frac{\partial g_{k2}}{\partial y_{2k}} = -\frac{h}{2}, \end{aligned} \tag{9.31}$$

$$\begin{aligned} \frac{\partial g_{k3}}{\partial x_{1k}} = \frac{h}{2(m_1+m_2)l_1} \{ & (m_1+m_2)\, g \cos[\tfrac{1}{2}(x_{1(k-1)} + x_{1k})] \\ & - Q_2 \cos \Omega (t_{k-1} + \tfrac{1}{2}h) \sin[\tfrac{1}{2}(x_{1(k-1)} + x_{1k}) - \tfrac{1}{2}(x_{2(k-1)} + x_{2k})] \\ & - \tfrac{1}{h} m_2 l_2 (y_{2k} - y_{2(k-1)}) \sin[\tfrac{1}{2}(x_{1(k-1)} + x_{1k}) - \tfrac{1}{2}(x_{2(k-1)} + x_{2k})] \\ & + m_2 l_2 [\tfrac{1}{2}(y_{2(k-1)} + y_{2k})]^2 \cos[\tfrac{1}{2}(x_{1(k-1)} + x_{1k}) - \tfrac{1}{2}(x_{2(k-1)} + x_{2k})] \\ & - \tfrac{1}{2}\delta_2 l_2 (y_{2(k-1)} + y_{2k}) \sin[\tfrac{1}{2}(x_{1(k-1)} + x_{1k}) - \tfrac{1}{2}(x_{2(k-1)} + x_{2k})]\}, \end{aligned} \tag{9.32}$$

$$\begin{aligned} \frac{\partial g_{k3}}{\partial x_{2k}} = \frac{h}{2(m_1+m_2)l_1} \{ & \tfrac{1}{2}\delta_2 l_2 (y_{2(k-1)} + y_{2k}) \sin[\tfrac{1}{2}(x_{1(k-1)} + x_{1k}) \\ & - \tfrac{1}{2}(x_{2(k-1)} + x_{2k})] + Q_2 \cos \Omega (t_{k-1} + \tfrac{1}{2}h) \sin[\tfrac{1}{2}(x_{1(k-1)} + x_{1k}) \\ & - \tfrac{1}{2}(x_{2(k-1)} + x_{2k})] + \tfrac{1}{h} m_2 l_2 (y_{2k} - y_{2(k-1)}) \sin[\tfrac{1}{2}(x_{1(k-1)} + x_{1k}) \\ & - \tfrac{1}{2}(x_{2(k-1)} + x_{2k})] - m_2 l_2 [\tfrac{1}{2}(y_{2(k-1)} + y_{2k})]^2 \cos[\tfrac{1}{2}(x_{1(k-1)} + x_{1k}) \\ & - \tfrac{1}{2}(x_{2(k-1)} + x_{2k})]\}, \end{aligned} \tag{9.33}$$

$$\frac{\partial g_{k3}}{\partial y_{1k}} = 1 + \frac{h(\delta_1 + \delta_2)}{2(m_1+m_2)}, \tag{9.34}$$

$$\begin{aligned}\frac{\partial g_{k3}}{\partial y_{2k}} =& \frac{h}{2\left(m_1+m_2\right)l_1}\{\delta_2 l_2 \cos[\frac{1}{2}(x_{1(k-1)}+x_{1k})-\frac{1}{2}(x_{2(k-1)}+x_{2k})]\\&+\frac{2}{h}m_2 l_2 \cos[\frac{1}{2}(x_{1(k-1)}+x_{1k})-\frac{1}{2}(x_{2(k-1)}+x_{2k})]\\&+m_2 l_2(y_{2(k-1)}+y_{2k})\sin[\frac{1}{2}(x_{1(k-1)}+x_{1k})-\frac{1}{2}(x_{2(k-1)}+x_{2k})]\},\end{aligned} \tag{9.35}$$

$$\begin{aligned}\frac{\partial g_{k4}}{\partial x_{1k}} =& \frac{h}{2m_2 l_2}\{-\frac{1}{2}\delta_2 l_1(y_{1(k-1)}+y_{1k})\sin[\frac{1}{2}(x_{1(k-1)}+x_{1k})\\&-\frac{1}{2}(x_{2(k-1)}+x_{2k})]-\frac{1}{h}m_2 l_1(y_{1k}-y_{1(k-1)})\\&\times\sin[\frac{1}{2}(x_{1(k-1)}+x_{1k})-\frac{1}{2}(x_{2(k-1)}+x_{2k})]\\&-m_2 l_1[\frac{1}{2}(y_{1(k-1)}+y_{1k})]^2\\&\times\cos[\frac{1}{2}(x_{1(k-1)}+x_{1k})-\frac{1}{2}(x_{2(k-1)}+x_{2k})]\},\end{aligned} \tag{9.36}$$

$$\begin{aligned}\frac{\partial g_{k4}}{\partial x_{2k}} =& \frac{h}{2m_2 l_2}\{m_2 g\cos[\frac{1}{2}(x_{2(k-1)}+x_{2k})]\\&+\frac{1}{h}m_2 l_1(y_{1k}-y_{1(k-1)})\sin(\frac{1}{2}(x_{1(k-1)}+x_{1k})-\frac{1}{2}(x_{2(k-1)}+x_{2k}))\\&+m_2 l_1[\frac{1}{2}(y_{1(k-1)}+y_{1k})]^2\cos[\frac{1}{2}(x_{1(k-1)}+x_{1k})-\frac{1}{2}(x_{2(k-1)}+x_{2k})]\\&+\frac{1}{2}\delta_2 l_1(y_{1(k-1)}+y_{1k})\sin[\frac{1}{2}(x_{1(k-1)}+x_{1k})-\frac{1}{2}(x_{2(k-1)}+x_{2k})]\},\end{aligned} \tag{9.37}$$

$$\begin{aligned}\frac{\partial g_{k4}}{\partial y_{1k}} =& \frac{h}{2m_2 l_2}\{\delta_2 l_1 \cos[\frac{1}{2}(x_{1(k-1)}+x_{1k})-\frac{1}{2}(x_{2(k-1)}+x_{2k})]\\&+\frac{2m_2 l_1}{h}\cos[\frac{1}{2}(x_{1(k-1)}+x_{1k})-\frac{1}{2}(x_{2(k-1)}+x_{2k})]\\&-m_2 l_1(y_{1(k-1)}+y_{1k})\sin[\frac{1}{2}(x_{1(k-1)}+x_{1k})-\frac{1}{2}(x_{2(k-1)}+x_{2k})]\},\end{aligned} \tag{9.38}$$

$$\frac{\partial g_{k4}}{\partial y_{2k}} = 1+\frac{h\delta_2}{2m_2}. \tag{9.39}$$

References

1. Luo, A.C.J., and C. Guo. 2019. A period-1 motion to chaos in a periodically forced, damped double pendulum. *Journal of Vibration Testing and System Dynamics* 3 (3): 250–280.
2. Herrmann, G., and I.C. Jong. 1966. On nonconservative stability problems of elastic systems with slight damping. *Journal of Applied Mechanics* 33: 125–133.
3. Roorda, J., and S. Nemat-Nasser. 1967. An energy method for stability analysis of nonlinear, nonconservative systems. *American Institute of Aeronautics and Astronautics* 5: 1262–1268.
4. Jin, J.-D., and Y. Matsuzaki. 1988. Bifurcations in a two-degree-of-freedom elastic system with follower forces. *Journal of Sound and Vibration* 126: 265–277.
5. Jin, J.-D., and Y. Matsuzaki. 1992. Bifurcation analysis of double pendulum with a follower force. *Journal of Sound and Vibration* 154: 191–204.
6. Thomsen, J.J. 1995. Chaotic dynamics of the partially follower-loader elastic double pendulum. *Journal of Sound and Vibration* 188: 385–405.
7. De Paula, A.S., M.A. Savi, and F.H.I. Pereira-Pinto. 2006. Chaos and transient chaos in an experimental nonlinear pendulum. *Journal of Sound and Vibration* 294: 585–595.
8. Luo, A.C.J. 2015. Periodic flows in nonlinear dynamical systems based on discrete implicit maps. *International Journal of Bifurcation and Chaos* 25 (3): 1550044.
9. Luo, A.C.J. 2015. *Discretization and implicit mapping dynamics*. Beijing/Dordrecht: HEP/Springer.
10. Guo, Y., and A.C.J. Luo. 2017. Routes of periodic motions to chaos in a periodically forced pendulum. *International Journal of Dynamics and Control* 5 (3): 551–569.

Chapter 10
Analytical Periodic Motions for a First-Order Nonlinear Circuit System Under Different Excitations

Yan Liu, Kai Ma, Hao He, and Jun Xiao

10.1 Introduction

The research on the characteristics of nonlinear circuits has become one hotspot in electronic circuit design. In general, the response, stability and control of nonlinear circuits are the main studying for the characteristic of nonlinear circuit. Some methods, such as wavelet equilibrium method, fast wavelet configuration technique, inverse operator method and algorithm of congruent transformation, are used to analyze the response, stability and control of nonlinear circuits and some satisfactory results have been obtained [1–4]. In recent years, researchers have paid more attention to the analytical solution of nonlinear systems [5–8]. At the same time, many new methods have been found in the analytical solutions of nonlinear systems, such as mixed numerical, differentiation matrix, and homotopy analysis method [9–14]. In particular, Luo proposed a generalized harmonic balance method for solving chaotic periodic motion in nonlinear dynamic systems [15]. In the following years, Luo and his groups further improved this analytical method, by means of truncating high order resonant amplitude and controlling calculation precision [16–18]. Therefore, the method could be introduced to study the characteristic of nonlinear circuit.

In this chapter, the periodic motion of a first-order nonlinear circuit at different excitation forces is studied in detail on the basis of the generalized harmonic balance method and the analytical solution of periodic motion in the system is obtained

Y. Liu (✉) · K. Ma · H. He
School of Mechanical Engineering, Northwestern Polytechnical University, Xi'an, Shaanxi Province, P. R. China
e-mail: liuyan@nwpu.edu.cn

J. Xiao
State Key Laboratory of Compressor Technology, Hefei General Machinery Research Institute Co., Ltd., Hefei, Anhui, P. R. China

J. Zhang (ed.), *Dynamics and Fault Diagnosis of Nonlinear Rotors and Impellers*, Nonlinear Systems and Complexity 34, https://doi.org/10.1007/978-3-030-94301-1_10

by the finite Fourier series. Then, the stability and bifurcation are analyzed by the eigenvalue of the coefficient matrix of the nonlinear circuit. Finally, the first order nonlinear circuit simulation model and circuit further are simulated to verify the effectiveness of the generalized harmonic balance method in the analysis of first-order nonlinear circuit.

10.2 First-Order Nonlinear Circuit Simulation Model

The first-order nonlinear circuit simulation model is designed to satisfy Eq. (10.1) and shown in Fig. 10.1. It mainly consists of amplifier, multiplier, resistor, capacitance, oscilloscope, power supply and some electronic elements. In the nonlinear circuit, the power supply voltage is set as $u_i = Q_0\ sin\ (\omega t)$, the amplitude of the power supply voltage is Q_0 and the frequency is f $(f = \omega/2\pi)$. The output voltage of the circuit is u_o, and the coefficients of the multiplier unit are K_1 and K_2.

$$\dot{u}_o + \alpha_1 u_o + \alpha_2 {u_o}^3 = Q_0 \sin(\omega t) \tag{10.1}$$

The relationships of parameters between Eq. (10.1) and the elements in the circuit simulation model can be presented as the following

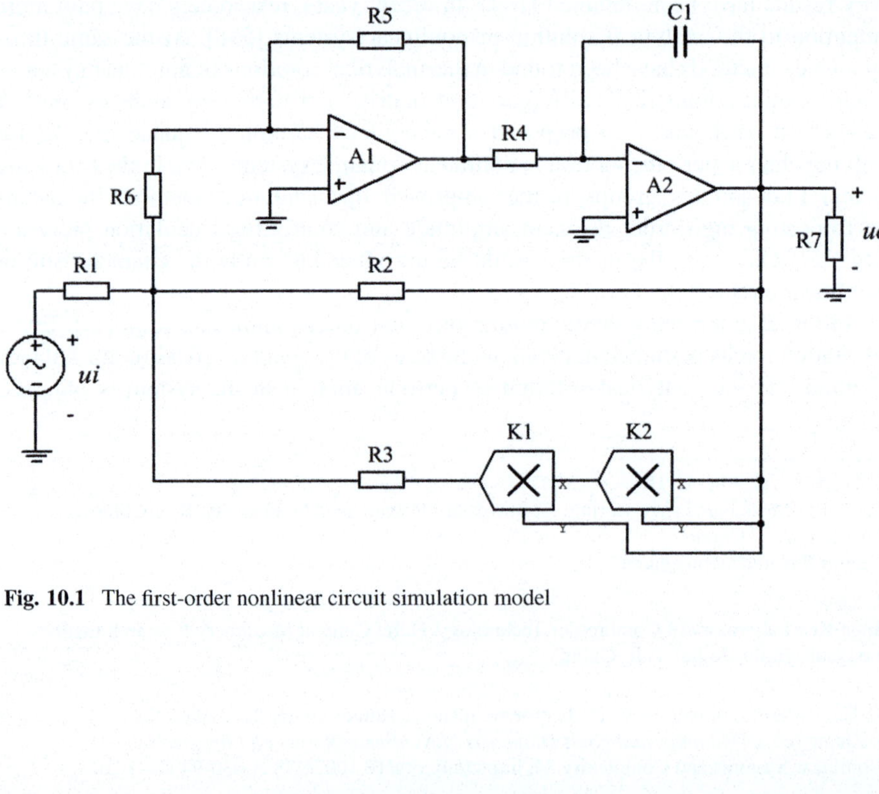

Fig. 10.1 The first-order nonlinear circuit simulation model

$$\alpha_1 = -\frac{R_1}{R_2}, \alpha_2 = -K_1K_2\frac{R_1}{R_3}, R_1R_4C_1 = R, \frac{1}{R} = \frac{1}{R_1} + \frac{1}{R_2} + \frac{1}{R_3}$$

Then, Eq. (10.1) can be rewritten as

$$R_1R_4C_1\left(\frac{1}{R_1} + \frac{1}{R_2} + \frac{1}{R_3}\right)\dot{u}_o - \frac{R_1}{R_2}u_o - K_1K_2\frac{R_1}{R_3}{u_o}^3 = u_i \tag{10.2}$$

10.3 Analytical Solution of the First-Order Nonlinear Circuit

In the section, the generalized harmonic balance method is used to analyze the periodic motion of the first-order nonlinear circuit, determine its analytical solution, and study the stability of the system.

At first, Eq. (10.1) can be transformed into the following form

$$\dot{u}_o = -\alpha_1 u_o - \alpha_3 {u_o}^3 + Q_0 \sin(\omega t) \tag{10.3}$$

Then, it can be written further as following,

$$\dot{u}_o = f(u_o, t) \tag{10.4}$$

According to Luo [15], there exists a periodic solution in the first-order nonlinear dynamical system, so the periodic motion can be expressed by Fourier series, and then the solution of the first-order nonlinear circuit can be written as

$$u_o = a_0(t) + \sum_{k=1}^{N}[b_k(t)\cos(k\omega t) + c_k(t)\sin(k\omega t)] \tag{10.5}$$

Moreover, the simplified Eq. (10.5) can be obtained

$$u_o = a_0 + \sum_{k=1}^{N}[b_k\cos(k\omega t) + c_k\sin(k\omega t)] \tag{10.6}$$

Then, the derivative of Eq. (10.6) is

$$\dot{u}_o = \dot{a}_0 + \sum_{k=1}^{N}\left[\left(\dot{b}_k + k\omega c_k\right)\cos(k\omega t) + (\dot{c}_k - k\omega b_k)\sin(k\omega t)\right] \tag{10.7}$$

Substitute Eqs. (10.6) and (10.7) into Eq. (10.3), and yields,

$$\dot{a}_0 = A_0$$

$$\dot{b}_k = -k\omega c_k + B_k \tag{10.8}$$

$$\dot{c}_k = k\omega b_k + C_k$$

Solve the above equations and get

$$A_0 = \frac{1}{T}\int_0^T f(u_o, t)\, dt$$

$$B_k = \frac{2}{T}\int_0^T f(u_o, t)\cos(k\omega t)\, dt \tag{10.9}$$

$$C_k = \frac{2}{T}\int_0^T f(u_o, t)\sin(k\omega t)\, dt$$

with

$$\begin{aligned} A_0 = -\alpha_1 a_0 - \alpha_2 \Big[a_0^3 + \tfrac{3a_0}{2} \sum_{i=1}^{N} \left(b_i^2 + c_i^2\right) \\ + \tfrac{1}{4} \sum_{i=1}^{N}\sum_{j=1}^{N}\sum_{l=1}^{N} b_i b_j b_l \left(\delta^0_{|i-j+l|} + \delta^0_{|i+j-l|} + \delta^0_{|i-j-l|}\right) \\ + \tfrac{3}{4} \sum_{i=1}^{N}\sum_{j=1}^{N}\sum_{l=1}^{N} b_i c_j c_l \left(\delta^0_{|i+j-l|} + \delta^0_{|i-j+l|} - \delta^0_{|i-j-l|}\right) \Big] \end{aligned}$$

$$B_k = -\alpha_1 b_k - \alpha_2 \left[3a_0^2 b_k + f_1^c + f_2^c + f_3^c + f_4^c\right] \tag{10.10}$$

$$C_k = -\alpha_1 c_k - \alpha_2 \left[3a_0^2 c_k + f_1^s + f_2^s + f_3^s + f_4^s\right] + Q_0 \delta_k^1$$

where

$$f_1^c = \frac{3a_0}{2} \sum_{i=1}^{N}\sum_{j=1}^{N} b_i b_j \left(\delta^k_{|i+j|} + \delta^k_{|i-j|}\right)$$

$$f_2^c = \frac{3a_0}{2} \sum_{i=1}^{N}\sum_{j=1}^{N} c_i c_j \left(\delta^k_{|i-j|} - \delta^k_{|i+j|}\right)$$

$$f_3^c = \frac{1}{4}\sum_{i=1}^{N}\sum_{j=1}^{N}\sum_{l=1}^{N} b_i b_j b_l \left(\delta_{|i+j+l|}^k + \delta_{|i-j+l|}^k + \delta_{|i+j-l|}^k + \delta_{|i-j-l|}^k\right)$$

$$f_4^c = \frac{3}{4}\sum_{i=1}^{N}\sum_{j=1}^{N}\sum_{l=1}^{N} b_i c_j c_l \left(\delta_{|i+j-l|}^k + \delta_{|i-j+l|}^k - \delta_{|i+j+l|}^k - \delta_{|i-j-l|}^k\right)$$

$$f_1^s = \frac{3a_0}{2}\sum_{i=1}^{N}\sum_{j=1}^{N} b_i c_j \left(\delta_{|i+j|}^k - \operatorname{sgn}(i-j)\,\delta_{|i-j|}^k\right) \tag{10.11}$$

$$f_2^s = \frac{3a_0}{2}\sum_{i=1}^{N}\sum_{j=1}^{N} c_i b_j \left(\delta_{|i+j|}^k + \operatorname{sgn}(i-j)\,\delta_{|i-j|}^k\right)$$

$$\begin{aligned} f_3^s = \tfrac{3}{4}\sum_{i=1}^{N}\sum_{j=1}^{N}\sum_{l=1}^{N} b_i b_j c_l \Big(&\delta_{|i+j+l|}^k + \operatorname{sgn}(i-j+l)\,\delta_{|i-j+l|}^k \\ &- \operatorname{sgn}(i+j-l)\,\delta_{|i+j-l|}^k - \operatorname{sgn}(i-j-l)\,\delta_{|i-j-l|}^k\Big) \end{aligned}$$

$$\begin{aligned} f_4^s = \tfrac{1}{4}\sum_{i=1}^{N}\sum_{j=1}^{N}\sum_{l=1}^{N} c_i c_j c_l \Big(&\operatorname{sgn}(i+j-l)\,\delta_{|i+j-l|}^k \\ &+ \operatorname{sgn}(i-j+l)\,\delta_{|i-j+l|}^k - \delta_{|i+j+l|}^k - \operatorname{sgn}(i-j-l)\,\delta_{|i-j-l|}^k\Big) \end{aligned}$$

and

$$\delta_l^k = \begin{cases} 1 & l = k \\ 0 & l \neq k \end{cases} \tag{10.12}$$

$$\operatorname{sgn}(y) = \begin{cases} 1 & y > 0 \\ -1 & y < 0 \end{cases} \tag{10.13}$$

So, the nonlinear system governed by Eq. (10.1) could be transformed into Fourier series coefficient system. Here define

$$\mathbf{Z} = (a_0, \mathbf{b}, \mathbf{c})^{\mathrm{T}} \tag{10.14}$$

$$f = (A_0, \mathbf{B}, \mathbf{C})^{\mathrm{T}} \tag{10.15}$$

where

$$\mathbf{b} = (b_1, b_2, b_3, \ldots, b_N)^T$$

$$\mathbf{c} = (c_1, c_2, c_3, \ldots, c_N)^T \tag{10.16}$$

$$\mathbf{B} = (-\omega c_1 + B_1, -2\omega c_2 + B_2, \cdots, -N\omega c_N + B_N)^{\mathrm{T}}$$

$$\mathbf{C} = (\omega b_1 + C_1, 2\omega b_2 + C_2, \cdots, N\omega b_N + C_N)^{\mathrm{T}}$$

So

$$\dot{\mathbf{Z}} = f(\mathbf{Z}) = (\mathrm{A}_0, \mathbf{B}, \mathbf{C})^{\mathrm{T}} \tag{10.17}$$

If $Z^{(*)}$ is the equilibrium point, then $f(Z^{(*)}) = \mathbf{0}$, and let $\mathbf{Z} = \mathbf{Z}^{(*)} + \Delta\mathbf{Z}$, we can get

$$Df\left(\mathbf{Z}^{(*)}\right) = \partial f(\mathbf{Z}) / \partial \mathbf{Z}\,|_{\mathbf{Z}^{(*)}} \tag{10.18}$$

a_0, b_k and c_k are governed by Eq. (10.14). u_o and $\dot{u}_o$ can be expressed by Eqs. (10.6) and (10.7), implying the analytical solution of the first-order nonlinear system could be obtained finally.On the other hand, the eigenvalues of $Df(\mathbf{Z}^{(*)})$, used in stability analysis, could be computed by the following formula

$$\left|Df\left(\mathbf{Z}^{(*)}\right) - \lambda \mathbf{I}_{n(2N+1)\times n(2N+1)}\right| = 0 \tag{10.19}$$

With the analytical solution, the Jacobian matrix is exactly given by

$$Df = \begin{bmatrix} \frac{\partial A_0}{\partial a_0} & \frac{\partial A_0}{\partial \mathbf{b}} & \frac{\partial A_0}{\partial \mathbf{c}} \\ \frac{\partial \mathbf{B}}{\partial a_0} & \frac{\partial \mathbf{B}}{\partial \mathbf{b}} & \frac{\partial \mathbf{B}}{\partial \mathbf{c}} \\ \frac{\partial \mathbf{C}}{\partial a_0} & \frac{\partial \mathbf{C}}{\partial \mathbf{b}} & \frac{\partial \mathbf{C}}{\partial \mathbf{c}} \end{bmatrix} \tag{10.20}$$

In Luo [15], the eigenvalues of $\mathrm{D}f(\mathbf{Z}^{(*)})$ are classified as $(n_1, n_2, n_3|n_4, n_5, n_6)$, where n_1 is the total number of negative real eigenvalues, n_2 the total number of positive real eigenvalues, n_3 the total number of zero eigenvalues, n_4 the total pair of complex eigenvalues with negative real parts, n_5 the total pair of complex eigenvalues with positive real parts, and n_6 the total pair of complex eigenvalues with zero real parts.

1. If all eigenvalues of the equilibrium possess negative real parts, the approximate analytical solution with truncated harmonic terms is stable.
2. If one of eigenvalues of the equilibrium possesses positive real part, the approximate analytical solution is unstable.
3. The boundary between stable and unstable solutions is determined by bifurcation point, including saddle-node and Hopf bifurcation.

10.4 Characteristics of Numerical Model and Circuit Simulation Model

Equation (10.6) can be written in terms of harmonic amplitudes and phases as following,

$$u_0 = a_0 + \sum_{k=1}^{N} a_k \sin(k\omega t + \varphi_k) \tag{10.21}$$

and the harmonic amplitude and phase are given as

$$a_k = \sqrt{b_k^2 + c_k^2},\ \varphi_k = \arctan\frac{b_k}{c_k}$$

The parameters in the first-order circuit simulation model are shown in Table 10.1.

According to the parameters and working state of the circuit simulation model, the parameters in the first-order differential equation could be obtained as listed in Table 10.2.

In Fig. 10.2, the direct current a_0 and harmonic amplitudes a_k of analytical solution in the periodic motion were calculated in sequences with the change of excitation amplitude as different excitation frequencies $\omega(\omega = 5.6$ rad/s, 6.6 rad/s, 7.6 rad/s, 8.6 rad/s, 9.6 rad/s, 10.6 rad/s) are selected in the circuit simulation model. The solid and dashed lines represent stable and unstable periodic motions

Table 10.1 Parameters in the first-model circuit simulation model

R_1	R_2	R_3	R_4	R_5	R_6	R_7	C_1	K_1	K_2
1 kΩ	1 kΩ	1 Ω	1 kΩ	1 kΩ	1 kΩ	1 kΩ	1 μF	−0.1	0.1

Table 10.2 Parameters in the first-order differential equation

α_1	α_2	$\omega(\text{rad}\cdot\text{s}^{-1})$	Q_0(V)
−1	10	5.6, 6.6, 7.6, 8.6, 9.6, 10.6	0–3.3

respectively. It is clear that there exist stable and unstable periodic motions in the first-order system. Generally, there are stable and unstable symmetric branches in such first order dynamic systems. The points of saddle-node bifurcation between the stable and unstable symmetric periodic motion are varied with different excitation force. The direct current a_0 of analytical solution in the periodic motions is shown in Fig. 10.2(a). From Fig. 10.2(a), the periodic motions are symmetrical with $a_0 = 0$. Moreover, for $\omega = 5.6$ rad/s, the symmetric periodic motion is unstable when $Q_0 < 1.46$ V, and the symmetric periodic motion is stable as $Q_0 \in$ (1.46 V, 2.52 V). For the excitation frequency ω at 6.6 rad/s, 7.6 rad/s, 8.6 rad/s, 9.6 rad/s and 10.6 rad/s, the corresponding points of saddle-node bifurcation are (0, 1.72), (0, 1.97), (0, 2.23), (0, 2.49) and (0, 2.80), and the corresponding stable ranges of excitation amplitude Q_0 are (1.72–2.19), (1.97–2.52), (2.23–2.85), (2.49–3.97) and (2.80–5.60). The harmonic amplitudes a_1-a_{10} of periodic motions are shown in Fig. 10.2(b)-(i). As a result, the similar conclusion can be drawn in Fig. 10.2(b)-(i) about the corresponding stable range and the points of saddle-node bifurcation between unstable and stable symmetric periodic motions as in Fig. 10.2(a). Moreover, the excitation amplitudes at bifurcation point increase with the increase of excitation frequency.

The stable periodic motion of the first-order nonlinear circuit numerical model is presented under the excitation amplitude $Q_0 = 1$ V and excitation frequency $\omega = 5.6$ rad/s. In this numerical model, the waveform and state space of the output voltage in the state are shown in Fig. 10.3, and it is clear that there exists periodic motion in Fig. 10.3(a). Also, it can be seen that voltage peak is about 0.41 V and voltage valley is about 0.05 V. Further, it is also shown clearly from Fig. 10.3(b) that the first-order circuit system behaves as stable periodic motion at the excitation frequency and excitation amplitude.

Both circuit simulation model and numerical model of the first-order nonlinear system are simulated at $\omega = 5.6$ rad/s and $Q_0 = 1$ V as the system is in a stable state. The compared results on harmonic amplitude and harmonic phase are listed in Table 10.3 and shown in Fig. 10.4. In the compared results, k is the harmonic order, a_{nk} and φ_{nk} are the harmonic amplitude and phase of the numerical model, and a_{ck} and φ_{ck} are the harmonic amplitude and phase of the circuit simulation model. Further, Eqs. (10.22) and (10.23) are defined to describe the difference clearly,

$$\delta a_k = \left| \frac{a_{nk} - a_{ck}}{a_{nk}} \right| \times 100\% \tag{10.22}$$

$$\delta \varphi_k = \left| \frac{\varphi_{nk} - \varphi_{ck}}{\varphi_{nk}} \right| \times 100\% \tag{10.23}$$

In Table 10.3 and Fig. 10.4, the harmonic amplitudes of the numerical model and the circuit simulation model of the system match very well in the low order. However, the difference becomes obvious as the order increases, because of the limited resolution in the experimental data collection during sampling for circuit simulation model. Moreover, the difference will be increased as the order increases in harmonic phase. It is due to the frequency conversion errors of the exaction in circuit simulation model.

In order to verify the availability of the generalized harmonic balance method, another example is present under excitation frequency $\omega = 7.6$ rad/s and excitation amplitude $Q_0 = 1$ V. The results with comparison on harmonic amplitude and harmonic phase are shown in Fig. 10.5. From Fig. 10.5, it can be seen that the harmonic amplitudes continuously decrease for numerical model as the harmonic order increases, while the harmonic amplitudes of circuit simulation model keep at a level about 10^{-8}. Moreover, the difference becomes more obvious as the increase of the order in harmonic phase at $\omega = 7.6$ rad/s. Overall, the results are in agreement with those from Fig. 10.4.

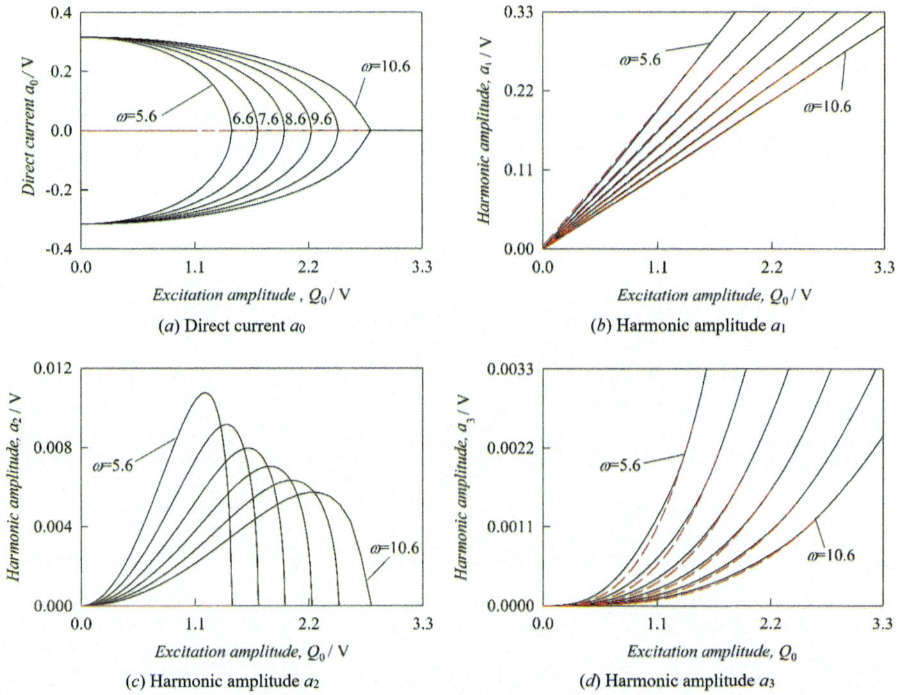

Fig. 10.2 Harmonic amplitudes in analytical solution of the first-order numerical model ((**a**)-(**k**) $a_k(k = 0\text{–}10)$, $Q_0 \in (0\text{ V}, 3.3\text{ V})$, $\alpha_1 = -1$, $\alpha_2 = 10$, $\omega = 5.6$ rad/s, 6.6 rad/s, 7.6 rad/s, 8.6 rad/s, 9.6 rad/s, 10.6 rad/s)

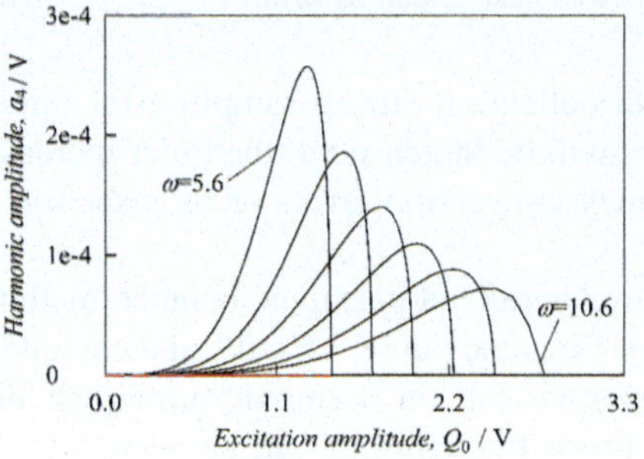

(*e*) Harmonic amplitude a_4

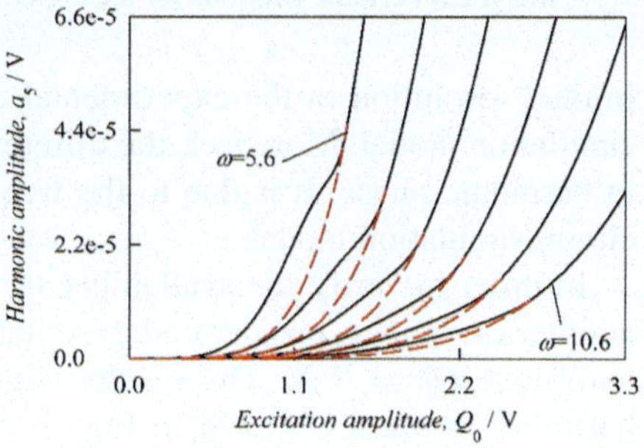

(*f*) Harmonic amplitude a_5

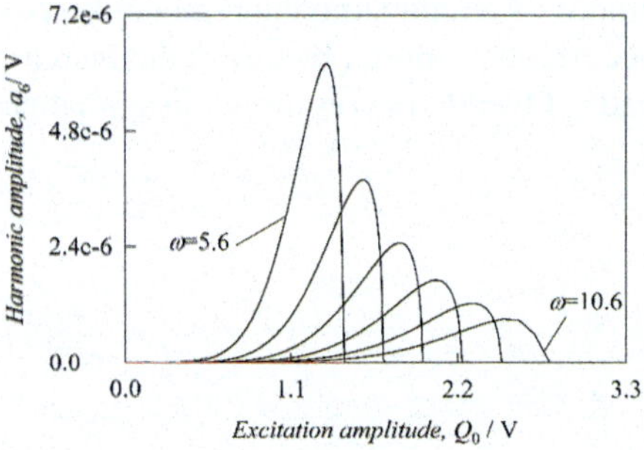

(*g*) Harmonic amplitude a_6

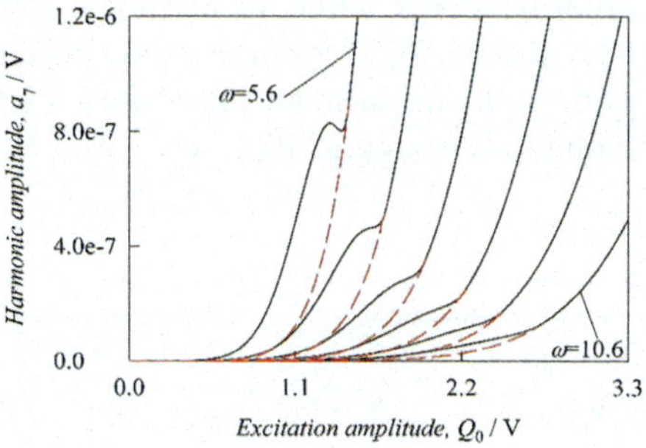

(*h*) Harmonic amplitude a_7

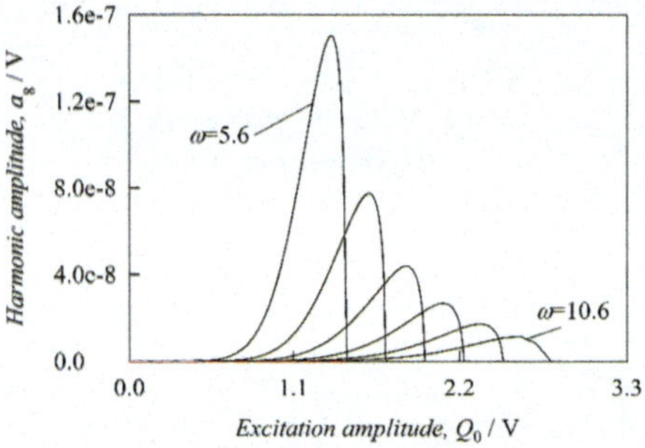

(*i*) Harmonic amplitude a_8

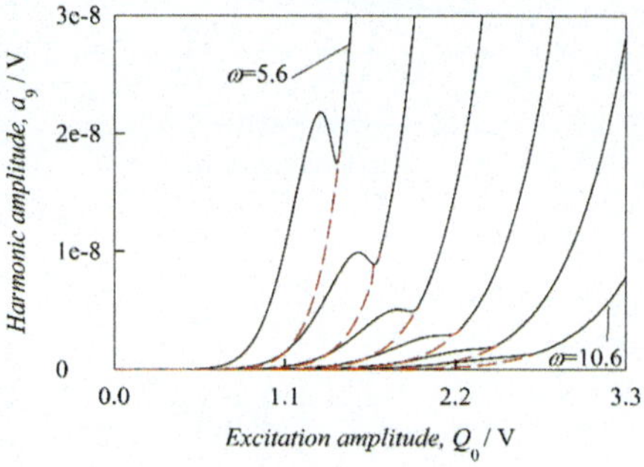

(*j*) Harmonic amplitude a_9

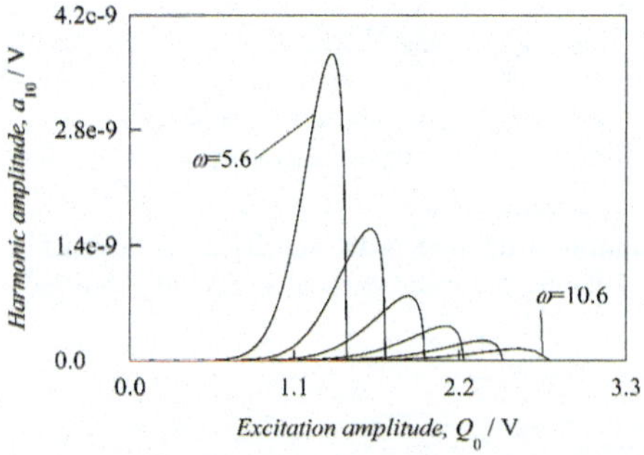

(*k*) Harmonic amplitude a_{10}

Fig. 10.2 (continued)

Fig. 10.3 Waveform and state space of the output voltage in stable periodic motion of the numerical model ($Q_0 = 1$ V, $\omega = 5.6$ rad/s, $\alpha_1 = -1$, $\alpha_2 = 10$)

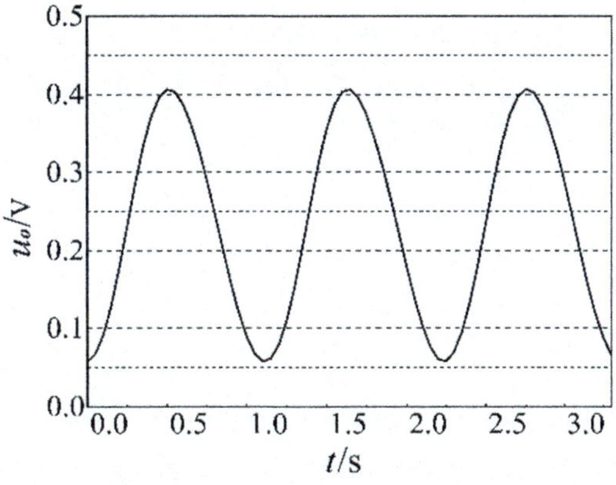

(*a*) Waveform of the output voltage of numerical model

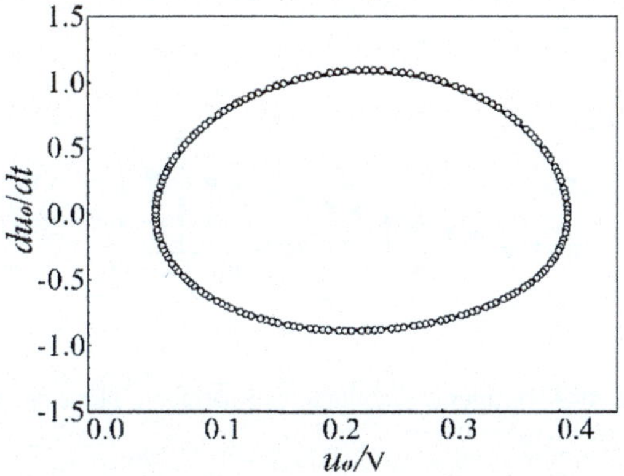

(*b*) State space of the output voltage

10.5 Output Response of the First-Order Nonlinear Circuit

According the circuit simulation model, the first-order nonlinear circuit is designed with excitation frequency $\omega = 5.6$ rad/s ($f = 0.89$ Hz) and excitation amplitude $Q_0 = 1$ V, as shown in Fig. 10.6. From Fig. 10.6, the circuit is composed of amplifier circuit, multiplication circuit, power supply, input circuit, output circuit and some auxiliary elements. The parameters in the circuit are chosen carefully to match the differential equation. In the circuit, AD835, which is a four-quadrant, voltage output, analog multiplier whose output is the linear product of input voltage X and Y, is selected as multiplier with power supply ± 5 V, and uA741 is used as amplifier with power supply ± 12 V.

Figure 10.7 shows the output voltage waveforms of the numerical model, circuit simulation model and circuit. It is clearly that the three waveforms are similar to each other. Specifically, the waveforms of numerical model and circuit simulation model are almost coincident with each other. For circuit, however, the waveform and

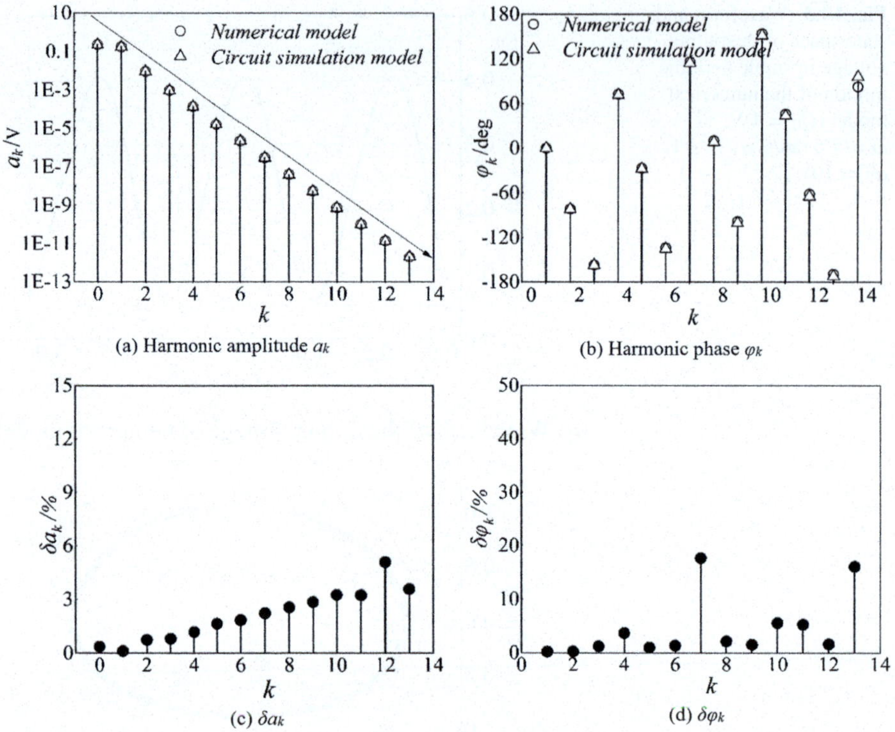

Fig. 10.4 Harmonic amplitude and harmonic phase for the system ($Q_0 = 1$ V, $\omega = 5.6$ rad/s, $\alpha_1 = -1, \alpha_2 = 10$)

voltage peak are all slightly lower than the results of the two other models, and the valley is a little higher than the two models. For the phase, the circuit is consistent with the other two. The main reasons for the errors are the numerical model and circuit simulation model are designed based on the ideal math models. In circuit, however, there are energy storage elements and intrinsic voltage attenuation in the chips. Moreover, changing parameters of the excitation as $\omega = 7.6$ rad/s ($f =$ Hz) and $Q_0 = 1$ V, the three waveforms are shown in Fig. 10.8. As shown in Fig. 10.8, the waveforms indicate the similar conclusion with that in Fig. 10.7.

10.6 Conclusions

In the chapter, the analytical solutions of periodic motion for the first-order nonlinear circuit are analyzed by using the generalized harmonic balance method, and the stability of the system are studied by analyzing the eigenvalues of the system response coefficients matrix in numerical model. Further, the numerical model,

Table 10.3 Harmonic amplitude and phase of numerical model and circuit simulation model for the system ($Q_0 = 1$ V, $\omega = 5.6$ rad/s, $\alpha_1 = -1$, $\alpha_2 = 10$)

k	a_{nk}/V	a_{ck}/V	δa_k/%	φ_{nk}/deg	φ_{ck}/deg	$\delta\varphi_k$/%
0	0.23295	0.23209	0.36875	0	0	
1	0.17453	0.17431	0.12204	−81.4664	−81.651	0.2266
2	0.00938	0.00931	0.75125	−156.956	−157.38	0.27014
3	9.70718E-4	9.62747E-4	0.82114	72.9033	72.0137	1.22025
4	1.40971E-4	1.39291E-4	1.19173	−26.6622	−27.65	3.70487
5	1.63943E-5	1.61231E-5	1.65423	−133.412	−134.75	1.00291
6	2.2406E-6	2.19908E-6	1.85308	116.44	114.848	1.36723
7	3.00364E-7	2.93632E-7	2.24128	10.3471	8.51804	17.67703
8	4.01785E-8	3.91493E-8	2.56157	−97.9965	−100.12	2.16691
9	5.5205E-9	5.36266E-9	2.85916	154.224	151.857	1.53478
10	7.55096E-10	7.3048E-10	3.26065	46.5968	43.9948	5.58407
11	1.0412E-10	1.0075E-10	3.23845	−61.32	−64.556	5.27723
12	1.4427E-11	1.3691E-11	5.09822	−169.072	−171.84	1.63717
13	2.0032E-12	2.0751E-12	3.58615	83.1206	96.5032	16.10022

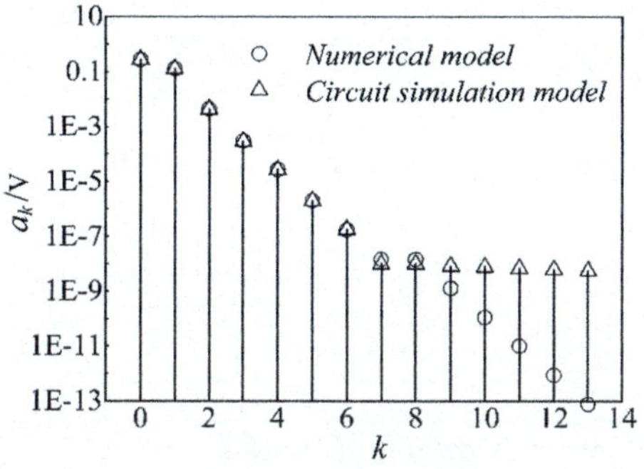

(a) Harmonic amplitude a_k

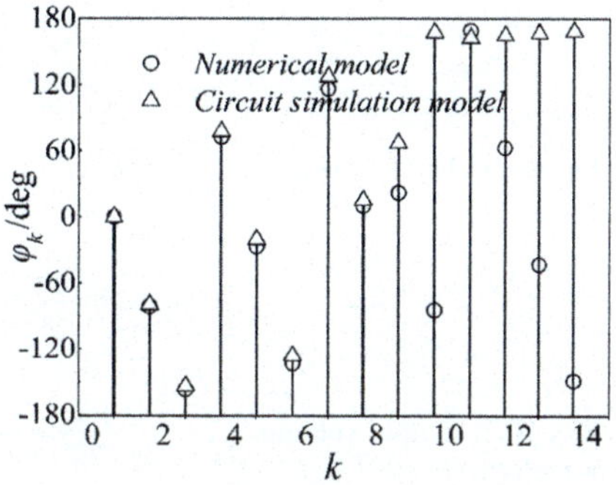

(b) Harmonic phase φ_k

Fig. 10.5 Harmonic amplitude and harmonic phase for the system ($Q_0 = 1$ V, $\omega = 5.6$ rad/s, $\alpha_1 = -1$, $\alpha_2 = 10$)

circuit simulation model and circuit of a first-order nonlinear system are studied by continuously changing the excitation amplitude at a certain excitation frequency. The numerical and simulated results show that the output voltage waveforms of the numerical model and circuit simulation model and circuit are closed to each other. In addition, there are small errors in amplitude and phase in the spectrum characteristics, and lower amplitude in the output waveforms of circuit.

As a result, the generalized harmonic balance method can well describe the periodic motion of the first-order nonlinear circuit at stable and unstable conditions, and it is of great significance to the analytical solution of nonlinear circuits. In addition, the method is practical for the accurate design and fault diagnosis of nonlinear circuits.

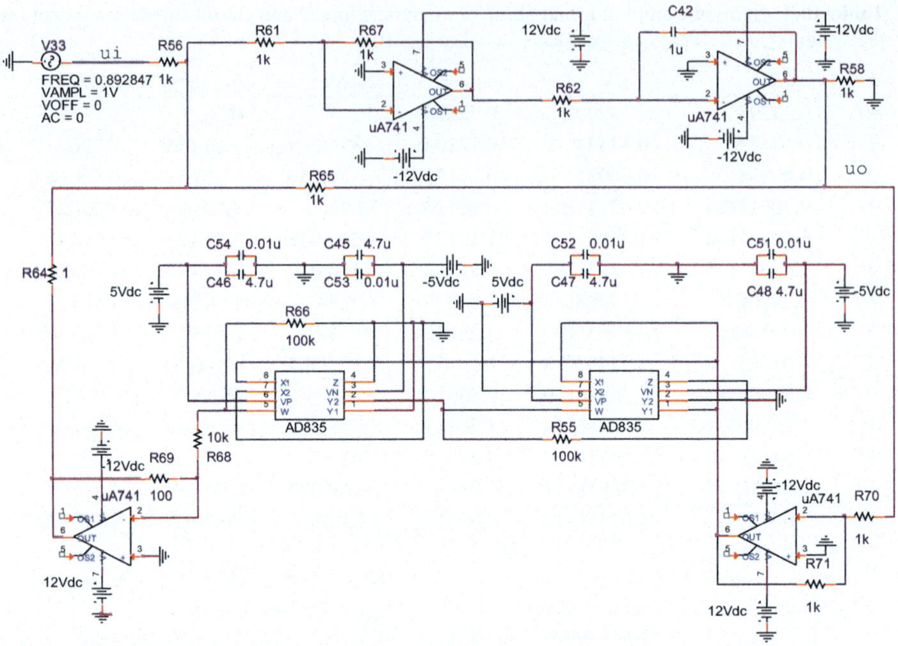

Fig. 10.6 The first-order nonlinear circuit ($Q_0 = 1$ V, $\omega = 5.6$ rad/s, $\alpha_1 = -1$, $\alpha_2 = 10$)

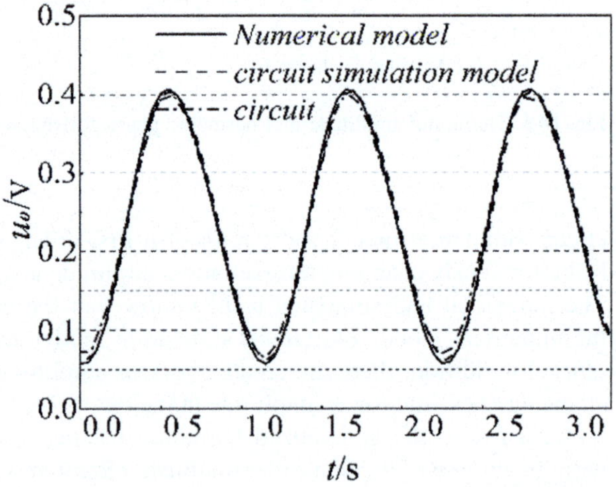

Fig. 10.7 Output voltage waveform ($Q_0 = 1$ V, $\omega = 5.6$ rad/s, $\alpha_1 = -1$, $\alpha_2 = 10$)

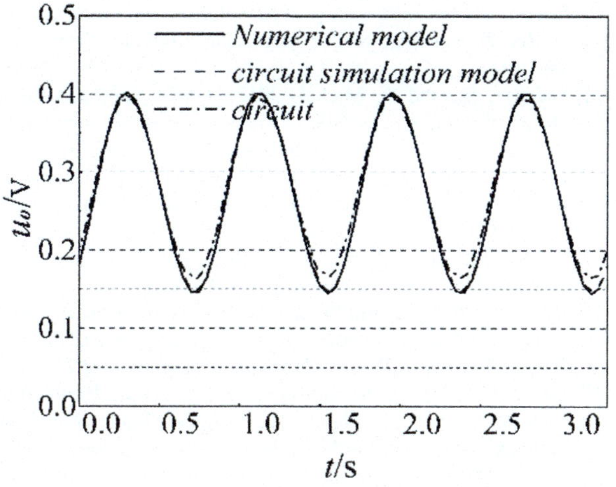

Fig. 10.8 Output voltage waveform ($Q_0 = 1$ V, $\omega = 7.6$ rad/s, $\alpha_1 = -1$, $\alpha_2 = 10$)

Acknowledgments This research is supported by National Natural Science Foundation of China (No. 51775437 and No. 51305355), Open Foundation of State Key Laboratory of Compressor Technology (No. SKL-YSJ201902) and National Key Research and Development Plan (No. 2019YFB1504601).

References

1. Li, X., B. Hu, X.T. Ling, and X. Zeng. 2002. A wavelet balance approach for steady-state analysis in nonlinear circuits. *IEEE Transactions on Circuits and Systems I: Fundamental Theory and Applications* 49 (5): 689–694.
2. Zhou, D., X.Q. Li, and W. Zhang. 1997. Nonlinear circuit simulation based on adaptive wavelet method, Proceedings of 1997 IEEE International Symposium on Circuits and Systems. *Circuits and Systems in the Information Age ISCAS 97* 3: 9–12.
3. Zhou, D., W. Cai, and W. Zhang. 1999. An adaptive wavelet method for nonlinear circuit simulation. *IEEE Transactions on Circuits and Systems I: Fundamental Theory and Applications* 46 (8): 1057–7122.
4. Emad, G., K. Roni, and S.N. Michel. 2000. A circuit reduction technique for finding the steady-state solution of nonlinear circuits. *IEEE Transactions on Microwave Theory & Techniques* 48 (12): 2389–2396.
5. Gong, W.Y., R.X. Gong, and H. Li. 2008. An approximate analytical solution of nonlinear circuit by the inverse operator method. *Journal of Guangxi University* 33 (B06): 159–161.
6. Khan, H., S.J. Liao, R.N. Mohapatra, and K. Vajravelu. 2009. An analytical solution for a nonlinear time-delay model in biology. *Communications in Nonlinear Science and Numerical Simulation* 14 (7): 3141–3148.
7. Wieslaw, M., and T. Zdzislaw. 2014. Mixed-mode oscillations and chaotic solutions of jerk (Newtonian) equations. *Journal of Computational and Applied Mathematics* 262: 373–383.
8. ———. 2010. Mixed-mode oscillations in a modified Chua's circuit. *Circuits, Systems and Signal Processing* 29 (6): 1075–1087.
9. ———. 2014. Mixed numerical and analytical analysis of nonlinear circuits with nonsmooth inputs: A hyperbolic algebra approach. *Midwest Symposium on Circuits and Systems* 23: 426–430.

10. Garcia, N. Periodic steady-state solutions of nonlinear circuits based on a differentiation matrix. *Proceedings of 2010 IEEE International Symposium on Circuits and Systems* 2010: 141–144.
11. Tadeusz, S., and R. Michal. 2017. A new approach to steady state analysis of nonlinear electrical circuits. *COMPEL - The international journal for computation and mathematics in electrical and electronic engineering* 36 (3): 716–728.
12. Farid, T., Y. Mohammad, and N. Honssein. 2017. Nonlinear dynamics of MEMS/NEMS resonators: analytical solution by the homotopy analysis method. *Microsystem Technologies* 23 (6): 1913–1926.
13. Bayat, M., and I. Pakar. 2013. On the approximate analytical solution to non-linear oscillation systems. *Shock and Vibration* 20 (1): 43–52.
14. Bayat, M., I. Pakar, and M. Bayat. 2016. Approximate analytical solution of nonlinear systems using homotopy perturbation method. *Proceedings of the Institution of Mechanical Engineers, Part E: Journal of Process Mechanical Engineering* 230 (1): 10–17.
15. Luo, A.C.J. 2012. *Continuous dynamical systems*. Beijing/Glen Carbon: HEP/L&H Scientific.
16. Luo, A.C.J., and J.Z. Huang. 2012. Approximate solutions of periodic motions in nonlinear systems via a generalized harmonic balance. *Journal of Vibration and Control* 18: 1661–1874.
17. ———. 2012. Analytical dynamics of period-m flows and chaos in nonlinear systems. *International Journal of Bifurcation and Chaos* 2 (04): 1250093.
18. Xu, Y.Y., A.C.J. Luo, and Z.B. Chen. 2017. Analytical solutions of periodic motions in 1-dimensional nonlinear systems. *Chaos, Solitons and Fractals* 97: 1–10.

Chapter 11
Model Reduction on Approximate Inertial Manifolds for NS Equations through Multilevel Finite Element Method and Hierarchical Basis

M. Nauman Aslam, Jiazhong Zhang, Nannan Dang, and Riaz Ahmad

11.1 Introduction

In computational analysis, mathematical models which exhibit complex physical behavior along high fidelity and strong nonlinearity has been the focus of many researchers recently. In engineering, nonlinear continuous dynamical systems describe the vast majority of phenomena, such as complex fluid flows, fluid structure interactions. For fluid flow problems, the motion of flow exhibits complex behavior and behind this complexity is the fact that the dynamics of the systems may be the product of multiple different interacting forces. It is known that some dynamic systems with continuum mechanics, are governed by a set of nonlinear partial differential equations, and some nonlinear behaviors, which can be captured by numerical solution, occur in such kind of systems. Despite the complexity of the flow topology, the entire behavior of most fluid flows is described by the so-called Navier-Stokes equations. Since in most cases, these equations do not provide the known analytical solutions, many numerical methods have been developed over the years to solve them [1–5]. For such systems, when applying discretization methods, the cost of computing time for the resulting equations is considerably expansive due to high number of degrees of freedom. Normally, finite element method is applied to approach the solution to such governing equations. Consequently, the resulting equations are mostly nonlinear dissipative evolution equation with a lot of degrees-of-freedom. In order to analyze the dynamics of the equations, the

M. Nauman Aslam (✉)
Department of Mathematics and Statistics, University of Lahore, Lahore, Punjab, Pakistan

J. Zhang · N. Dang
School of Energy and Power Engineering, Xi'an Jiaotong University Xi'an, Xi'an, China

R. Ahmad
Faculty of Science, Yibin University, Yibin, China

J. Zhang (ed.), *Dynamics and Fault Diagnosis of Nonlinear Rotors and Impellers*, Nonlinear Systems and Complexity 34, https://doi.org/10.1007/978-3-030-94301-1_11

system is changed into phase-space. However, in finite dimensional phase space with higher dimension, many difficulties appear from analyzing the nonlinear dynamics both qualitatively and quantitatively. Indeed, the disadvantage of the above mentioned numerical methods is that they require considerable computing time with great difficulties, due to large number of degrees-of-freedom, and the long term behaviors of the systems have great influences from the numerical computational errors [6, 7]. For such problem, there are some numerical methods. For example, the systems with local nonlinearity have been analyzed by IRS and balanced realization methods by Friswell et al. [8]. In structural dynamics with nonlinearities, there are some reduction methods mostly having numerical algorithms based on component synthesis techniques which can be efficiently used for linear dynamic systems and the solution to these problems are obtained through many numerical experiments and computational analysis [9, 10]. While Slaats et al. [11] introduced a reduction method based on three modes for nonlinear dynamical systems using finite element discretization. Therefore, model reduction for high dimensional or infinite dimensional fluid dynamic systems are required to overcome such difficulties.

On the other hand, in the study of long term behaviors of dissipative nonlinear evolution systems, one encounters with the global attractors, which is compact, invariant set with finite fractal dimensions attracting all the orbits of the systems uniformly, such attractors have complicated and dynamic structures [12]. A theoretical approach was shown that there is a approximate inertial manifold for the long term behavior of some dissipative partial differential equations in Titi [13]. Consequently, it has been proved that infinite-dimensional dissipative systems can be reduced to finite dimensional systems by reduction technique. Thus, a number of methods have been applied to construct a finite system exhibiting asymptotic dynamic behavior in the original dynamic system [14, 15]. More, an important feature of nonlinear dynamics is given related to model reduction, which explains the asymptotic properties based on spectral theory and decomposition process of the dynamical systems [16]. Therefore, developing a feasible model reduction method is very urgent as continuum dynamic systems are studied numerically.

For decades, the concept of inertial manifold is an important development in the study of systems with complicated attractor, since it reduces an infinite dimensional problem into a finite dimensional one without introducing errors [17]. Infinite dimensional dynamics systems are converted into finite dimensional ones by applying the theory of inertial manifolds. It also reduces the computing time and space keeping the topology of the systems intact. The presence of inertial manifold has been verified and shown reducing the degree of freedom involving in the dynamic systems. This provides a strong mathematical basis for model reduction in fluid dynamical systems [18]. An inertial manifold is defined to be a finite dimensional Lipschitz manifold, which is invariant for a semi-group defined on a phase space and attracts all the orbits exponentially and uniformly for the given initial conditions on the bounded set. In fact, the study of integral manifolds and center manifolds lead to methods for the construction of IMs theoretically. Since then, some model or dimension reduction method are proposed. Among them, the approach to model reduction by the theory of Inertial Manifolds (IMs) is more powerful and has been carried out for different periods of time [19, 20].

Theoretically, the presence of IMs usually sustains by very definitive spectral gap conditions. Therefore, practically the notion of Approximate Inertial Manifolds (AIMs) was introduced by Titi [13] and Foias et al. [21]. Haller and Sten [22] developed a model reduction method for the nonlinear mechanical systems by establishing certain conditions which can reduce the higher degrees of freedom to lower degrees of freedom. For this decomposition, they analyzed that there exists an exponential convergence defined on the slow manifold for the reduced model. For interaction between low and high modes, Schmidtmann [23] proposed a method using Approximate Inertial Manifolds by approximating the solution of MHD equation examining the behaviour of incompressible fluids. A Nonlinear Galerkin Method for coupling between lower and higher modes using proper orthogonal decomposition approached by Approximate Inertial Manifold was used by Kang et al. [24]. Laing et al. [25] applied AIMs for post processing calculation to lift the modes of the solution. On the other hand, some studies have been carried out for time-dependent second order autonomous dynamical systems which are nonlinear dissipative systems in nature by applying IMs and AIMs.

Further, some results have obtained by using AIMs approach for partial differential equations of second order depending upon time with delay [26]. So reduction of time-dependent second order autonomous dynamical systems, which are nonlinear dissipative systems with many degrees-of-freedom, has been developed for infinite dynamical systems together with mode synthesis analysis, which are used mostly in engineering and has induced effects of model reduction on long term behaviors of dynamical systems. Also, the concept of inertial manifold with time-delay along with traditional Galerkin method exhibiting the dynamic behavior of nonlinear phenomena such as buckling of shallow arches under the load impact is introduced using AIMs [27].

Recently, AIMs approach was used for reduction of model for Navier-Stokes Equation using multilevel finite element method, and some numerical examples are given to verify the method by Zhang et al. [18]. For such problem, the numerical approach to the AIMs is the main point. In finite element method, the hierarchical element or basis can describe the addition of further nodal elements and has no effect on the preceding ones in a major way. Particularly, hierarchical representation is the series for the approximation of a function. The terms which are added in the series does not affect the previous ones. To attain a hierarchical shape function, every component is required for optimal shape function, variables, physical description and discretization used must be organized hierarchically, this concept has been impressively demonstrated in Whiting and Jansen [28]. Application of mesh refinement through adaptive finite element approximation has been applied, giving uniform refined results in a hierarchy of approximation spaces [29]. Derivation and analysis for solving boundary value problems using hierarchical basis multigrid method on rectangular regions is used in Bank et al. [30]. Following that, the multilevel finite element method with hierarchical basis can be available to approach the AIMs, and that is the main idea in this paper.

In this study, numerical method (Euler Scheme) is used to approach the unknown terms involved in the system of governing equations. The pressure-correction technique is applied numerically for the time derivative. The hierarchical basis is applied to the improvement in approaching AIMs, showing a broad set of hierarchical modes, which are decomposed into two subspaces, a finite dimensional space spanned by low modes and high modes. The flow dynamics is initially carried out by low modes rapidly converging onto AIMs characterized by high modes. The hierarchical basis functions are based on the construction for the specified variable k-order nodal meshes. Then, the finite element mesh is defined based on topological hierarchy of mesh entities. The decomposition of the solution space spanned by IMs is based on refinement of mesh domain by hierarchical finite element basis. Finally, to investigate the numerical method, dynamics of flow around an airfoil is carried out to show the accuracy and reduced computing time in comparison with the experimental results and traditional commercial software ANSYS(Fluent), showing a close agreement setting a benchmark for the problem.

11.2 Mathematical Modelling

11.2.1 Governing Equations

The Incompressible Navier-Stokes Equations with elementary variables are

$$\frac{\partial u_i}{\partial t} + u_j \frac{\partial u_i}{\partial x_j} = -\frac{1}{\rho}\frac{\partial p}{\partial x_i} + \nu \frac{\partial^2 u_i}{\partial x_j \partial x_j} \tag{11.1}$$

$$\frac{\partial u_j}{\partial x_j} = 0 \tag{11.2}$$

where u is the flow velocity, p is the pressure, ν is the kinemetic viscosity, ρ is the fluid density. The characteristics of flow depends on the non-dimensional parameters.

The initial conditions
$u(x_i, 0) = u_0(x_i)$

11.2.2 The Boundary Conditions

We introduce the following system defining the set of boundary conditions $\Gamma = \Gamma_1 \cup \Gamma_2$

where Γ_1 is boundary condition for velocity, Γ_2 boundary condition for pressure and on boundary Γ_1 and $u_i = \tilde{u}_i,\ i = 1, 2$ on boundary Γ_2 . Also $p_{ij}n_j = \tilde{p}_i\ (i, j = 1, 2)$, n_j is outward unit normal vector component and tensor $p_{ij} = -\frac{p}{\rho}\delta_{ij} + \upsilon(\frac{\partial u_i}{\partial x_j} + \frac{\partial u_j}{\partial x_i})$.

11.2.3 Existence of Inertial Manifolds

A brief introduction to Inertial Manifolds and its existence is given in the following as we study the dynamical system generated by the evolution equation (11.3). To approach inertial manifold, spectral method is normally used, and the governing equations are projected onto the space spanned by eigenfunctions of positive definite operator in terms of two components, namely small and large components.

Let u be the solution to governing equation in an approximate Hilbert space. For model reduction, a mapping from higher modes to lower modes in the system is constructed.

$$\frac{du}{dt} + Au + E(u) = f,\ \ t > 0 \tag{11.3}$$

$$u(0) = u_0 \tag{11.4}$$

In Eq. (11.3), A is the space of self-adjoint linear operator H, $E(u) = E(u, u)$ corresponds to linear and low-order, $E(u, u)$ is also termed as bilinear operator. The Navier-Stokes Equations subject to boundary conditions can be written as abstract dissipative evolution equation given by (11.3), where A is positive definite operator. As spectral method is used to approach Inertial Manifolds, the components involved for the projection are approached by eigenfunctions of the definite operator of the governing equations. Now if $\omega_j, (j = 1, 2, \ldots)$, denotes the eigenfunctions of the operator for the governing equations, gives $A\omega_j = \lambda\omega_j$, which leads to the construction of Inertial Manifolds in terms of components P and Q. Since the existence of an IMs for Navier-Stokes Equations has not been proven, therefore it is obliged to use manifolds which are close to global attractors to approximate the permanent regime of dynamical systems. In this paper we provide AIMs approach to NSE, where Inertial Manifold for dissipative equations is obtained by modifying the Navier-Stokes Equations.

The subspace P is spanned by the eigenfunctions or low modes and the subspace Q is spanned by high modes. So, if inertial manifolds exists under the spectral gap condition, then we have a graph defined as $\phi : P \longrightarrow Q$

Then

$$\phi : P_m H \longrightarrow Q_m H$$

exists, where P_m and Q_m denote the orthogonal projections, i.e.,

$$P_m : H \longrightarrow span\{\omega_1, \omega_2, \ldots, \omega_m\},\ Q_m = I - P_m \tag{11.5}$$

So the projection is obtained as

$$u(t) = p_m(t) + q_m(t) \tag{11.6}$$

where $p_m = P_m u$ and $q_m = Q_m u$, p_m and q_m are large components and small components. Equation (11.3) is projected onto the space $P_m H$ and $Q_m H$, respectively, obtaining the following two equations,

$$\frac{dp_m}{dt} + Ap_m + P_m R(p_m + q_m) = P_m f,\ \ p_m(0) = p_m^0 = P_m u_0 \tag{11.7}$$

$$\frac{dq_m}{dt} + Aq_m + Q_m R(p_m + q_m) = Q_m f,\ \ q_m(0) = q_m^0 = Q_m u_0 \tag{11.8}$$

where m is large enough, the time derivative term $\frac{dq_m}{dt} \approx 0$, and $Q_m R(p_m + q_m) \approx Q_m R(p_m)$. Finally, Eq. (11.8) becomes

$$Aq_m + Q_m R(p_m) = Q_m f \tag{11.9}$$

Thus obtaining

$$q_m = \Phi(p_m) = A^{-1}[Q_m f - Q_m R(p_m)],\ \ p_m \in P_m H \tag{11.10}$$

Eq. (11.10) results in H the finite dimensional manifold.

11.2.4 Multilevel Finite Element Method and Its Basis

Numerically, Inertial Manifold is implemented by using finite element method in such a way that spannings of both subspaces are given as coarse grid element space and incremental fine grid element space, which are finite. Meshing of the domain is carried out in terms of quadrangle elements, especially for the coarse grid finite element space given by Fig. 11.1. The initial element is represented as quadrangle elements having four nodes as initial mesh shown in Fig. 11.2, whereas the incremental refined mesh by addition of further 27 nodes in the quadrangle elements is shown in Fig. 11.3. Then, the fundamental finite element space is spanned by construction of hierarchical shape functions as basis functions in the initial element space accordingly.

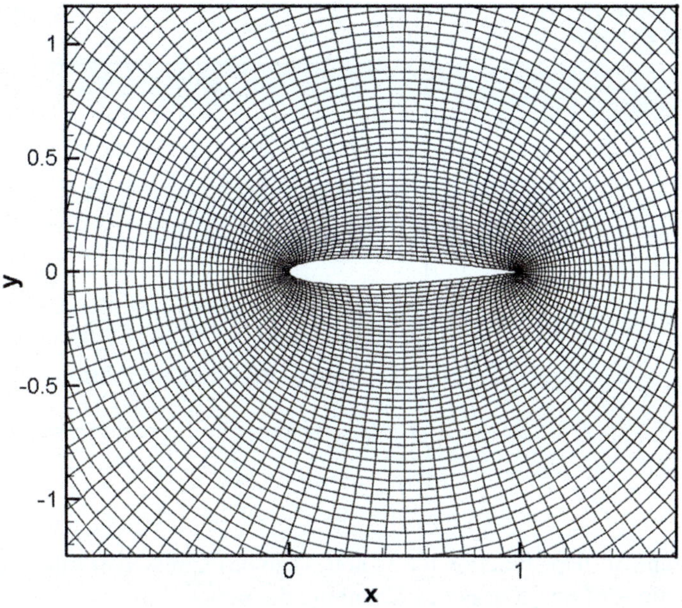

Fig. 11.1 The mesh of the domain

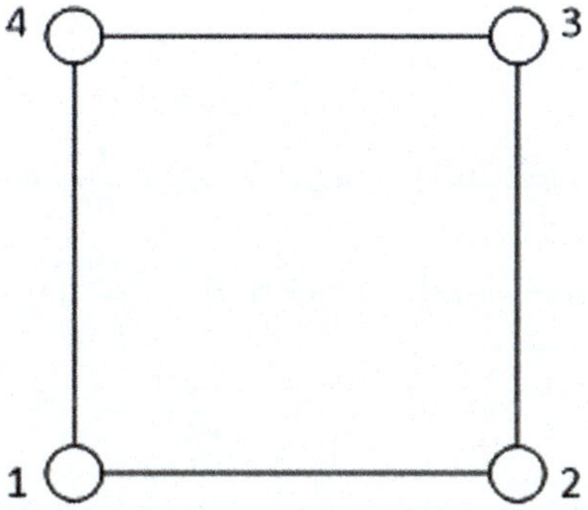

Fig. 11.2 Quadrangle element with 4-nodes

The quadrangle elements with four nodes have the following shape functions,

$$N_1 = \frac{1}{4}(1-\xi)(1-\eta),\ N_2 = \frac{1}{4}(1+\xi)(1-\eta) \tag{11.11}$$

$$N_3 = \frac{1}{4}(1+\xi)(1+\eta),\ N_4 = \frac{1}{4}(1-\xi)(1+\eta) \tag{11.12}$$

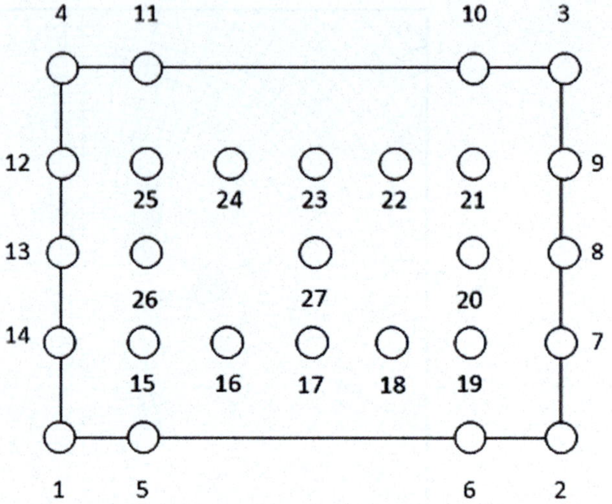

Fig. 11.3 Quadrangle element with 27-nodes

The hierarchical shape function for 27 nodes can be expressed as following with the shape functions of first four nodes remaining the same,

$$H_1 = N_1,\ H_2 = N_2,\ H_3 = N_3,\ H_4 = N_4, \tag{11.13}$$

$$\bar{H}_5 = \frac{9}{120}(\xi^2-1)(3\xi-2)(\eta^2-\eta)(4\eta^2-1),\ \bar{H}_6 = \frac{-9}{120}(\xi^2-1)(3\xi+2)(\eta^2-\eta)(4\eta^2-1), \tag{11.14}$$

$$\bar{H}_7 = \frac{-1}{60}(\xi^2+\xi)(9\xi^2-1)(9\xi^2-4)(\eta^3-\eta)(2\eta-1),\ \bar{H}_8 = \frac{1}{10}(\xi+1)(9\xi^2-4)(\eta^2-1)(4\eta^2-1), \tag{11.15}$$

$$\bar{H}_9 = \frac{-1}{60}(\xi^2+\xi)(9\xi^2-1)(9\xi^2-4)(\eta^3-\eta)(2\eta+1),\ \bar{H}_{10} = \frac{-9}{120}(3\xi+2)(\xi^2-1)(\eta^2+\eta)(4\eta^2-1), \tag{11.16}$$

$$\bar{H}_{11} = \frac{9}{120}(\xi^2-\xi)(3\xi^2-2)(\eta^2+\eta)(4\eta^2-1),\ \bar{H}_{12} = \frac{-1}{60}(\xi^2-\xi)(9\xi^2-1)(9\xi^2-4)(\eta^3-\eta)(2\eta+1), \tag{11.17}$$

$$\bar{H}_{13} = \frac{-1}{10}(\xi-1)(9\xi^2-4)(\eta^2-1)(4\eta^2-1),\ \bar{H}_{14} = \frac{-1}{60}(\xi^2-\xi)(9\xi^2-1)(9\xi^2-4)(\eta^3-\eta)(2\eta-1), \tag{11.18}$$

$$\bar{H}_{15} = \frac{9}{10}(\xi^2-\xi)(3\xi^2-2)(9\xi^2-1)(\eta^3-\eta)(2\eta-1),\ \bar{H}_{16} = \frac{9}{32}(\xi^3-\xi)(3\xi-1)(9\xi^2-4)(2\eta-1), \tag{11.19}$$

$$\bar{H}_{17} = \frac{1}{8}(\xi^2-\xi)(9\xi^2-1)(9\xi^2-4)(2\eta-1),\ \bar{H}_{18} = \frac{9}{32}(\xi^3-\xi)(3\xi+1)(9\xi^2-4)(2\eta-1), \tag{11.20}$$

$$\bar{H}_{19} = \frac{9}{10}(\xi^2-\xi)(3\xi+2)(9\xi^2-1)(\eta^3-\eta)(2\eta-1),\ \bar{H}_{20} = \frac{-9}{20}(\xi^2-1)(3\xi+2)(\eta^2-\eta)(4\eta^2-1), \tag{11.21}$$

$$\bar{H}_{21} = \frac{3}{10}(\xi^2-\xi)(3\xi+2)(9\xi^2-1)(\eta^3-\eta)(2\eta+1),\ \bar{H}_{22} = \frac{9}{32}(\xi^3-\xi)(3\xi+1)(9\xi^2-4)(2\eta+1), \tag{11.22}$$

$$\bar{H}_{23} = \frac{-1}{8}(\xi^2 - 1)(9\xi^2 - 4)(9\xi^2 - 1)(2\eta + 1),\ \bar{H}_{24} = \frac{9}{32}(\xi^3 - \xi)(3\xi - 1)(9\xi^2 - 4)(2\eta + 1), \tag{11.23}$$

$$\bar{H}_{25} = \frac{3}{10}(\xi^2 - \xi)(3\xi - 2)(9\xi^2 - 1)(\eta^3 - \eta)(2\eta + 1),\ \bar{H}_{26} = \frac{-9}{20}(\xi^2 - \xi)(3\xi - 2)(\eta^2 - 1)(4\eta^2 - 1), \tag{11.24}$$

$$\bar{H}_{27} = \frac{-1}{4}(\xi^2 - 1)(9\xi^2 - 4)(4\eta^2 - 1) \tag{11.25}$$

In terms of Inertial Manifolds, the elements of velocity $u^{(e)}$ are decomposed into two components $y^{(e)}$ and $z^{(e)}$, which are large component and small component, respectively, reads,

$$u^{(e)} = y^{(e)} + z^{(e)} \tag{11.26}$$

In summation notation, the above expression can be written as

$$u^{(e)} = \sum_{1}^{4} H_i y_i^{(e)},\ z^{(e)} = \sum_{5}^{27} \bar{H}_i z_i^{(e)} \tag{11.27}$$

11.3 Numerical Scheme for Navier-Stokes Equations

The governing equations for the incompressible flow are

$$\frac{\partial u_j}{\partial x_j} = 0 \tag{11.28}$$

$$\frac{\partial u_i}{\partial t} + u_j \frac{\partial u_i}{\partial x_j} = -\frac{1}{\rho}\frac{\partial p}{\partial x_i} + \nu \frac{\partial^2 u_i}{\partial x_j \partial x_j} \tag{11.29}$$

Euler method is used to approach the time derivative as pressure-correction technique is implemented numerically for the governing equations, where $i = 1, 2$; n is the step time

$$\frac{\partial \hat{u}_j^{(n+1)}}{\partial x_j} = 0 \tag{11.30}$$

$$\frac{\hat{u}_i^{(n+1)} - u_i^{(n)}}{\Delta t} + u_j^{(n)} \frac{\partial u_j^{(n)}}{\partial x_j} = -\frac{1}{\rho}\frac{\partial p^{(n+1)}}{\partial x_i} + \nu \frac{\partial^2 u_i^{(n)}}{\partial x_j \partial x_j} \tag{11.31}$$

An intermediate velocity $\hat{u}$ is introduced in the implicit method, and then the momentum equation can be rewritten as

$$\frac{\hat{u}_i - u_i^{(n)}}{\Delta t} + u_j^{(n)} \frac{\partial u_j^{(n)}}{\partial x_j} = -\frac{1}{\rho} \frac{\partial p^{(n)}}{\partial x_i} + \nu \frac{\partial^2 u_i^{(n)}}{\partial x_j \partial x_j} \tag{11.32}$$

Subtracting Eq. (11.31) from Eq. (11.32) as the quantity p is explicit in the method, yields

$$u_i^{n+1} - u_i = -\frac{\Delta t}{\rho} \left(\frac{\partial p^{(n+1)}}{\partial x_i} - \frac{\partial p^{(n)}}{\partial x_i} \right) \tag{11.33}$$

Introducing

$$u_i^{n+1} - u_i = -\frac{\partial \phi}{\partial x_i} \tag{11.34}$$

$$-\frac{\partial \phi}{\partial x_i} = -\frac{\Delta t}{\rho} \left(\frac{\partial p^{(n+1)}}{\partial x_i} - \frac{\partial p^{(n)}}{\partial x_i} \right) \tag{11.35}$$

$$\frac{\partial \phi}{\partial x_i} = \frac{\Delta t}{\rho} \frac{\partial}{\partial x_i} (p^{(n+1)} - p^{(n)}) \tag{11.36}$$

$$\frac{\rho}{\Delta t} \phi = (p^{(n+1)} - p^{(n)}) \tag{11.37}$$

Finally, pressure can be obtained as

$$p^{(n+1)} = \frac{\rho}{\Delta t} \phi + p^{(n)} \tag{11.38}$$

Also, in order to obtain ϕ, we take the divergence of Eq. (11.34) and applying continuity equation,

$$\nabla (u_i^{n+1} - u_i) = \nabla \left(-\frac{\partial \phi}{\partial x_i} \right) \tag{11.39}$$

$$\frac{\partial \hat{u_1}}{\partial x_1} + \frac{\partial \hat{u_2}}{\partial x_2} = \frac{\partial}{\partial x_1} \frac{\partial \phi}{\partial x_1} + \frac{\partial}{\partial x_2} \frac{\partial \phi}{\partial x_2} \tag{11.40}$$

$$\frac{\partial \hat{u_1}}{\partial x_1} + \frac{\partial \hat{u_2}}{\partial x_2} = \frac{\partial^2 \phi}{\partial x_1} + \frac{\partial^2 \phi}{\partial x_2} \tag{11.41}$$

which is relevant to the poisson equation. So when substituting the value of ϕ into Eqs. (11.38) and (11.37), both u and p can be obtained finally.

11.4 Weak Form of Navier-Stokes Equation

Before applying Multilevel Finite Element method to the governing equations, the weak form of governing equations should be obtained, i.e.,

$$\int_{\Omega} \frac{\partial u_j}{\partial x_j} \delta p \, d\Omega = 0 \tag{11.42}$$

$$\int_{\Omega} \left[\frac{\partial u_i}{\partial t} + u_j \frac{\partial u_i}{\partial x_j} + \frac{1}{\rho} \frac{\partial p}{\partial x_i} - \nu \frac{\partial^2 u_i}{\partial x_j \partial x_j} \right] \delta u_i \, d\Omega = 0 \tag{11.43}$$

Using Greens Formulas both Eqs. (11.42) and (11.43) can be rewritten as

$$\int_{\Omega} u_j \frac{\partial}{\partial x_j} (\delta p) \, d\Omega = \int_{\Gamma_1} u_n \delta p \, d\Gamma \tag{11.44}$$

$$\int_{\Omega} \Big\{ \big(\frac{\partial u_i}{\partial t} + u_j \frac{\partial u_i}{\partial x_j} \big) \delta u_i + \Big[\frac{-p}{\rho} \delta_{ij} + \nu \big(\frac{\partial u_i}{\partial x_j} + \frac{\partial u_j}{\partial x_i} \big) \Big] \frac{\partial}{\partial x_j} (\delta u_i) \Big\} d\Omega = \int_{\Gamma_2} \bar{p}_i \delta u_i d\Gamma \tag{11.45}$$

We first solve Eq. (11.45) for $i = 1, 2$

$$\int_{\Omega} \Big\{ \big(\frac{\partial u_1}{\partial t} + u_1 \frac{\partial u_1}{\partial x_1} + u_2 \frac{\partial u_1}{\partial x_2} \big) \delta u_i + \Big[- \frac{p}{\rho} \frac{\partial}{\partial x_1} (\delta u_i) + \nu \Big(2 \frac{\partial u_1}{\partial x_1} \frac{\partial}{\partial x_1} (\delta u_i) + \frac{\partial u_1}{\partial x_2} \frac{\partial}{\partial x_2} (\delta u_i) + \frac{\partial u_2}{\partial x_1} \frac{\partial}{\partial x_2} (\delta u_i) \Big) \Big] \Big\} \, d\Omega = \int_{\Gamma_2} \bar{p}_i \delta u_i \, d\Gamma \tag{11.46}$$

$$\int_{\Omega} \Big\{ \big(\frac{\partial u_2}{\partial t} + u_1 \frac{\partial u_2}{\partial x_1} + u_2 \frac{\partial u_2}{\partial x_2} \big) \delta u_i + \Big[- \frac{p}{\rho} \frac{\partial}{\partial x_2} (\delta u_i) + \nu \Big(2 \frac{\partial u_2}{\partial x_2} \frac{\partial}{\partial x_2} (\delta u_i) + \frac{\partial u_2}{\partial x_1} \frac{\partial}{\partial x_1} (\delta u_i) + \frac{\partial u_1}{\partial x_2} \frac{\partial}{\partial x_1} (\delta u_i) \Big) \Big] \Big\} \, d\Omega = \int_{\Gamma_2} \bar{p}_i \delta u_i \, d\Gamma \tag{11.47}$$

By Inertial Manifold, the elements of velocity and pressure are decomposed mainly into two components, and applying the hierarchical basis functions, the above equations can be written as,

$$\int_{\Omega} \frac{\partial \bar{u}_1}{\partial t} \bar{H}_i \bar{H}_j d\Omega + \int_{\Omega} u_1 \frac{\partial u_1}{\partial x_1} \bar{H}_i d\Omega + \Big(\int_{\Omega} u_1 \bar{H}_i \frac{\partial \bar{H}_j}{\partial x_1} d\Omega \Big) \bar{u}_1 + \Big(\int_{\Omega} \frac{\partial u_1}{\partial x_1} \bar{H}_i \bar{H}_j d\Omega \Big) \bar{u}_1$$
$$+ \Big(\int_{\Omega} \bar{H}_i \frac{\partial \bar{H}_j}{\partial x_1} \bar{H}_l d\Omega \Big) \bar{u}_1 \bar{u}_1 + \int_{\Omega} u_2 \frac{\partial u_1}{\partial x_2} \bar{H}_i d\Omega + \Big(\int_{\Omega} u_2 \bar{H}_i \frac{\partial \bar{H}_j}{\partial x_2} d\Omega \Big) \bar{u}_1$$

$$+\Big(\int_{\Omega}\frac{\partial u_1}{\partial x_2}\bar{H}_i\bar{H}_j d\Omega\Big)\bar{u}_2+\Big(\int_{\Omega}\bar{H}_i\frac{\partial \bar{H}_j}{\partial x_2}\bar{H}_l d\Omega\Big)\bar{u}_2\bar{u}_1-\frac{1}{\rho}\int_{\Omega}p\frac{\partial \bar{H}_i}{\partial x_1}d\Omega$$
$$+\int_{\Omega}\nu\Big[2\frac{\partial u_1}{\partial x_1}\frac{\partial \bar{H}_i}{\partial x_1}+\frac{\partial u_1}{\partial x_2}\frac{\partial \bar{H}_i}{\partial x_2}+\frac{\partial u_1}{\partial x_1}\frac{\partial \bar{H}_i}{\partial x_2}\Big]d\Omega+\Big(\int_{\Omega}\nu\Big[2\frac{\partial \bar{H}_i}{\partial x_1}\frac{\partial \bar{H}_j}{\partial x_1}$$
$$+\frac{\partial \bar{H}_i}{\partial x_2}\frac{\partial \bar{H}_j}{\partial x_2}\Big]d\Omega\Big)\bar{u}_1+\Big(\int_{\Omega}\nu\frac{\partial \bar{H}_i}{\partial x_2}\frac{\partial \bar{H}_j}{\partial x_1}d\Omega\Big)\bar{u_2}=\int_{\Gamma_2}\bar{p}_{1m}\psi_m\bar{H}_i d\Gamma \quad (11.48)$$

$$\int_{\Omega}\frac{\partial \bar{u}_2}{\partial t}\bar{H}_i\bar{H}_j d\Omega+\int_{\Omega}u_1\frac{\partial u_2}{\partial x_1}\bar{H}_i d\Omega+\Big(\int_{\Omega}u_1\bar{H}_i\frac{\partial \bar{H}_j}{\partial x_1}d\Omega\Big)\bar{u}_2+\Big(\int_{\Omega}\frac{\partial u_2}{\partial x_1}\bar{H}_i\bar{H}_j d\Omega\Big)\bar{u}_1$$
$$+\Big(\int_{\Omega}\bar{H}_i\frac{\partial \bar{H}_j}{\partial x_1}\bar{H}_l d\Omega\Big)\bar{u}_1\bar{u}_2+\int_{\Omega}u_2\frac{\partial u_2}{\partial x_2}\bar{H}_i d\Omega+\Big(\int_{\Omega}u_2\bar{H}_i\frac{\partial \bar{H}_j}{\partial x_2}d\Omega\Big)\bar{u}_2$$
$$+\Big(\int_{\Omega}\frac{\partial u_2}{\partial x_2}\bar{H}_i\bar{H}_j d\Omega\Big)\bar{u}_2+\Big(\int_{\Omega}\bar{H}_i\frac{\partial \bar{H}_j}{\partial x_2}\bar{H}_l d\Omega\Big)\bar{u}_2\bar{u}_2-\frac{1}{\rho}\int_{\Omega}p\frac{\partial \bar{H}_i}{\partial x_2}d\Omega$$
$$+\int_{\Omega}\nu\Big[2\frac{\partial u_2}{\partial x_2}\frac{\partial \bar{H}_i}{\partial x_2}+\frac{\partial u_2}{\partial x_1}\frac{\partial \bar{H}_i}{\partial x_1}+\frac{\partial u_1}{\partial x_2}\frac{\partial \bar{H}_i}{\partial x_1}\Big]d\Omega+\Big(\int_{\Omega}\nu\Big[2\frac{\partial \bar{H}_i}{\partial x_2}\frac{\partial \bar{H}_j}{\partial x_2}$$
$$+\frac{\partial \bar{H}_i}{\partial x_1}\frac{\partial \bar{H}_j}{\partial x_1}\Big]d\Omega\Big)\bar{u}_2+\Big(\int_{\Omega}\nu\frac{\partial \bar{H}_i}{\partial x_1}\frac{\partial \bar{H}_j}{\partial x_2}d\Omega\Big)\bar{u}_1=\int_{\Gamma_2}\bar{p}_{2m}\psi_m\bar{H}_i d\Gamma \quad (11.49)$$

From Eqs. (11.48) and (11.49), re-writing the terms as:

$$\mathbf{A}=\int_{\Omega}\bar{H}_i\bar{H}_j d\Omega,\quad \mathbf{B}=\int_{\Omega}\bar{H}_i\frac{\partial \bar{H}_j}{\partial x_1}\bar{H}_l d\Omega,\quad \mathbf{C}=\int_{\Omega}\bar{H}_i\frac{\partial \bar{H}_j}{\partial x_2}\bar{H}_l d\Omega$$

$$\mathbf{F}_1=\int_{\Omega}u_1\bar{H}_i\frac{\partial \bar{H}_j}{\partial x_1}d\Omega+\int_{\Omega}\frac{\partial u_1}{\partial x_1}\bar{H}_i\bar{H}_j d\Omega+\int_{\Omega}u_2\bar{H}_i\frac{\partial \bar{H}_j}{\partial x_2}d\Omega+\int_{\Omega}\nu\Big[2\frac{\partial \bar{H}_i}{\partial x_1}\frac{\partial \bar{H}_j}{\partial x_1}+\frac{\partial \bar{H}_i}{\partial x_2}\frac{\partial \bar{H}_j}{\partial x_2}\Big]d\Omega,$$

$$\mathbf{F}_2=\int_{\Omega}\frac{\partial u_1}{\partial x_2}\bar{H}_i\bar{H}_j d\Omega+\int_{\Omega}\nu\frac{\partial \bar{H}_i}{\partial x_1}\frac{\partial \bar{H}_j}{\partial x_2}d\Omega \qquad \mathbf{F}_3=\int_{\Omega}\frac{\partial u_2}{\partial x_1}\bar{H}_i\bar{H}_j d\Omega+\int_{\Omega}\nu\frac{\partial \bar{H}_i}{\partial x_2}\frac{\partial \bar{H}_j}{\partial x_1}d\Omega,$$

$$\mathbf{F}_4=\int_{\Omega}u_1\bar{H}_i\frac{\partial \bar{H}_j}{\partial x_1}d\Omega+\int_{\Omega}u_2\bar{H}_i\frac{\partial \bar{H}_j}{\partial x_2}d\Omega+\int_{\Omega}\frac{\partial u_1}{\partial x_1}\bar{H}_i bar H_j d\Omega+\int_{\Omega}\nu\Big[2\frac{\partial \bar{H}_i}{\partial x_2}\frac{\partial \bar{H}_j}{\partial x_2}+\frac{\partial \bar{H}_i}{\partial x_1}\frac{\partial \bar{H}_j}{\partial x_1}\Big]d\Omega$$

$$\mathbf{E}_1=-\int_{\Omega}u_1\frac{\partial u_1}{\partial x_1}\bar{H}_i d\Omega-\int_{\Omega}u_2\frac{\partial u_1}{\partial x_2}\bar{H}_i d\Omega+\frac{1}{\rho}\int_{\Omega}p\frac{\partial \bar{H}_i}{\partial x_1}d\Omega-\int_{\Omega}\nu\Big[2\frac{\partial u_1}{\partial x_1}\frac{\partial \bar{H}_i}{\partial x_1}+\frac{\partial u_1}{\partial x_2}\frac{\partial \bar{H}_i}{\partial x_2}$$
$$+\frac{\partial u_1}{\partial x_1}\frac{\partial \bar{H}_i}{\partial x_2}\Big]d\Omega+\int_{\Gamma_2}\bar{p}_{1m}\psi_m\bar{H}_i d\Gamma$$

$$\mathbf{E}_2 = -\int_\Omega u_1 \frac{\partial u_2}{\partial x_1} \bar{H}_i d\Omega - \int_\Omega u_2 \frac{\partial u_2}{\partial x_2} \bar{H}_i d\Omega + \frac{1}{\rho}\int_\Omega p \frac{\partial \bar{H}_i}{\partial x_2} d\Omega - \int_\Omega \nu\Big[2\frac{\partial u_2}{\partial x_2}\frac{\partial \bar{H}_i}{\partial x_2} + \frac{\partial u_2}{\partial x_1}\frac{\partial \bar{H}_i}{\partial x_1} + \frac{\partial u_1}{\partial x_2}\frac{\partial \bar{H}_i}{\partial x_1}\Big] d\Omega + \int_{\Gamma_2} \bar{p}_{2m}\psi_m \bar{H}_i d\Gamma$$

Now solving Eq. (11.44) for $i = 1, 2$

$$\int_\Omega u_1 \frac{\partial}{\partial x_1}(\delta p) d\Omega + \int_\Omega u_2 \frac{\partial}{\partial x_2}(\delta p) d\Omega = \int_{\Gamma_1} u_n \delta p d\Gamma$$

$$\int_\Omega \Big[(u_{1j} + \bar{u}_{1j}\bar{H}_j)\frac{\partial}{\partial x_1}\psi_k + (u_{2j} + \bar{u}_{2j}\bar{H}_j)\frac{\partial}{\partial x_2}\psi_k\Big]\bar{H}_j d\Omega = \int_{\Gamma_1} \bar{u}_{ni} H_i \psi_k d\Gamma$$

$$\int_\Omega u_1 \bar{H}_i \frac{\partial}{\partial x_1}\psi_k d\Omega + \Big(\int_\Omega \bar{H}_i \bar{H}_j \frac{\partial}{\partial x_1}\psi_k d\Omega\Big)\bar{u}_1 + \int_\Omega u_2 \bar{H}_i \frac{\partial}{\partial x_2}\psi_k d\Omega + \Big(\int_\Omega \bar{H}_i \bar{H}_j \frac{\partial}{\partial x_1}\psi_k d\Omega\Big)\bar{u}_2 = \int_{\Gamma_1} \bar{u}_{ni} H_i \psi_k d\Gamma$$

Re-writing the above equations as;

$$\mathbf{M}_1 = \int_\Omega \bar{H}_i \bar{H}_j \frac{\partial}{\partial x_1}\psi_k d\Omega, \quad \mathbf{M}_2 = \int_\Omega \bar{H}_i \bar{H}_j \frac{\partial}{\partial x_2}\psi_k d\Omega$$

$$\mathbf{G} = \int_\Gamma \bar{u}_{ni} H_i \psi_k d\Gamma - \int_\Omega u_1 \bar{H}_i \frac{\partial}{\partial x_1}\psi_k d\Omega - \int_\Omega u_2 \bar{H}_i \frac{\partial}{\partial x_2}\psi_k d\Omega$$

Since the relationship between small and large components of the velocity in element can be expressed as

$$u^{(e)} = y^{(e)} + z^{(e)} = \sum_1^4 H_i y_i^{(e)} + \sum_5^{27} \bar{H}_i z_i^{(e)} \tag{11.50}$$

Also the pressure in the element is given as,

$$P^{(e)} = \sum_1^4 \psi_i P_i^{(e)} + \sum_5^{27} \bar{\psi}_i P_i^{(e)} \tag{11.51}$$

We choose the weight function in Eq. (11.45) as $\delta u_i = \bar{H}_m (m = 5, 6, \ldots, 27)$, and for Eq. (11.44) we set $\delta p = \bar{\psi}_n (n = 5, 6, \ldots, 27)$. So Eqs. (11.44) and (11.45) can be rewritten as

$$\mathbf{A}\dot{\bar{u}}_1 + \mathbf{B}\bar{u}_1\bar{u}_1 + \mathbf{C}\bar{u}_2\bar{u}_1 + \mathbf{D}_1 p + \mathbf{F}_1\bar{u}_1 + \mathbf{F}_2\bar{u}_2 = \mathbf{E}_1 \tag{11.52}$$

$$\mathbf{A}\dot{\bar{u}}_2 + \mathbf{B}\bar{u}_1\bar{u}_2 + \mathbf{C}\bar{u}_2\bar{u}_2 + \mathbf{D}_2 p + \mathbf{F}_3\bar{u}_1 + \mathbf{F}_4\bar{u}_2 = \mathbf{E}_2 \tag{11.53}$$

$$\mathbf{M}_1\bar{u}_1 + \mathbf{M}_2\bar{u}_2 = \mathbf{G} \tag{11.54}$$

The relationship between y_i and z_i is clearly expressed by Eqs. (11.50) and (11.51). We can get the values of y_i since z_i is known by the relationship. To approach the time derivative, Euler method is used,

$$\dot{u}_k = \frac{u_k^{(i+1)} - u_k^{(i)}}{\Delta t} \quad (k = 1, 2) \tag{11.55}$$

For expressing explicitly at $t = t_{i+1}$, we establish the following equations,

$$\begin{aligned} \mathbf{A}u_1^{(i+1)} + \Delta t(\mathbf{B}u_1^{(i+1)}u_1^{(i+1)} + \mathbf{C}u_2^{(i+1)}u_1^{(i+1)} + \mathbf{D}_1\bar{p}^{(i+1)} \\ + \mathbf{F}_1 u_1^{(i+1)} + \mathbf{F}_2 u_2^{(i+1)}) = \Delta t\mathbf{E}_1 + \mathbf{A}u_1^{(i)} \end{aligned} \tag{11.56}$$

$$\begin{aligned} \mathbf{A}u_2^{(i+1)} + \Delta t(\mathbf{B}u_1^{(i+1)}u_2^{(i+1)} + \mathbf{C}u_2^{(i+1)}u_2^{(i+1)} + \mathbf{D}_2\bar{p}^{(i+1)} \\ + \mathbf{F}_3 u_1^{(i+1)} + \mathbf{F}_4 u_2^{(i+1)}) = \Delta t\mathbf{E}_2 + \mathbf{A}u_2^{(i)} \end{aligned} \tag{11.57}$$

Newtons iterative formula is used for the $u^{(i+1)}$ step, since $u^{(i)}$ is known at the ith step,

$$J^{(i+1),(k)}v^{(i+1),(k+1)} = J^{(i+1),(k)}v^{(i+1),(k)} - R^{(i+1),(k)} \tag{11.58}$$

where J is Jacobean Matrix, $R = [R_1 \;\; R_2 \;\; R_3]^T$ is the residual matrix and $v = [u_1 \;\; u_2 \;\; p]^T$.

11.4.1 Verification of Numerical Method

The flow around NACA 0012 airfoil, which is a typical problem in computational fluid dynamics, is studied numerically. Verification of the above method will be analyzed by simulating flow over NACA 0012 airfoil at various angles with $Re = 7 \times 10^4$. The angles of attack were ranged $\alpha = 0° - 7°$, obtaining lift coefficients. Figure 11.4 shows a good agreement of the lift coefficients calculated with present method to the experimental data available in literature [31]. So we can conclude that the method used is suitable for studying the dynamics of flow over an airfoil. Figure 11.5 presents the drag coefficients, computed at similar angles of attack as for lift coefficients, and compared with the experimental results [31]. It is observed that for laminar flow, skin friction drag is dominant as flow remains attached to the airfoil upto certain angles of attack showing less skin drag friction as compared to turbulent flow from experimental results shown in literature [31]. At high angle of attack, boundary layer separation starts when pressure drag becomes dominant and

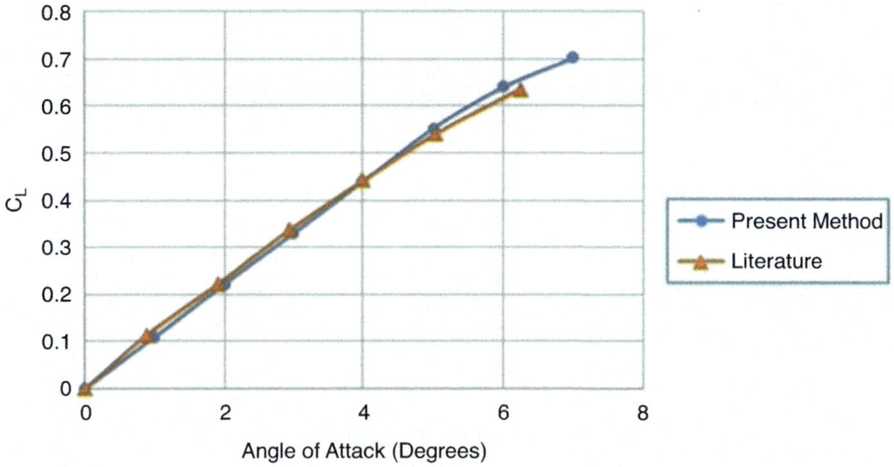

Fig. 11.4 Comparisons between experiments and numerical results for Lift Coefficients

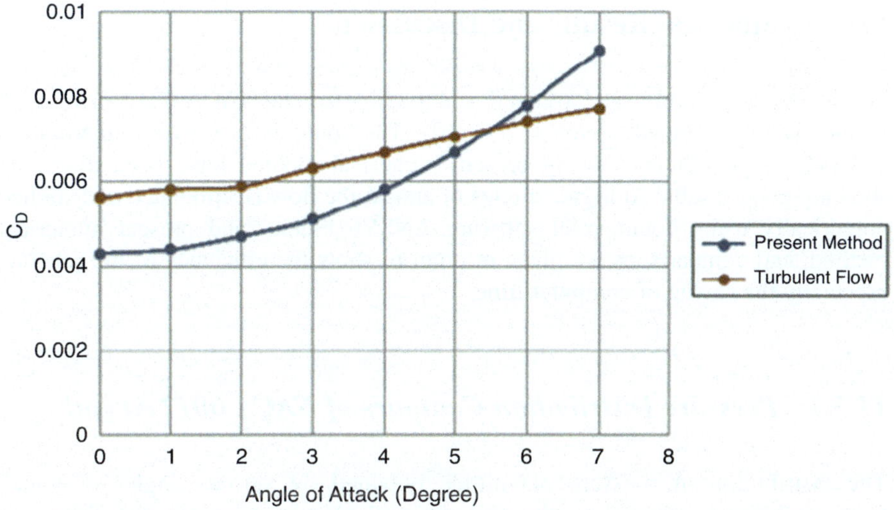

Fig. 11.5 Comparisons between turbulent model and numerical results for Drag Coefficients

is considered less for turbulent flows in comparison with laminar flows. So, from Fig. 11.5, we can observe a sharp rise of curve at an angle of 6° as separation starts for laminar flow. In many lift generating methods, the important quantity is the ratio of lift to drag presented in terms of Cl/Cd verses angle of attack. Figure 11.6 shows the most efficient angle of attack at 5° for attaining maximum efficiency of NACA 0012 airfoil. After that the lift to drag ratio decreases rapidly as we reach separation.

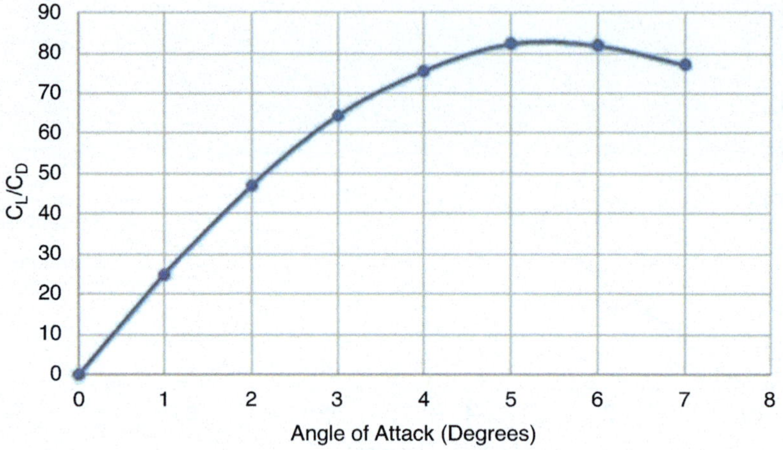

Fig. 11.6 Ratio of Cl/Cd for various angles of attack

11.5 Numerical Results and Discussion

For numerical example, we have analyzed the flow around NACA 0012 airfoil with various angles of attack at $Re = 7 \times 10^4$. The contours of pressure distribution and pressure coefficients using present numerical method have been discussed showing good results. At higher angles of attack, the flow is stimulated and studied numerically using commercial software ANSYS(Fluent) and present numerical method, and comparisons are given in order to show the efficiency of the scheme, especially the saving of computer time.

11.5.1 Pressure Distribution Contours of NACA 0012 Airfoil

The distribution of pressure magnitude obtained by various angles of attack from simulations results by the present method are shown in the following Figs. 11.7, 11.8, 11.9 and 11.10. We can see there is a region of high pressure at the stagnation point (leading edge) since the airfoil is symmetrical. As the flow accelerates with increase in angle of attack we observe the pressure on the upper surface is less and the stagnation point changes its position where the pressure increases thus creating lift since velocity on the upper surface increases rapidly. Similar effects can be observed as the pressure profile increase on the lower surface of the airfoil upon increasing angles of attack until stall occurs. The distribution of flow created on the upper surface show the boundary layer separation with a smooth transition of flow on the upper surface of the airfoil.

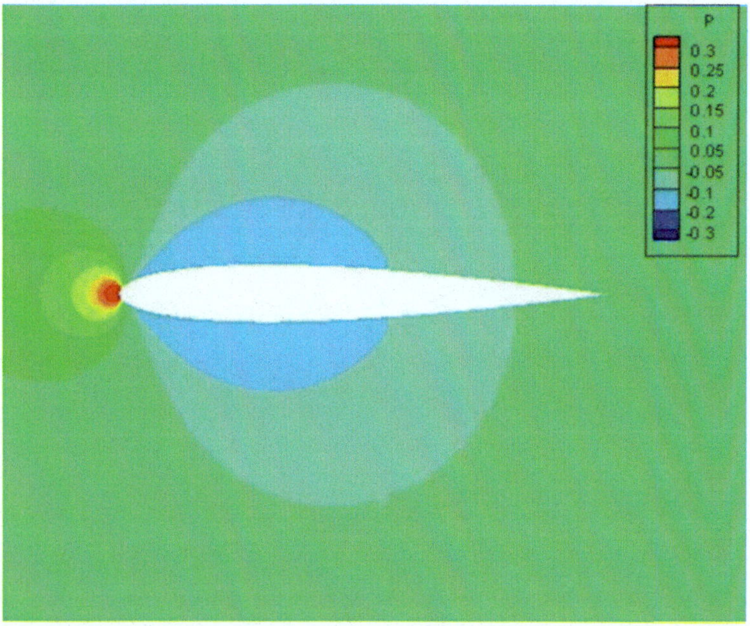

Fig. 11.7 Pressure Distribution Contour at 0°

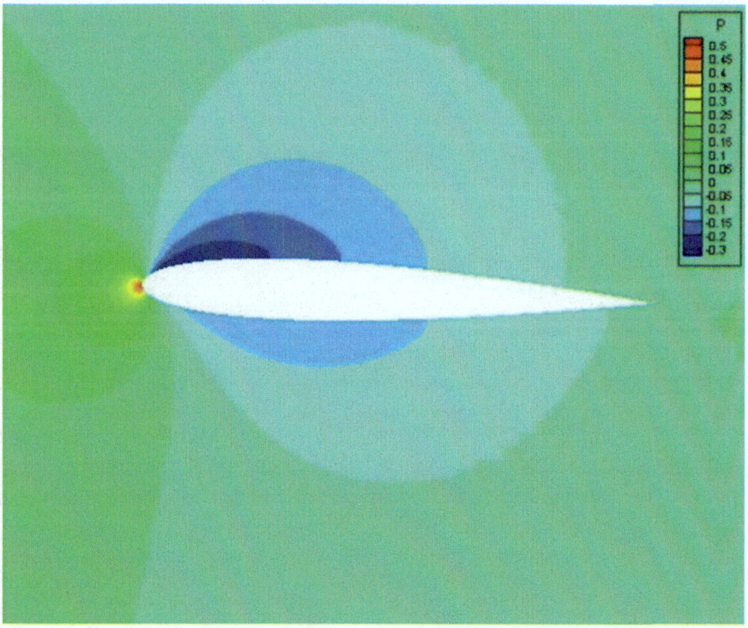

Fig. 11.8 Pressure Distribution Contour at 2°

Fig. 11.9 Pressure Distribution Contour at 4°

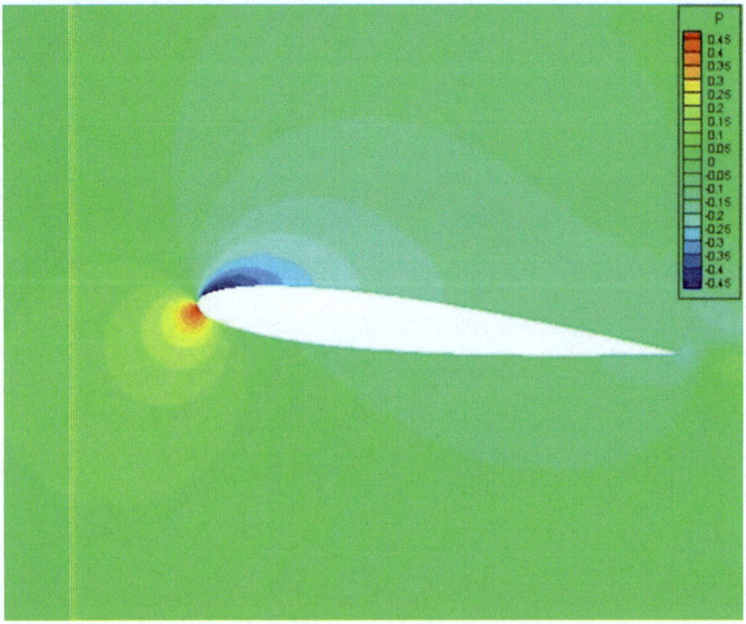

Fig. 11.10 Pressure Distribution Contour at 6°

11.5.2 Distributions of Pressure Coefficients of NACA 0012 Airfoil

The distribution of pressure coefficient of NACA 0012 airfoil under different angles of attack is shown in the following Figs. 11.11, 11.12, 11.13, and 11.14. When $\alpha = 0°$, the pressure coefficient at the leading edge of the airfoil is maximum giving symmetrical variation of distribution with net zero velocity. When $\alpha = 2°$, the pressure coefficient on the lower surface increases as compared with the upper surface. When $\alpha = 4°$ pressure coefficient increase rapidly on the lower surface and low pressure on the upper surface contribute to increase in lift. When $\alpha = 6°$, the pressure coefficient at the trailing edge increase subsequently while on the upper surface low pressure induces smooth boundary layer separation before stall occurs at high angles of attack. With increase in angle of attack, greater is the difference of pressure coefficient between the lower and upper surface. We can also see that the coefficient of pressure difference is much larger on the front edge as compared with pressure on the rear edge of airfoil.

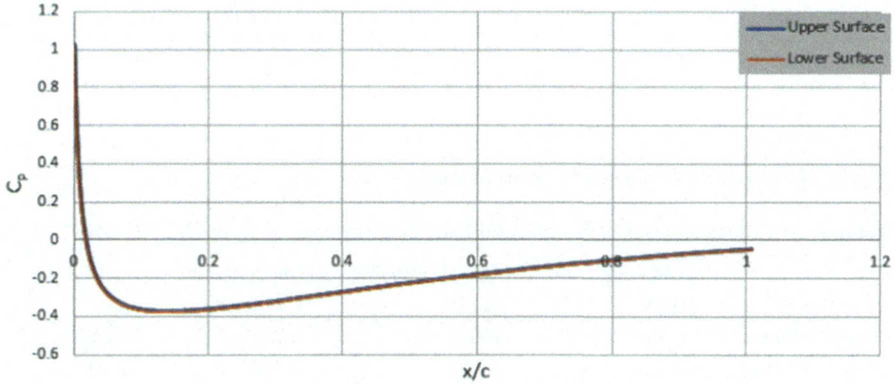

Fig. 11.11 Pressure coefficient at 0° angle of attack

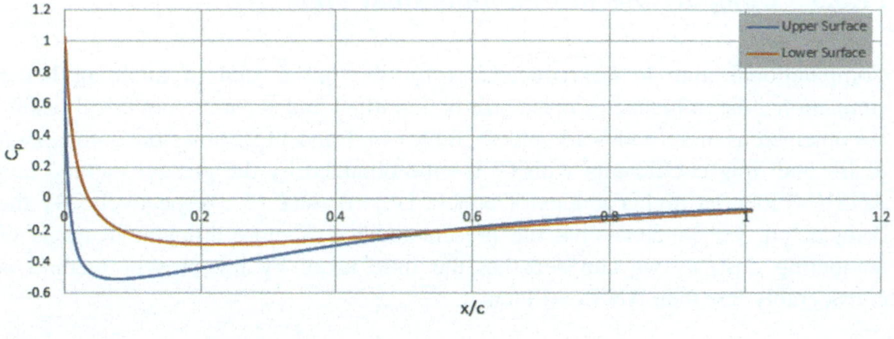

Fig. 11.12 Pressure coefficient at 2° angle of attack

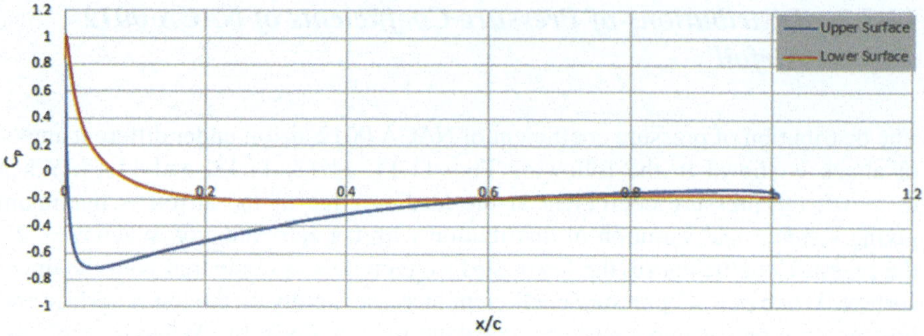

Fig. 11.13 Pressure coefficient at 4° angle of attack

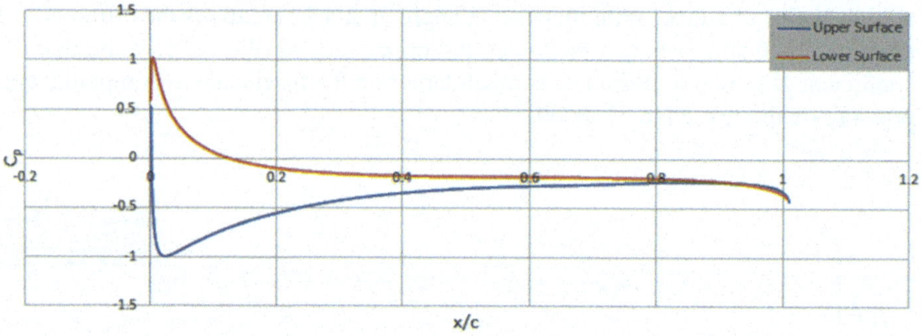

Fig. 11.14 Pressure coefficient at 6° angle of attack

Table 11.1 Computational Coefficients of lift and drag at 10° angle of attack-NACA0012

	Lift coefficient (C_l)	Drag coefficient (C_d)	Max.pressure coefficient	CPU-time (s)
ANSYS(Fluent)	0.815	0.013	0.777	16280
Present method	0.891	0.002	0.734	12816

11.5.3 Comparisons of Computational Time

Computational time to be reduced, as for the initial step of attaining better simulations, the influence of mesh size is important and accurate numerical results are obtained as more nodes are added. Tables 11.1 and 11.2 shows the comparison of lift and drag coefficients with CPU-time obtained by the present method and ANSYS(Fluent)at higher angles of attack, i.e., 10° and 12°, respectively. By the comparison, we conclude that the present method is more efficient in terms of computing cost, as we can see that the time taken by the present method is considerably less then ANSYS(Fluent).

Table 11.2 Computational Coefficients of lift and drag at 12° angle of attack-NACA0012

	Lift coefficient (C_l)	Drag coefficient (C_d)	Max.pressure coefficient	CPU-time (s)
ANSYS(Fluent)	0.913	0.054	0.892	13656
Present method	0.999	0.012	0.834	9594

11.6 Conclusion

Following Inertial Manifold, a dimension reduction method, based on multilevel finite element method with hierarchical basis, is presented for the Navier-Stokes Equation. The couplings between low modes and high modes have been considered and included by refining mesh domain using hierarchical finite element basis functions. The numerical simulation results indicate that the present method is feasible and efficient for the numerical analysis of complex flows with less computational time. The current work gives an efficient contribution towards model reduction method for the nonlinear dynamic systems with continuum mechanics, and reduces the original system to a system with less degrees-of-freedom. As for further work, this method could be applied to flow with complicated domains.

Acknowledgments The research is supported by the National Basic Research Program of China (973 Program, Grant No. 2012CB026002) and the National Natural Science Foundation of China (Grant No.51305355).

References

1. Kamakoti, R., and W. Shyy. 2004. Fluid-structure interaction for aeroelastic applications. *Progress in Aerospace Sciences* 40(8): 535–558.
2. Bendiksen, O., and G. Seber. 2008. Fluid-structure interactions with both structural and fluid nonlinearities. *Journal of Sound and Vibration* 315(3): 664–684.
3. Gordnier, R. 2009. High fidelity computational simulation of a membrane wing airfoil. *Journal of Sound and Vibration* 25(5): 897–917.
4. Rojratsirikul, P., Z. Wang, and I. Gursul. 2009. Unsteady fluid-structure interactions of membrane airfoils at low Reynolds numbers. *Experiments in Fluids* 46(5): 859–872.
5. Rega, G., and H. Troger. 2005. Dimension reduction of dynamical systems: Methods, models, applications. *Nonlinear Dynamics* 41(1): 1–15.
6. Steindl, A., and H. Troger. 2001. Methods for dimension reduction and their application in nonlinear dynamics. *International Journal of Solids and Structures* 38: 2131–2147.
7. Zhang, J.Z., Y. Liu, and D.M. Chen. 2005. Error estimate for influence of model reduction of nonlinear dissipative autonomous dynamical system on long-term behaviours. *Applied Mathematics and Mechanics* 26: 938–943.
8. Friswell, M.I., J.E.T. Penny, and S.D. Garvey. 1996. The application of the IRS and balanced realization methods to obtain reduced models of structures with local nonlinearities. *Journal of Sound and Vibration* 196: 453–468.
9. Fey, R.H.B., D.H. Von Campen, and A. De Kraker. 1996. Long-term structural dynamics of mechanical system with local nonlinearities. *ASME Journal of Vibration and Acoustics* 118: 147–163.

10. Kordt, M., and H. Lusebrink. 2001. Nonlinear order reduction of structural dynamic aircraft models. *Aerospace Science and Technology* 5: 55–68.
11. Slaats, P.M.A., J. De Jongh, and A.A.H.J. Sauren. 1995. Model reduction tools for nonlinear structural dynamics. *Computers and Structures* 54: 1155–1171.
12. Temam, R. 1997. *Infinite-dimensional dynamical system in mechanics and physics*. New York: Springer.
13. Titi, E.S. 1990. On approximate inertial manifolds to the Navier-Stokes equations. *Journal of Mathematical Analysis and Applications* 149: 540–557.
14. Jauberteau, F., C. Rosier, and R. Temam. 1990. A nonlinear Galerkin method for the Navier-Stokes equations. *Computer Methods in Applied Mechanics and Engineering* 80: 245–260.
15. Chueshov, I.D. 1996. On a construction of approximate inertial manifolds for second order in time evolution equations. *Nonlinear Analysis, Theory, Methods and Applications* 26: 1007–1021.
16. Mezic, I. 2005. Spectral properties of dynamical systems, model reduction and decomposition. *Nonlinear Dynamics* 41: 309–325.
17. Rezounenko, A.V. 2002. Inertial manifolds for retarded second order in time evolution equations. *Nonlinear Analysis* 51: 1045–1054.
18. Zhang, J.Z., R. Shen, and M. Guanhua. 2011. Model Reduction on inertial manifolds for N-S equations approached by multilevel finite element method. *Communications in Nonlinear Science and Numerical Simulation* 16: 195–205.
19. Chow, S.N., and K. Lu. 2001. Invariant manifolds for flows in Banach space. *Journal of Differential Equations* 74: 285–317.
20. Foias, C., G.R. Sell, and E.S. Titi. 1989. Exponential tracking and approximation of inertial manifolds for dissipative nonlinear equations. *Journal of Dynamics and Differential Equations* 1: 199–244.
21. Foias, C., O. Manley, and R. Temam. 1988. Modelling of the interaction of small and large eddies in two dimensional turbulent flows. *Mathematical Modelling and Numerical Analysis* 22: 93–118.
22. Haller, G., and S. Ponsioen. 2017. Exact model reduction by a slow-fast decomposition of nonlinear mechanical systems. *Nonlinear Dynamics* 90: 617–647.
23. Schmidtmann, O. 1996. Modelling of the interaction of lower and higher modes in two-dimensional MHD-equations. *Nonlinear Analysis, Theory, Methods and Applications* 26: 41–54.
24. Kang, W., J.Z. Zhang, R. Shen, and L. Penfei. 2015. Nonlinear Galerkin method for low-dimensional modeling of fluid dynamic system using POD modes. *Communications in Nonlinear Science and Numerical Simulation* 22: 943–952.
25. Laing, C.R., A. McRobie, and J.M.T. Thompson. 1999. The post-processed Galerkin method applied to non-linear shell vibrations. *Dynamical Stability and Systems* 14: 163–181.
26. Rezounenko, A.V. 2002. Inertial manifolds for retarded second order in time evolution equations. *Nonlinear Analysis* 51: 1045–1054.
27. Zhang, J.Z., Y. Liu, P.F. Lei, and X. Sun. 2007. Dynamic snap-through buckling analysis of shallow arches under impact load based on approximate inertial manifolds. *Dynamics of Continuous, Discrete and Impulsive Systems Series-B(DCDIS-B)* 14: 287–291.
28. Whiting, C.H., and K.E. Jansen. 2001. A Stabilized finite element methods for incompressible Navier-Stokes equation using a hierarchical basis. *International Journal for Numerical Methods in Fluids* 35: 93–116.
29. Krysl, P., E. Grinspun, and P. Schroder. 2003. Natural hierarchical refinement for finite element methods. *International Journal for Numerical Methods in Engineering* 56: 1109–1124.
30. Bank, E., F. Dupont, and H. Yserentant. 1988. The hierarchical basis multigrid method. *Numerische Mathematik Springer* 52: 427–456.
31. Abbott, I.H., and A.E. Von Doenhoff. 1959. *Theory of wing sections*. New York: Dover Publishing.

Index

J. Zhang (ed.), *Dynamics and Fault Diagnosis of Nonlinear Rotors and Impellers*, Nonlinear Systems and Complexity 34, https://doi.org/10.1007/978-3-030-94301-1

The manufacturer's authorised representative in the EU is Springer Nature Customer Service Centre GmbH, Europaplatz 3, 69115 Heidelberg, Germany. If you have any concerns regarding our products, please contact ProductSafety@springernature.com

Printed and bound by CPI Group (UK) Ltd, Croydon, CR0 4YY
01/10/2026
02225872-0005